FRED HOYLE'S UNIVERSE

Proceedings of a Conference
Celebrating Fred Hoyle's Extraordinary Contributions to Science
25–26 June 2002
Cardiff University, United Kingdom

Editors:

CHANDRA WICKRAMASINGHE
Cardiff Centre for Astrobiology
Cardiff University
Cardiff, United Kingdom

GEOFFREY BURBIDGE
Centre for Astrophysics and Space Sciences (CASS)
University of California – San Diego
La Jolla, USA

and

JAYANT NARLIKAR
Inter-University Centre for Astronomy and Astrophysics
Pune, India

Reprinted from *Astrophysics and Space Science*
Volume 285, No. 2, 2003

SPRINGER-SCIENCE+BUSINESS MEDIA, B.V.

Library of Congress Cataloging-in-Publication Data

Fred Hoyle's universe / edited by Chandra Wickramasinghe, Geoffrey Burbidge, and Jayant Narlikar.

p.cm.

ISBN 978-90-481-6339-7 ISBN 978-94-017-1605-5 (eBook)

DOI 10.1007/978-94-017-1605-5

1. Hoyle, Fred, Sir – Congresses. 2. Astronomy-Congresses. I. Wickramasinghe, N.C. (Nalin Chandra), 1939 – II. Burbidge, Geoffrey R. III. Narlikar, Jayant Vishnu, 1938 –

2003058909

Printed on acid-free paper

TABLE OF CONTENTS

Foreword from Vice-Chancellor Dr David Grant, CBE 1

G. BURBIDGE /Fred Hoyle's Universe 3

PART I: PERSONAL REMINISCENCE

SIR H. BONDI / Working with Fred (1942–49) 9

C. DOMB / Fred Hoyle and naval radar 1941–5 13

M. BURBIDGE / Fred Hoyle and the Anglo-Australian telescope 23

R.D. DAVIES / Fred Hoyle and Manchester 29

R. MEYERS / Meeting with a remarkable man: Sir Fred Hoyle Memorial Conference 41

PART II: STELLAR STRUCTURES AND EVOLUTION

L. MESTEL / Fred's contributions to stellar evolution 47

J. FAULKNER / Fred Hoyle, Red giants and beyond 59

D. GOUGH / Some remarks on solar neutrinos 61

D.D. CLAYTON / Novae as thermonuclear laboratories 73

I. ROXBURGH / The *Eddington* mission 83

S.J. AARSETH / Black hole binary dynamics 87

PART III: COSMOLOGY

M.J. REES / Numerical coincidences and 'Tuning' in cosmology 95

J.V. NARLIKAR / Working with Fred on action at a distance 109

T. PADMANABHAN / Gravity from spacetime thermodynamics 127

W.M. NAPIER / A statistical evaluation of anomalous redshift claims 139

W.G. TIFFT / Redshift periodicities, The galaxy-quasar connection 149

H. ARP / Research with Fred 171

D.F. ROSCOE / The discovery of major new phenomenology in spiral discs and its theoretical meaning 179

R. TEMPLE / The Prehistory of the steady state theory 191

PART IV: INTERSTELLAR MATTER

P.M. SOLOMON / Interstellar matter and star formation 201

C. WICKRAMASINGHE / Elevating the status of dust 217

PART V: COMETS

W.M. NAPIER / Giant comets and human culture 233

S.V.M. CLUBE / An exceptional cosmic influence and its bearing on the evolution of human culture as evident in the apparent early development of mathematics and astronomy 241

PART VI: PANSPERMIA

C. WICKRAMASINGHE / Panspermia according to Hoyle 255

F. HOYLE and C. WICKRAMASINGHE / Astronomy or biology? 259

J.V. NARLIKAR, D. LLOYD, N.C. WICKRAMASINGHE, M.J. HARRIS, M.P. TURNER, S. AL-MUFTI, M.K. WALLIS, M. WAINWRIGHT, P. RAJARATNAM, S. SHIVAJI, G.S.N. REDDY, S. RAMADURAI and F. HOYLE / A balloon experiment to detect microorganisms in the outer space 275

M. WAINWRIGHT / A microbiologist looks at Panspermia 283

A.K. CAMPBELL / What Darwin missed 291

M.K. WALLIS / Cosmic genes in the Cretaceous-tertiary transition 307

J. FAULKNER / Remembering Fred Hoyle 313

FOREWORD FROM VICE-CHANCELLOR DR DAVID GRANT, CBE

Fred Hoyle will go down in history as one of the most remarkable men of science of the 20th century. He challenged our habitual ways of thinking on several important scientific issues and in the process contributed greatly to our understanding of the world around us. The programme for this meeting illustrates the creativity and versatility of this great man and the influence he had in many areas of astronomy.

Cardiff University was privileged to be closely associated with Fred Hoyle for over a quarter of a century. The association began in 1973 with the appointment of the theoretical astronomer Chandra Wickramasinghe to the Chair of Applied Mathematics, which marked the revival of astronomy as an academic discipline in this University (it was last taught in the 19th Century). In the subsequent years Sir Fred was closely involved, through consultations, in the appointment of several senior astronomers currently in post and thus contributed to the making of a highly successful astronomical research centre in Cardiff.

As an Honorary Professor, Fred Hoyle was a regular visitor to Cardiff for his collaborations with Professor Chandra Wickramasinghe and also to deliver a memorable series of public lectures.

I am pleased to welcome participants to the conference and I wish you all a very productive and enjoyable visit to the University and to the capital city of Wales.

FRED HOYLE'S UNIVERSE

Introductory Remarks

Soon after Fred Hoyle died on August 20, 2001, Chandra Wickramasinghe, Jayant Narlikar and I decided that we would organize a meeting to celebrate Fred, all of his works, and the tremendous influence that he has had on all of us, and on many aspects of science over the last sixty years.

We are very happy that this meeting is being hosted by Cardiff University where Chandra and his group play such an important role.

There has already been one meeting in Cambridge, on April 16, to celebrate Fred's life and work. This was centered about the Institute of Astronomy, which he founded, and his college, St. John's College. This was a marvelous affair and some of the speakers who were at Cambridge are also here. The proceedings of that meeting will be published (Cambridge University Press, D. Gough Editor). At this meeting we have tried to cover as wide a range as possible of Fred's many interests and creative ideas. We start with a discussion of Fred Hoyle's earliest work on stars and with some of those who worked with him on it. His first collaborator in Cambridge was Ray Lyttleton who died in 1995, but it was after Fred had moved into radar research at the Admiralty in 1940 that he met Hermann Bondi, Cyril Domb, Tom Gold and others. Both Hermann Bondi and Cyril Domb are here and will speak this afternoon. Tom Gold is not able to be here. He has sent his regrets.

The first papers describing Fred's important work with Lyttleton and Bondi on accretion and on the structure of main sequence stars, and on red giants and stellar evolution (some of it with Martin Schwarzschild) will be given by Leon Mestel and John Faulkner. Ian Roxburgh will cover other topics on stars, and Douglas Gough will talk about solar structure and the resolution of the solar neutrino problem. Sverre Aarseth will speak on black hole binary dynamics and Martin Rees will talk about the problem of fine-tuning in the constants of physics and cosmology.

There is a general consensus that the most important contribution that Fred made to astrophysics was his initial proposal and later demonstration that all of the chemical elements, apart from the lightest isotopes, have been synthesized in stars. He first proposed this in 1946, and by 1957 when the work by him, together with Willy Fowler and Margaret and me was published, the basic case had been made, and it soon became generally accepted. This afternoon Dave Arnett and Donald Clayton will describe some of the more recent work on stellar nucleosynthesis. Partly because Fred's contribution is so well known and accepted, we have not invited more speakers to talk about this.

But, in anticipating the later discussions, I would like to add two points. Whatever sort of a universe we live in, the stellar nucleosynthesis proposal must be correct. At the same time, as a cosmologist, from 1948 onward, Fred hardly ever wavered

Astrophysics and Space Science **285:** 283–286, 2003.

from the view that some form of the steady state universe model must be correct. This being the case, the lightest isotopes D, ^3He, ^4He, and possibly some ^7Li must also be synthesized in stars. Fred was always very skeptical about a big bang origin, though he did do fundamental work on primordial nucleosynthesis. He was skeptical in large part because he was perfectly well aware that to make it work, the correct initial value of photons to baryons had to be *chosen*. Thus there is no real theory behind the conventional picture. In fact in some of the most recent work Fred found another initial starting point that would give the correct abundances. This was discussed in detail in our book (HBN 2000)[1]. However, the fact that the energy density of the microwave background, is exactly what is expected if all of the helium in the universe was produced in hydrogen burning in stars, and then thermalized, also suggests that in a quasi-steady state universe helium is of stellar origin. If this is true then it is fairly clear that deuterium must be made in the outside of stars, in stellar flares. Fred and I finally got this view published in 1998. While this part of the element synthesis picture is not yet accepted by the community, it does now seem likely that *all* of the elements in the periodic table were made, as Fred originally surmised, in stars.

On Tuesday we turn to field theory and cosmology. Jayant Narlikar will talk about the work that he did over many years with Fred on action at a distance. Later on Wednesday morning J.N. Islam will talk on the Hoyle-Narlikar theory of gravitation. Also on Tuesday morning T. Padmanabhan will give a talk on gravity from the thermodynamics of space time.

Fred Hoyle was not only a great scientist, but he also did a great deal to advance British astronomy on all fronts. His most important achievement was undoubtedly the creation of the Institute of Theoretical Astronomy in Cambridge which was started in the middle 1960s. There was a considerable amount of opposition to this in Cambridge, but Fred finally prevailed, and he was the first director until 1972 when there was an upheaval and he left Cambridge for good. Of course, the Institute, renamed the Institute of Astronomy, and incorporating the Observatories, has gone from strength to strength, and today it is one of the few great international research centers for astrophysics in the world.

Fred also played a major role in the planning and building of the Anglo-Australian Observatory. Other major players on the British side were originally Sir Hermann Bondi, Sir Richard Woolley (while he was Astronomer Royal), Margaret Burbidge (while she was director of the Royal Greenwich Observatory) and Jim Hosie, probably the most able Civil Servant in the UK or the USA that I have ever met. On the Australian side there was Olin Eggen, Director of Mount Stromlo, E.G. (Taffy) Bowen from CSIRO and Ken Jones representing the Australian government. Fred was the chairman of the steering committee over the critical period, succeeding Taffy Bowen. Margaret Burbidge is going to give us some flavour of Fred's work in this period.

[1] Hoyle, F., Burbidge, G. and Narlikar, J.V. 2000 (Cambridge University Press).

After Fred left Cambridge in 1972 he lived first in Cumbria, and finally in Bournemouth, but he accepted honorary professorships at Manchester and Cardiff. As many of us know Fred always preferred to work at home, and he never spent much time in Manchester or Cardiff. But he had close connections to both. Fred always realized the importance of radio astronomy at Manchester and Cambridge and supported both Martin Ryle and Bernard Lovell (despite his differences with Ryle). Sir Bernard Lovell was one of his close friends both socially and politically, and we are very sorry that Sir Bernard was not able to come to talk. However, this afternoon Rod Davies is going to give an account of Fred's Manchester connections.

One of Fred Hoyle's greatest strengths as a theorist was his belief that the observations are nearly always paramount. Hermann Bondi is often quoted as having said that if there is a direct conflict between observation and theory it is the observation that is likely to be in error! I believe that Hermann would only use that argument today in situations in which the observations are not repeatable, or are flawed in other ways. At least over the period that I knew Fred, more than 40 years, my experience was that be would examine observations with great care, but once he believed them he would conclude that if there was conflict between theory and the observations it was the theory that should be modified. In my opinion it is this approach together with his excellent intuition that led to Fred's moving away from mainstream opinion in cosmology and extragalactic astronomy over the last thirty years or more. Some of the evidence which he found so compelling and which led to this was described in our book published in 2000. One of the key questions is indeed whether or not all redshifts can be attributed to the expansion of the universe. We have three contributions on this topic by distinguished observers and statisticians whose work continued to influence Fred greatly, H.C. (Chip) Arp, W. Tifft, and W. Napier who will be speaking on Tuesday and Wednesday.

On the cosmological front John Maddox is going to argue against the big bang, and there is going to be some discussion of the pre-history of the steady state by R. Temple.

Another major field of astrophysics on which Fred Hoyle had a major impact was the study of the nature and the origin of the interstellar gas and dust. When he began this work in 1962 with his student Chandra Wickramasinghe, it was thought that the interstellar dust was a form of dirty ice and its origin was quite obscure. Hoyle and Wickramasinghe showed the importance of carbon, and first proposed that the dust was produced in the outer layer of giant stars. Over many years, they and an increasing number of others turned the study of the composition and origin of interstellar dust into a very active field of research. Tomorrow and on Wednesday P. Solomon, C. Wickramasinghe and M. Edmunds are going to cover different aspects of this endeavor, and describe the current state of affairs.

Hoyle and Wickramasinghe suggested that organic molecules were contained in the dust, and in cometary material, so there is a connection with the solar system and the earth. This latter work though clearly now acknowledged to be correct, was not well received. Their suggestion that life itself may have been transported this way (cf. "Life Cloud - The Origin of Life in the Universe" with C. Wickramasinghe, Dent 1978) - in other words, that the earlier idea of panspermia is correct, has been heavily and unfairly criticized.

We shall have a discussion of this whole area of research on Wednesday by W. Napier, V. Clube and C. Wickramasinghe. Also Narlikar and Wickramasinghe who with their colleagues have been flying a balloon to try to detect micro-organisms in the earth's upper atmosphere will give an account of this investigation.

This completes my summary of practically the whole of the scientific program. But even now so much has been left out. There will be no discussion of Fred's popular writings in astronomy and in other fields, although we shall hear a tribute to him from Arthur C. Clarke directly from Sri Lanka. Also in the summer before Fred died, R. Mayer interviewed him on film as part of a scientific documentary about radical views in astronomy that he is making. We shall see a short segment of that film, and an account of it by its maker appears in this volume.

Also no one is going to talk about Fred's science fiction, particularly, 'A for Andromeda' and the 'Black Cloud', or about his children's plays or his opera lyrics. Fred provided the libretti for two of the composer Leo Smit's compositions, the opera *Alchemy of Love* (1969) and *Copernicus: Narrative and Credo* (1973) for chorus, narrator and chamber ensemble. For many years Fred was very close to the actor and producer Bernard Miles (later Lord Miles) at the Mermaid Theatre in London. And then there were Fred's investigations of Stonehenge (*From Stonehenge to Modern Cosmology*, 1973, W. H. Freeman), and his extended essay on Western Religion 'The Origin of the Universe and the Origin of Religion' (Moyer; Bell 1993). And that is not all. Thus, in introducing this meeting I simply want to stress that many of Fred's great achievements will be described, but still much has been omitted.

Geoffrey Burbidge

PART I: PERSONAL REMINISCENCE

WORKING WITH FRED (1942–49)

SIR HERMANN BONDI

Abstract. How working together in the radar research establishment of the Royal Navy led to Fred, Tommy Gold and I spending many evenings together discussing astronomical problems brought up by Fred. The accretion question was the subject of my first serious paper and led to my Trinity Fellowship. Continuation of the joint discussions after our return to Cambridge. The solar corona. Fred's generosity.

At the beginning of April 1942 I joined the radar research establishment of the Royal Navy, the main part of which was then located in Portsmouth. Initially I was working there with Maurice Pryce whom I had met at Trinity College, Cambridge where his brother Jacques was in the same year as I. Soon I heard that at our aerial research branch station at Nutbourne near Chichester there was an unusual scientist, a man of ideas but impatient with committee work and with bureaucratic delays in getting equipment and improved procedures to the fleet. (It must be remembered that at this desperate stage of the war anything that could help in the critical Battle of the Atlantic was urgently needed.) Soon Fred and I met and appreciated each other.

In June 1942 our whole establishment was moved to one site (Witley in Surrey) and reorganised. A Theory Division was set up with Fred as its head, me as his deputy and a very few other members (amongst them Cyril Domb). It is perhaps worth mentioning that 5 of the members of this small Division subsequently became Fellows of the Royal Society. The move to Surrey posed no difficulties for a bachelor like me. I simply exchanged digs (bed, breakfast and dinner) in Portsmouth for being billeted on a well-to-do family near Witley. But for Fred it made life much more difficult. He had a house near Nutbourne where he lived with his wife and little son. Houses near Witley, in the Surrey commuter belt, were very expensive. So Fred cycled 10 hilly miles from his house to a railway station on the Portsmouth-London line and then had 20 minutes on a train to get to Witley. In summer this was a waste of time and energy; in winter, with both journeys in the dark, it was a strain even for somebody as tough as Fred.

In our work for the Admiralty Fred, as head of the Division, handled the links with the upper echelons while I looked at technical questions he referred to me. I learned a great deal about noise and its relation to signal/noise ratios, so important for the users and designers of our radars. Some of what I did may even have been useful to the Navy! Fred and I began to make aerial design a little more systematic. Fred also encouraged me to work on magnetrons, the vital transmitting valves that allowed us to use centimetric wavelengths, giving us a great lead in the naval domain. Our magnetrons certainly worked but one did not understand why.

The next point to mention is that I asked Fred to recruit Tommy Gold to our small Division and he agreed to do so. I had met Gold in internment and then

Astrophysics and Space Science **285:** 289–292, 2003.

spent half a year at Trinity in close contact with him. He impressed me greatly with his understanding of physics, his practical outlook and his general ability. Unfortunately the undergraduate engineering course he took did not inspire him and he performed very badly in his final exams in June 1942. Against this paper fact Fred had to fight the system, guided by nothing except my assessment. I am still amazed that he placed so much confidence in my judgement after only a few months acquaintance. Fred was successful and Tommy joined us in October 1942. He was put into the same billet as I, but this did not suit him at all. He was very keen to have his independence and I concurred. So we rented a house in the delightful village of Dunsfold from the beginning of 1943. The one awkward point about living there was that travelling to and from our work at Witley by public transport was lengthy. One could go by bike, but in bad weather this was tiresome. Before long we suggested to Fred (who had meanwhile acquired an old car) to spend some nights during the week with us. This arrangement was of advantage to Tommy and me (getting to and from work by car) and to Fred (who avoided a long journey). Thus the three of us had many evenings together. Later Fred's wife and little son joined us as well on many occasions.

Fred's enthusiasm for astrophysics was soon conveyed to Tommy and me. Every evening we spent together we had an exciting scientific discussion, occasionally about a question related to our daytime work, but mostly about some astronomical problem explained to us by Fred. [It is perhaps worth mentioning here that he was not only our senior by 4 years according to the calendar, but was an established scientist who had published a number of significant papers and had been a Junior Research Fellow at St. John's College, Cambridge. By contrast, I had been a Research Student for a bare 6 months and had published one very minor paper while Tommy had nothing behind him but his undergraduate years in engineering].

Fred's marvellous ability to expound a problem and to sketch in existing work meant that I was soon busy with the accretion question that he and Ray Lyttleton had put on the agenda a few years earlier. My work on the theory of magnetron transmitting valves had made me familiar with multistream states, which, it turned out, were of great significance in the accretion problem. I wrote up my work on accretion as a Fellowship thesis for my College (Trinity) and was elected a Junior Research Fellow in October 1943, a concrete result of our fortuitous housing arrangements and of Fred's outgoing and stimulating personality. He and I carried the accretion effort further forward by evaluating the braking effect of the cloud on the star and by considering non-steady states. The whole work was then published as a joint paper (H. Bondi and F. Hoyle: 1944, *M.N.R.A.S.* **104**, 273), my first substantial contribution.

But the immediate output of astronomical papers is not an adequate measure of the value of these evening discussions among three people, each of whom was in full-time non-astronomical employment of great importance. We all gained greatly from these tough discussions, to which we brought our individual attitudes and skills. But Fred's personality, generosity, wide knowledge and insight were at the root of it all.

While 1943 was full of these discussions, 1944 was rather less so, since our Admiralty work made each of us travel a great deal. But the mutual education continued until, with the end of the war, Fred and I returned to Cambridge in summer 1945. Tommy followed somewhat later. Then history repeated itself: Fred, with his wife and, by then, two children required a sizeable house. In and near Cambridge such properties commanded a price hard to contemplate for an academic of Fred's then position. So he bought a house at Quendon, nearly 20 miles away. Therefore he needed a base in Cambridge, where he could spend the time between lectures and supervisions in congenial company. To my delight and advantage, my Fellow's set of rooms in Trinity served this purpose admirably. In 1947 I married Christine Stockman, who had been his Research Student. I had to move out of College, but we moved into a College flat just outside, which served his needs equally well. Tommy had acquired a house a couple of miles away and our central position was convenient for him too. So our scientific discussions continued as before and now included questions arising out of our teaching. But gradually our interests diverged and when in April 1949 Christine and I bought a house one-and-a half miles away, the almost daily discussions came to an end.

But these sessions had been long and intense. Fred once remarked that at the end of such a day we were likely to be as far apart as at the beginning, but he had been persuaded of my point of view and I of his. As an education in the science of astronomy, these discussions could scarcely be bettered. The topics involved stellar structure and evolution [but Tommy and I could not contribute to Fred's special love, the origin of the elements, as neither of us had worked in nuclear physics, while Fred had.], the origin of the high temperature of the solar corona, and cosmology.

Though our work on the corona in which R.A. Lyttleton was actively involved (H.B., F.H. and R.A.L.: 1947, *M.N.R.A.S.* **107,** 184) is now only of historical interest, it taught me a lot, not least about Fred's generosity. During the early stages of our effort I had fumbled about the thermodynamics. So I was surprised when Fred had included my name as an author and suggested he should omit it. He responded by saying that in that case he needed to put in the text a remark about his indebtedness to me for having led him by the nose for two years. I gave in!

As regards cosmology, we had numerous discussions on this seductive topic over the years. Eventually all three of us were much impressed by Tommy's suggestion of a steady state with the attendant need for continual creation. Fred's attempt to base this on a field theory appealed as little to Tommy and me as our more philosophical approach appealed to him. Hence there were two papers, fortunately published in the same issue **108** of *M.N.R.A.S.* 1948, (H.B. and T.G., 252, F.H. 372).

As a model for joint work, our sometimes fierce discussions all evening long could scarcely be bettered, especially when the senior person is as generous as Fred invariably was. I am still aware how much I owe to the fortunate circumstances that made me a participant all these many years ago.

FRED HOYLE AND NAVAL RADAR 1941–5

CYRIL DOMB

Abstract. The author spent the years 1941–5 as a member of Fred Hoyle's Theoretical Group in radar research for the Admiralty. Fred was in his mid-twenties when we first met, and a few personal reminiscences are presented. A brief description follows of the composition of the theoretical group. The major aim of the paper is to outline three important contributions which Fred made to the use of radar at sea: (1) height estimation of aircraft at metre wavelengths; (2) anomalous propagation at centimeter wavelengths; (3) detection of aircraft in the presence of window jamming. Finally, an account is given of the determined manner in which Fred negotiated his release from the Admiralty in 1945 so that he could work full time on an idea of major importance to astrophysics which had recently occurred to him; this idea, which he referred to as 'the abundance of the elements', (later known as nucleosynthesis) is generally regarded as his most significant contribution to science.

1. Introduction

I first met Fred Hoyle in July 1941 when I joined the Admiralty Signal Establishment (ASE), the body responsible for radar research and development for the Navy. I was then 20 and had just graduated in mathematics from Cambridge; Fred was 26, a Junior Fellow of St. John's College. The electrical engineers who ran ASE did not quite know what to do with theoreticians; their attitude was scornful, and they assumed that the area in which theoreticians could best contribute was probably aerials and propagation. The aerials division was located at Nutbourne in fields and primitive huts in the countryside northeast of Portsmouth; headquarters of ASE were at Eastney near Portsmouth.

Fred had joined a few months before me, and had been assigned to Hut No. 2 with lots of literature about radar to think about. I was less prestigious and they decided to teach me radar from the beginning, starting with the soldering iron. When I proved completely hopeless at this they assigned me to Fred in Hut No. 2.

Fred greeted me warmly and we chatted about our experiences at Cambridge. Although he had also taken the Mathematics Tripos, he had no real bent for mathematics; for him mathematics was an invaluable tool for scientific exploration. Since I had an interest in mathematics *per se* he would gladly assign any pure mathematical problems to me. He also said that although he appreciated the importance of devoting time and thought to the war effort, he did not consider that we needed to stop thinking about fundamental problems in physics.

He then told me how he had moved from quantum electrodynamics to astrophysics, and outlined the ideas that he and Lyttleton had put forward on accretion, and the difficulty they had experienced in getting them accepted by the establishment. Turning to my personal situation, he assumed that I must be looking around for a field in which to do basic research, and he strongly recommended

Astrophysics and Space Science **285:** 293–302, 2003.

astrophysics. 'The leading British authorities in the field' he said, 'are Eddington, Jeans and Milne. Of these three Jeans and Milne have done nothing of real significance. Eddington may have done one or two significant things, but he is still pretty mediocre; in a field in which the top researchers are so mediocre you should have no difficulty in making your mark quickly'. A week or two later he came in one morning excitedly waving the current issue of the 'Monthly Notices'. 'Eddington has just made an elementary error which you can easily identify. If you send off a quick note pointing it out you will have started your research career'. It was not appropriate for Fred himself to do this since he was a party to the dispute. Since Eddington had been one of my Part III examiners only a few weeks previously, I did not feel that I was the person to engage in public criticism of his work.

During the next few months I saw quite a lot of Fred since we were the sole occupants of the hut, and I should like to devote a few minutes to my personal impressions. His mental energy was amazing. He never seemed to switch off completely but instead switched from one topic of interest to another. Towards the end of the day he would relax by reading science fiction. When I chided him for wasting time on such low grade literature, he replied 'I have a purpose in mind. These people don't know any real science and they make money by writing this stuff. I, who know some science, should be able to do much better.' In fact he did later on pioneer some first class science fiction, which paid very well, and dwarfed his academic salary; it enabled him to resign his Cambridge Chair and move to the Lake District, a crucial step in his unconventional career.

On one occasion he expressed a desire to learn what pure mathematics was all about, and I managed to get hold of a book by Littlewood. He read steadily for a few days and was impressed. Then he came upon a point which he was sure Littlewood had got wrong. I was certainly not capable of deciding, but it taught me that Fred could not devote attention to any topic without evoking a personal reaction, generally critical.

Fred was anti-establishment. He was a junior Fellow of St. Johns, and he would regale me with stories of how an *enfant terrible* in the Senior Common Room succeeded in deflating the ego, and ridiculing the conventional stuffiness of the senior dons who ran the College. Fred described himself to me as a committed atheist, and he knew that I was an observant Jew; our basic beliefs could not have been further apart, and yet I cannot recall any occasion on which this led to a clash or to ill-feelings. In fact, Fred was very considerate of other people's feelings, and he confided to me that, despite his atheism, when he married a couple of years previously, he agreed to his bride's request, on sentimental grounds, for a church wedding; but he told the vicar of his real views, and the latter agreed to keep the ceremony short.

Fred was also scornful of government honours and awards. 'If ever they offer me a knighthood, I'll turn it down; by accepting it you put yourself in the same social class as treacle merchants'. He quoted with relish the story of George Bernard Shaw who, when offered a knighthood, replied 'I wouldn't accept anything less than a Barony, and I couldn't afford that'. They then offered him the prestigious 'Order of Merit' (OM). 'Ah' he replied, 'I awarded that to myself some years ago'. In fact many years later Fred did accept a knighthood without demur; perhaps he mellowed as he did later in regard to atheism, but I suspect that consideration of his wife's feelings played a part in his decision.

2. Theoretical Group at ASE

Hermann Bondi joined our hut in April 1942. He came with prestigious backing from headquarters at Eastney, and, when we were later formally constituted into a Theoretical Group, he served as Deputy Head. One of his first important achievements was to persuade the authorities to let Tommy Gold join ASE. Tommy and he had been close friends at Cambridge, but Tommy had only a 4th class engineering degree to back his case. As a student he found the engineering courses boring, and he devoted his time and energy to climbing King's College Chapel. Hermann demonstrated his ingenuity and organizational ability in overcoming this handicap, and Tommy became the fourth occupant of our hut in November 1942. Needless to say Tommy soon exploited his rare combination of experimental versatility with wide theoretical understanding, and his reputation grew rapidly.

Fred, Hermann, Tommy and I formed the core of the theoretical group. We were later supplemented by Charles (E.T.) Goodwin, a Cambridge friend of Fred's whom he rescued from an unhappy situation in ballistics research, John Gillams, a young Oxford graduate, and S. Rosseland, a Norwegian astronomer, who made no contribution to radar research, but whose membership of our group made the free Norwegian government happy.

Three members of other groups were closely associated with us. Maurice (M.H.L.) Pryce, whom Fred knew well from Cambridge; Pryce was 2 years senior to Fred, and enjoyed a fabulous reputation as a theoretical physicist; he had recently been appointed Reader In Theoretical Physics at Liverpool University, and was much higher in the ASE hierarchy. Otto Bohm, a highly talented electrical engineer who had been Research Director of Telefunken in Germany before he had to leave because of Hitler; he designed the wave-guide fed 'cheese' aerials for the new generation of 10 cm wavelength radar sets. R.J. Pumphrey, a capable biologist with a good grasp of general scientific ideas was quite high up in the administration.

Fred served as a benevolent Head of the group, allowing everyone to follow his own path, and providing administrative help to the best of his ability. I should now like to describe three major contributions to Naval radar for which he was personally responsible. They well illustrate his thorough grasp of scientific principles, and his imaginative approach to new problems.

3. Height Finding

Shortly after joining ASE Fred was asked if he could help with a major problem which was troubling the Navy. Radar sets operating on a long wavelength of 7 metres were picking up aircraft at a satisfactory range. But there was no indication of their height. Could he suggest a method of height estimation?

Fred assumed that at the wavelengths and distances under consideration the curvature of the earth can be ignored; and with horizontal polarization over the sea reflection is almost perfect. It is therefore a simple matter to draw up a coverage diagram resulting from the interference of the incident and reflected waves. If h_o is the height of the transmitting aerial, and h the height of the aircraft at a distance d from the transmitter, the field strength at the aircraft is given by

$$\frac{E_o}{d} \sin \frac{4\pi h_o h}{\lambda d}, \tag{1}$$

where E_o represents the power of the transmitter. Hence the field strength of the reflected signal back at the transmitter is

$$\frac{E_o A}{d^2} \sin^2 \frac{4\pi h_o h}{\lambda d}, \tag{2}$$

where A is the equivalent reflecting area of the aircraft. The coverage diagram of the radar will be given by equating (2) to the minimum signal e_o which can be picked up by the receiver.

The form of the (d,h) coverage diagram is shown in Figures 1, 2 for a particular choice of h_o, and it is determined by three parameters E_o, A, e_o. Once these parameters have been fixed the height of the aircraft can be read off from the range of first detection. In his autobiography (*Home is Where the Wind Blows*, University Science Books 1994, pp. 177–179) Fred tells how he tried to take account of the various sources of error in determining the parameters that he assumed would combine randomly. But the curve that he drew as a result gave altitude estimates that were completely wrong.

After much cogitation he came to the conclusion that the errors were not random but systematic. But instead of trying to improve the accuracy of the estimates he noted that all the uncertainties could be lumped together into one unknown parameter. He calculated curves for 10 different values of this parameter. A ship's radar officer could use a plane at a known altitude as a calibration to decide which of the curves applied to his particular radar set. This curve could then be used afterwards for unknown aircraft. The point at which the signal disappeared at the other side of the lobe could be used as a check.

The method proved extremely successful and was used right up to the end of the war.

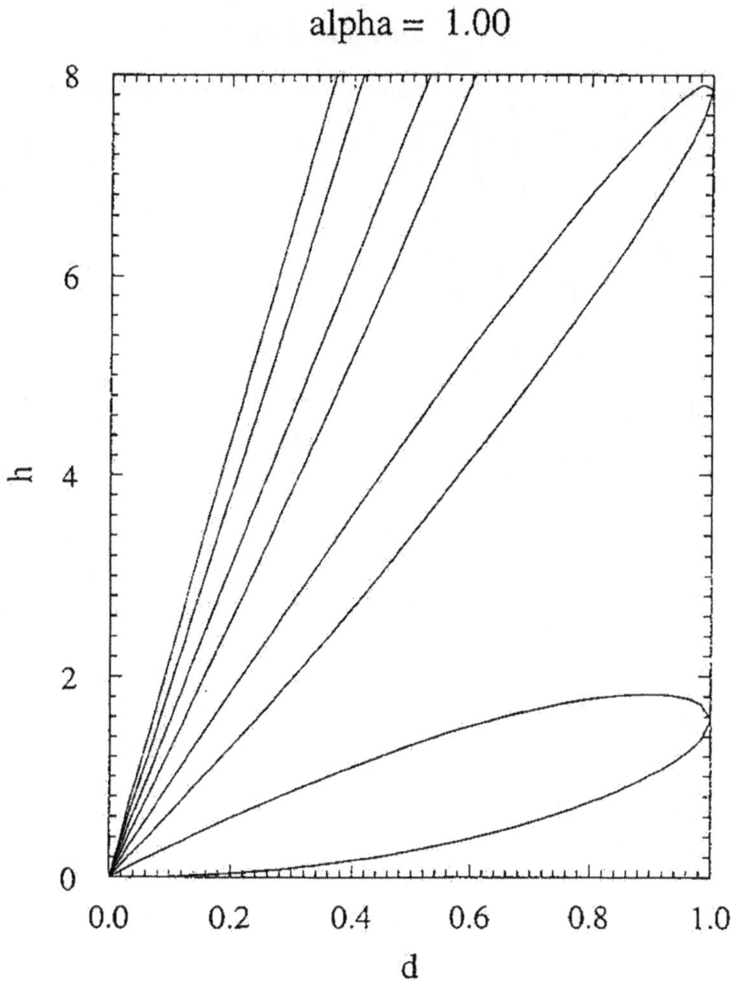

Figure 1. Interference pattern from radar aerial above the sea.

4. Anomalous Propagation

With the introduction of radar sets operating on a wavelength of 10cm a new phenomenon was rapidly discovered, termed anomalous propagation. There were days on which the range of detection of surface targets was extended to two or three times its normal value. This aspect was very beneficial, but it also meant that the radiation could be detected at ranges far beyond normal, so that extra care was needed.

A suggested explanation was soon forthcoming that certain weather conditions involving a temperature inversion would give rise to a rapid increase of density leading to a wave guide effect in the lower layers of the atmosphere. The

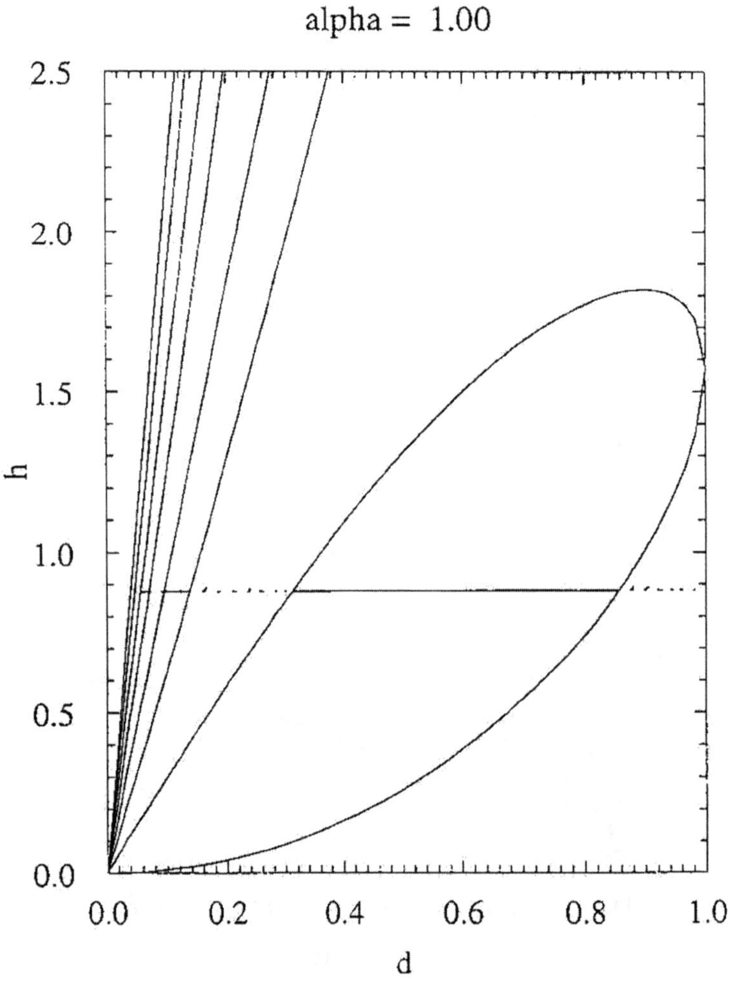

Figure 2. Use of coverage diagram for height finding.

phenomenon was considered a matter of major importance, and a special Radio-Meteorological Panel was established under the Chairmanship of Sir Edward Appleton on which representatives of the various government bodies concerned with the phenomenon could meet for discussions. Fred was a member of the panel.

A stream of theoretical papers flowed from TRE (the Air Force Radar Establishment) with propagation calculations for a variety of density profiles. But they did not impinge on the key problem of identifying in practice the conditions under which anomalous propagation would occur.

One morning a paper arrived form the meteorologists presenting a method of forecasting these conditions. It claimed successes in eighty percent of the cases investigated. Pumphrey had already noted that once the conditions for anomalous

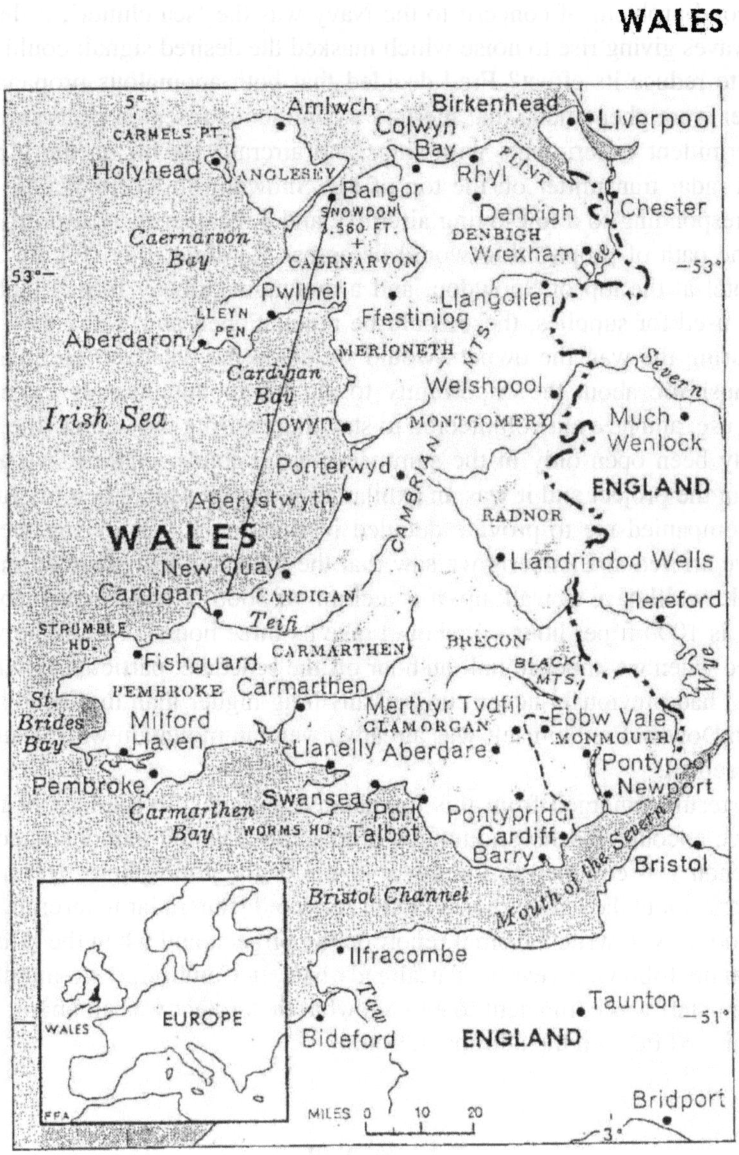

Figure 3. Propagation path from Snowdon to Aberporth.

propagation set in, they persisted for quite a period of time. He therefore hit on an alternative and obvious method of prediction that 'tomorrow is the same as today'. He found success in eighty five percent of the cases! Fred then pointed out that the real criterion for practical use was predicting the onset and termination of anomalous propagation; if this criterion were used they were correct in only fifteen percent of the cases.

A second problem of concern to the Navy was the 'sea-clutter', reflections of the sea-waves giving rise to noise which masked the desired signal; could anything be done to reduce its effect? Fred decided that both anomalous propagation and sea-clutter were phenomena that merited a continuous research programme rather than intermittent experiments with ships and aircraft. He hit on the idea of stationing a radar transmitter on the top of Mt. Snowdon in North Wales, (3,500 ft high corresponding to a low-flying aircraft), and a receiver in Aberporth in South Wales; the path of propagation would be across Cardigan Bay (Figure 3). There was a hotel at the top of Snowdon, and a mountain railway from Llanberis that could be used for supplies, these could be requisitioned and, since they were not in use during the war, the owners would welcome the takeover. Hermann Bondi was enthusiastic about the opportunity to put his mountaineering experience to practical use, and he even volunteered to stay there during the winter (the hotel had previously been open only in the summer). In the course of time, we all played our part in the project and it was an exhilarating experience. When my turn came, Fred accompanied me to provide detailed instruction. It was during the summer and as we arrived at Llanberis we saw that the train had just steamed out. 'O.K.' said Fred, 'we'll have to walk up, it is a climb of 3000 ft. and a reasonable rate of climbing is 1000 ft per hour – it should take us three hours'. I think he was quite impressed when we knocked half an hour off the schedule, particularly when I told him that I had previously never climbed anything higher than the Devil's Dyke in the South Downs. Fred himself was already a veteran mountain walker and always kept himself fit.

What results emerged from this imaginative project? To the best of my recollection we encountered *no* anomalous propagation, which demonstrated that the phenomenon was confined to the lower layers of the atmosphere. Hermann drew conclusions about the nature of sea-clutter reflected from radar in aircraft. But Fred was far too busy to write a formal report on the project, and when the bills came to be paid in the following year he was already back in Cambridge: no member of the permanent staff was competent to explain what the project was all about, and there were some red faces in the administration.

5. Window

Window was the code name for strips of aluminum foil released from an aircraft to mask genuine signals and confuse enemy radar. Anything connected with Window was 'Top Secret'. In fact both sides had thought of the idea, but neither had used it for fear that they might lose more than they gained when their opponent adopted it. Eventually the British decided to initiate its use in a number of crucial attacks. Anticipating the enemy response, Fred was asked by the Chief Scientist at ASE if there was any method of discriminating between the echoes from an aircraft and Window.

Fred decided that to make progress with this question it was vital to investigate the detailed character of echoes from aircraft at a wavelength of 10 cms. For this purpose a continuous supply of flying aircraft would be needed, and he hit on the idea of using the Coastal Command station at St. Erth on the North Cornwall coast, where Liberator aircraft flew round the clock. The radar set could be situated near Bedruthen Steps, a part of the coast he knew well. He teamed up with Tommy Gold, and in his autobiography (pp. 199–203) Fred describes the nasty headaches that they encountered in using the circuitry provided by the Admiralty. Tommy devised an ingenious method of separating echoes from individual pulses; experiments continued for some time in which other members of the group played their part. The surprising result was that a significant change occurred in the 1/500 of a second between successive pulses. The solution to the problem then became apparent since during this time there could be no significant change be in the echoes from the slow moving aluminum strips. Hence a subtraction of adjacent blocks of pulses would eliminate the Window but preserve the aircraft signal. The idea was simple but no technology was then available for preserving information for a few milliseconds. Tommy's ingenuity again saved the situation – he converted the electrical information into sound, and used mercury delay lines.

The idea worked perfectly. It was not ready for use in World War 2, but was used for many years thereafter.

Finally, may I tell of a request with a characteristic Fred solution. In one phase of the war against Japan the need arose for a safe code which could be deciphered quickly; Fred's reply 'Use two Welshman – I doubt if there are even any Japanese academics who know Welsh'.

6. Leaving ASE

The European war ended in May 1945. It seemed likely that the British would be less involved in the Japanese war, and there seemed to be a good chance that manpower restrictions would be eased. In fact scientists with teaching appointments were released immediately; Hermann Bondi managed to obtain such an appointment, and left immediately for Cambridge. But Fred had only a Research Fellowship with no teaching commitments and he was obliged to stay on. He let it be known that he had a new idea for explaining the abundance of the elements in the universe which he regarded as far more important than anything he had done before. Details of the tortuous path by which he was led to it are given in his autobiography (pp. 218–233).

As the weeks rolled on he became increasingly frustrated, until one morning he announced that he was leaving ASE to go to Cambridge to work on his idea. When he told this to the administration they were in a quandary since there was no mechanism by which they could release him officially. They therefore wrote 'Away on Duty' in the Attendance Book; but this meant that the Admiralty continued to pay Fred's salary. Fred was scrupulously honest, and would not dream of taking government money whilst working on his own problems. He decided to go over the heads of all the civilian hierarchy, and arranged an appointment with the Captain Superintendent at Haslemere, who was the Naval Officer in charge of ASE.

Fred explained that he had no objection to participating in the war against Japan, but he knew form previous experience that it would take 18 months for any contribution he made to be put to practical use and he was sure that the war would be over by then. In the meantime he had an idea of supreme scientific importance which he urgently needed to develop, and the literature he needed to consult was available at Cambridge. Surprisingly the Capt. Superintendent was sympathetic and said 'Perhaps you can spend some of your time at Cambridge thinking about our problems and then we can legitimately consider you Away on Duty at Cambridge'. Fred went back, but was not too happy with the arrangement. Within a short period the Admiralty came up with a formula by which he could be released.

A year or so later he published his first paper on nucleosynthesis which, after a very long haul, led to the seminal 1957 paper with the Burbidges and W.A. Fowler.

7. My Own Future

I got back to Cambridge some months later, and by then I had decided that my area of research interest was statistical mechanics. Fortunately Fred had also been involved with statistical mechanical calculations of nuclear reactions in stars, and when Sir Ralph Fowler died, his statistical mechanics course was assigned to Fred. It was natural, therefore, that I should choose him as my Ph.D. supervisor. He was happy to let me work on my own, and helped me greatly at the crucial early stage of my career to get my feet on the academic ladder.

Although I did no research in astrophysics, I followed his remarkable and unconventional research and academic career with enormous interest. We met from time to time and remained close friends for the whole of his life. It has been a privilege to pay my own humble tribute to an outstanding scientist and a great human being.

Acknowledgement

I am grateful to Prof. Dennis Rapaport for his generous help in preparing the diagrams.

FRED HOYLE AND THE ANGLO-AUSTRALIAN TELESCOPE

MARGARET BURBIDGE
Center for Astrophysics and Space Sciences University of California, San Diego

Abstract. The 150-inch telescope on Siding Spring mountain in New South Wales, Australia, stands as a reminder of Fred Hoyle's genius in aspects of astronomy that surprise those who know him only as a theoretician, as one who did not himself use the telescopes that produced the all-important data for his theoretical work. This article shows another side of Fred's abilities – that he could immerse himself in the techniques of telescope building, from mount to mirror, and overcome the many hurdles that had to be faced.

A 60-page account of the early history of the Anglo-Australian Telescope (AAT) by Sir Bernard Lovell is published in the Quarterly Journal of the Royal Astronomical Society (1985, Vol. 26, pp. 393–455). The idea for a large optical telescope in Australia came originally from Sir Richard Woolley when he was Director of the Mt. Stromlo Observatory in Australia. The suggestion for a large telescope in the Southern Hemisphere did not at that time specify Australia as its location. The UK had the Radcliffe Observatory in South Africa, and in the period 1960–62 there were discussions between the USA and the UK about the possibility of the UK joining the USA in plans for a large telescope in Chile. Woolley, as Astronomer Royal in the UK, opposed British membership in the European Southern Observatory because he knew this would kill any possibility for a UK – Australia partnership, and he felt that UK astronomers would get very little observing time in the ESO plan. As the account by Lovell (1985) tells us, 'the conflict between the ESO and Commonwealth projects remained a critical feature of the discussions until 1962'. This period, from the early 1960s until 1967, when a formal agreement between Australia and the UK was reached, might well be described as 'Pride and Prejudice in Astronomy'.

For in the 1960s radio astronomy was flourishing in both hemispheres, through Jodrell Bank and Cambridge in the UK and at the Mills Cross and Parkes with the Australian teams. The development of radar during World War II had provided experts in this field, while optical astronomy was languishing in England, its only major effort being under the so frequently present clouds over Cambridge. The disastrous placing of the Isaac Newton Telescope at Herstmonceux, some 300 meters above sea level, in an area known as Pevensey Marshes, where hopeful astronomers assigned time on the 100-inch would walk in the evening through dewy grass and a rising mist, most probably to dodge clouds in a sky with a transparency sadly diminished from that at telescopes elsewhere in the world, was depressing. A Southern Hemisphere Observatory, in as good a climate as possible, was a very attractive prospect.

Astrophysics and Space Science **285:** 303–307, 2003.

Fred Hoyle on the Anglo-Australian Telescope Board

Two books on the design, construction, and management of the Anglo-Australian Telescope have been published. 'The Creation of the Anglo-Australian Observatory', by S.C. Ben Gascoigne, Katrina M. Proust, and Malcolm O. Robins (Cambridge University Press, 1990), gives the definitive history of the design, construction, siting, and management history of the AAT. A personal account was also written by Fred Hoyle: 'The Anglo-Australian Telescope', published by the University College Cardiff Press (1982). In this article I describe only a few highlights from the full history of what became an all-important organization for astronomical research in the Southern Hemisphere.

The organization set up to plan the design and choose the contractors for the construction, was the Anglo-Australian Telescope Board. At the time I first became acquainted with it, it consisted of three Australian and three UK members: Sir Richard Woolley, as Astronomer Royal and Director of the Royal Greenwich Observatory at Herstmoneux, Sir Fred Hoyle, Plumian Professor of Astronomy and Experimental Philosophy at Cambridge University, and Jim F. Hosie, Director of the Astronomy, Space, and Radio Division of the Science Research Council, as the UK government representative, were the UK members; the radio astronomer E.G. (Taffy) Bowen, the optical astronomer Olin Eggen, then Director of Mount Stromlo and Siding Spring Observatories, and a faculty member at the Australian National University, and the Australian governmental representative Ken N. Jones were the Australian members. My role in this came after Sir Richard Woolley retired in 1971, when I was appointed to succeed him on the Anglo-Australian Telescope Board.

It had been decided that the telescope should be a 150-inch aperture and equatorially mounted telescope, using the basic design of the 4-m telescope at Kitt Peak National Observatory.

Now Fred Hoyle, through his visits to the US and his many friends and contacts there, knew of a design problem that had been experienced with the Kitt Peak 4-m declination bearing, and his advice was crucial in helping the AAT Board to get their contractors to face and deal with this possible design problem. The book by Gascoigne et al. describes this in detail. Fred Hoyle, the theoretical astrophysicist? No, Fred Hoyle, the man of many talents. The design problem was faced and dealt with.

When the design and construction were well underway, the question of management had to be faced. Olin Eggen, as the Mt. Stromlo astronomer at the Australian National University and former Chief Assistant at the Royal Greenwich Observatory, saw himself as the future director of the AAT. This was the situation when Sir Richard Woolley's time as Director of the RGO came to an end and I was appointed in his place. I joined Hoyle and Hosie in the belief that the AAT should be independent of Mt. Stromlo, with a Director sought world-wide and appointed on his merits as an astronomer familiar with telescopes and instrumentation.

As can be imagined, meetings of the AAT Board during this time grew increasingly stressful, while the design and construction of the telescope, dome, living quarters, and laboratories were going well. The book by Gascoigne et al. should be consulted for more details of the personal difficulties. But I can give just a brief flavour of the personal conflicts by a description of one evening in Canberra.

Fred, Jim Hosie, and I would stay together in a hotel in Canberra for the meetings. I recall Fred coming into the dining room for dinner one evening and saying: 'I think I may have committed an indiscretion'. Under our questioning, he admitted that he had been writing a science fiction book, with personality conflicts among the astronomers involved. He described some – and told of his account in his manuscript of an astronomical catastrophe that occurred: the center of our Galaxy became a Seyfert nucleus! The Southern Hemisphere on Earth was consumed by fire and flame, while Scotland, a favorite habitat of Fred's, was far enough north so that the conflagration destroying the Southern Hemisphere did not reach it. Jim and I persuaded Fred to tone down his descriptions of some of his actors in this melodrama.

The La Jolla Meeting

Most meetings of the AAT Board and the people involved in the construction of the telescope and the Observatory dome, the workshop and laboratory buildings, the dormitory, etc., had been held in Canberra. When it became necessary to tackle head-on the question of management of the Observatory, a meeting early in 1972 was arranged at the University of California's Institute for Geophysics and Planetary Physics, located on the cliffs just north of the UC Scripps Institute of Oceanography. This was a crucial meeting for Fred.

On the flight from England to the U.S., Fred had drafted his statement of resignation from Cambridge University and the Institute of Theoretical Astronomy which he had created and headed since 1967. Candidates to take over from Fred when his five years as Director were over had been under discussion in Cambridge, and Fred had a good candidate in mind, an ex-patriate Brit who was not only an expert observational astronomer, but also a good theoretician, and a good candidate for a Cambridge professorship.

But there were those in Cambridge University who were much more adept at the unpleasant side of University politics and manipulations than was Fred, and an important meeting at which Fred would have spoken and made the case for his candidate, was put off and not rescheduled until Fred was away. Another candidate was then selected. This was the bad news which Jim Hosie had imparted on the journey to California.

However, the important business of the La Jolla meeting was the question of management of the Anglo-Australian Observatory. Olin Eggen had just made a bad mistake, written up in the newspapers, by implying that he saw himself as

Figure 1. Fred Hoyle and Prince Charles, Prince of Wales, at the dedication of the Anglo-Australian Telescope on 16 October 1974, from Chapter 9 of 'The Creation of the Anglo-Australian Observatory', by S.C. Ben Gascoigne, Katrina M. Proust, and Malcolm O. Robins (Cambridge University Press, 1990) (*Photograph by M. Jensen*).

the obvious head of the AAO. This is all documented in Gascoigne et al. (1990). The serious question of the need for a Director who was independent of Canberra and the UK, who had a good reputation in observational astronomy as well as in instrument design and building, was tackled head on. A crucial vote was taken on the way the search for Director should be conducted. There were three UK and three Australians on the Board: would the vote be split 3 – 3? The Board Chairman, Taffy Bowen, who had always recognized what this excellent telescope would do for astronomy, did not vote with Olin Eggen. The vote was 4 – 2 for a real world-wide search, which resulted in the selection of Dr Joseph Wampler as first Director.

Dedication of the Anglo-Australian Telescope

The opening ceremony was held on 16 October 1974. Chapter 9 of Gascoigne et al. (1990) describes the ceremony of dedication of the Anglo-Australian Telescope at which Prince Charles, the Prince of Wales, performed the Dedication. The illustration shows Fred Hoyle, with the Prince of Wales, on the Cassegrain platform of the telescope. It was an occasion when those who knew how crucial Fred's role had been in the achievement of the whole Anglo-Australian Observatory, from design, construction, and management throughout, could utter a silent prayer of thanks. Photographs of this ceremony are published in Chapter 9 of the book by Gascoigne et al., with a quote from Fred Hoyle's own description of his role in carrying out his task of welcoming Prince Charles at the entrance of the telescope dome in a storm of high wind. Apparently the weather at Siding Spring produced a howling wind, blowing the participants as they arrived outside the dome, as described by Fred in his book, but obviously the ceremony itself was a moving and auspicious occasion. I close by quoting what Fred Hoyle said in his speech: 'A telescope is a good example of the things that our civilization does well'.

FRED HOYLE AND MANCHESTER

RODNEY D. DAVIES
University of Manchester, Jodrell Bank Observatory, Macclesfield, Cheshire SK11 9DL, UK

1. Introduction

Fred Hoyle has had a significant influence on my scientific journey at several pivotal points. It is a pleasure to acknowledge this and to describe his relationship over the years with a number of us at Jodrell Bank Observatory.

As undergraduates in the Physics Department of the University of Adelaide in South Australia, we were encouraged to read widely around our subject and I came upon Edwin Hubble's 'Realm of the Nebulae' with its exciting portrayal of cosmology. The University Library included the Monthly Notices of the Royal Astronomical Society; in 1949 I came across volume 108 containing the papers by Fred Hoyle and by Hermann Bondi and Tommy Gold proposing a fascinating cosmology involving the continuous creation of matter. This captured my imagination partly because it resolved the problem of how the apparently 1.5×10^9 year old expanding Universe of more conventional cosmology could contain a 3×10^9 (soon to rise to 4.5×10^9) year old solar system; these were the days of a Hubble constant of 500-600 km/sec/Mpc. The nuances of the perfect (or 'wide') cosmological principle and physical interpretation of the cosmological constant somewhat escaped me. My excitement in discovering a new way of thinking about the demonstrably expanding Universe encouraged me to write an article in the Adelaide University Science Journal describing the Steady State theory and the continuous creation of matter. This was in a sense my first scientific publication, although at the time I had no idea that my own career would lead into astronomy, let alone cosmology.

Our professor during my last undergraduate year in Adelaide, in the line of professors leading down from our foundation professor W.H. Bragg was the newly-appointed Leonard Huxley who was one of the network of men developing radar in the UK during World War II. At the time of graduation he encouraged me to apply to the Radiophysics Division in Sydney of the Commonwealth Scientific and Industrial Research Organization; RP as it was fondly called, was the war-time research group established to develop radar and radio communication for the Pacific theatre of operations. Its leader at the time was E.G. (Taffy) Bowen who made pivotal contributions to Watson Watts' pioneering radar team in pre-war UK and who was diversifying RP's activities into radio astronomy, rainmaking and navigational aids. During my first two weeks at RP in January 1951 I had two memorable and career-forming experiences. The first, which had more immediate

Astrophysics and Space Science **285:** 309–319, 2003.

consequences, was being encouraged to study the film, I think made at Mt Palomar and Mt Wilson Observatories, showing sequences in H-alpha light of activity on the surface of the Sun and the arches of the flare-related loops anchored to sunspots but projecting half a solar radius above the limb. This was incredible; it led me to work for the next two years with W.N. Christiansen mapping the closely related radio activity above the sun's surface.

The second land-mark experience was hearing Fred Hoyle's talks on 'The Nature of the Universe' re-broadcast by the Australian Broadcasting Company from the original BBC programme. Listening to Fred's enthusiastic discourse in an accent not all-that-familiar to my Australian ears, one was transported to a world that eclipsed the beauty of the Sydney Heads, Sydney Harbor bridge in the balmy night air. 1951 was before the cosmological contributions of radio astronomy were foreseen. It was some 10 years before the Steady State theory would impinge on my own research odyssey. Equally it was some 5–10 years before the lively controversies that resulted from the first measurements of the log N-log S relation for radio sources in Cambridge and Sydney and which were seen as definitive tests of the Steady State theory.

In 1953 I was appointed to an assistant lecturer post in the University of Manchester working in Bernard Lovell's research group at what was then known as the Jodrell Bank Experimental Station. Bernard had made major contributions to the development of radar; his team produced the successful H_2S centimetre radar used by the Royal Air Force in WW2. In the context of this article it is interesting that Fred Hoyle's group, which included Bondi and Gold, was working for the Royal Navy on the logistical and mathematical aspects of naval radar.

2. Hydrogen in the Universe

2.1. Introduction

The 21-cm hyperfine splitting line from neutral hydrogen was predicted by H.C. van der Hulst while still a graduate student in wartime Holland and published in 1945. Radio emission from this fundamental spectral line was detected almost simultaneously by radio astronomers in Sydney (RP), Harvard and Leiden in 1951. When I arrived at Jodrell Bank I joined the small group who were building a 21-cm receiver and were furnishing a drive system for the 30-ft telescope which had been used in WW2 as a monitoring dish on Beachy Head on England's south coast. As soon as we commissioned our hydrogen-line system, D.R.W. Williams and I began absorption-line measurements of the strongest extragalactic source, Cygnus A, and the strongest Galactic source, Cassiopeia A. It was evident that Cygnus A was extragalactic since it showed absorption at the velocity of each of the Galactic spiral arms in the line of sight. Cassiopeia A on the other hand showed absorption in the local Orion arm and in the Perseus arm but not in the Outer arm thus placing it in the Perseus arm some 3 kpc distant and certainly closer than the 5.5 kpc outer arm. This result rather confounded the optical pundits at the time who had determined a distance of 500 pc (Baade and Minkowski, 1954). I continued to use the 30-ft telescope in studies of neutral hydrogen (HI) emission from the Local System (Gould Belt) of our Galaxy. Fred Hoyle's Steady State predictions and the successful measurement of HI absorption against continuum radio sources kept going around in my mind which was also being assailed by images of the growing 250-ft telescope visible from my laboratory window. The last chapter of my PhD thesis (Davies, 1956) described a proposal to 'test' the Steady State theory by observations with the 250-ft telescope as soon as it was completed.

2.2. Testing Continuous Creation

The Hoyle (1948), Bondi and Gold (1948) Steady State theory required the continuous creation of matter to preserve a constant density in an expanding Universe. The created matter would be hydrogen and it was likely to be dispersed throughout intergalactic space. Hoyle estimated a smoothed-out density

$$\rho = \frac{1}{6\pi G} t^2$$

where t is the age of the Universe = H_o^{-1}. For the Hubble constant H_o = 180 km s^{-1} Mpc^{-1} offered by Mayall, Sandage and Humason (1955) at the time these H-line experiments were being planned, $\rho = 6 \times 10^{-29}$ g cm^{-3}. Bondi (1952) in fact suggested that the value might be 4–6 times greater. The argument that convinced me of the existence of an intergalactic medium was the optical observation by Stebbins and Whitford (1948) of an excess reddening of distant galaxies above and beyond the reddening due to recession. There was no other compelling observational evidence at that time for the presence of a pervading intergalactic gas. Bondi suggested that its temperature might be anywhere in the range from about zero to 6000K; it would be unionized.

A direct observation of intergalactic HI can in principle be made by measuring the absorption spectrum in a distant extragalactic source and combining this with an emission spectrum in a nearby direction to obtain both the density of hydrogen atoms and their spin temperature. Cygnus A at a redshift v = 16,800 km s^{-1} and the weaker Virgo A at v = 1200 km s^{-1} are suitable for this experiment. The absorption trough runs from v=0 to the redshift of the source. The emission step red-wards of v=0 sets an upper limit of $\rho < 2.6 \times 10^{-5}$ cm^{-3} corresponding to $\rho = 4.2 \times 10^{-29}$ g cm^{-3} for H_o = 75 km s^{-1} Mpc^{-1}.

Using absorption measurements this limit can be reduced in cosmologies which suggest that the intergalactic gas is predominantly neutral where the results can potentially indicate a value of 10^2 times less than predicted by Einstein-de Sitter theory.

The absorption data can be used to even lower limits to ρ in cosmologies where the hydrogen remains neutral and its spin temperature, T_s, can be estimated. George

Field (1959) showed how T_s could be calculated if the radiation field and the de-excitation conditions were known, so that given n_H/T_s from the Cygnus A observations, a value of ρ could be derived. The early Steady State papers suggested low kinetic temperatures in the intergalactic radiation field which lead to the following values (Davies and Jennison, 1964; Davies, 1964)

$$n_H < 9 \times 10^{-8} \text{cm}^{-3}$$

$$\rho < 1.4 \times 10^{-31} \text{g.cm}^{-3}$$

This value is two orders of magnitude below the Einstein-de Sitter value

$$\rho = 3H_o^2/8\pi G$$

$$= 1.1 \times 10^{-29} \text{ g.cm}^{-3} \text{ for } H_o = 75 \text{ km s}^{-1} \text{ Mpc}^{-1}$$

The Hoyle Steady State value is twice this. Two comments should be made about the assumptions which underlie this apparent refutation of the Einstein-de Sitter and the Hoyle predictions. Firstly the state of ionization was not clear at the time this work was done; the modern view is that the intergalactic medium is highly ionized and consequently ρ only applies to the neutral component. Secondly the spin temperature is probably $\sim 10^3$ to 10^4 K increasing the n_H upper limit by a factor of $\sim 10^2$ or more.

This experiment was my introduction to the complexities of cosmology.

2.3. Hydrogen around galaxies

One of the major programmes with the newly-completed 250-ft telescope was the investigation of HI in normal galaxies. This included the determination of the rotation curves and gas distribution in nearby spiral galaxies such as M31 and M33. Evidence began to appear in such observations for tidal effects in many of these galaxies; a tidal explanation of these effects was dramatically confirmed in the computer modelling by the Toomre twins. The question which then presented itself was whether such effects may occur on a larger scale in galaxy groups and clusters.

The first investigation of these matters made with the 250-ft telescope was an HI survey of the galaxies in the Virgo cluster. B.M. Lewis and I found that the Virgo cluster galaxies contained less HI than galaxies of the same type in the field. This was the first direct evidence for gas-stripping in cluster galaxies (Davies and Lewis, 1973). More recent investigations show that this phenomenon is widespread – atomic gas is stripped from the outer regions of galaxies although the more centrally concentrated molecular gas is unaffected. Where does this gas go? Presumably some fuels the ellipticals in the cluster, while some may become intra-cluster gas.

The most dramatic instance of tidal disruption of field galaxy groups was the M81/M82 system (Appleton, Davies, Stephenson, 1981). The 4 brightest galaxies in this group are embedded in a large dynamically complex cloud of HI; bridges of emission connect the 3 companion galaxies to M81. In addition the South-West cloud lying $\sim 1°$ from M81 has no optical counterpart. Tidal bridges and tails are relatively common in galaxy groups; they contribute to the HI reservoir in the intergalactic medium.

Some of this intergalactic gas may be captured by passing larger-mass galaxies. SO galaxies are sometimes culprits. One of the best examples is the SO galaxy NGC 1023 (Sancisi et al., 1984) which has extended HI features orbiting at 15 to 45 kpc from the galaxy centre. The kinematics suggest an intergalactic origin and a recent arrival of the HI features in the vicinity of NGC 1023.

An interesting question then arises. Where is the bulk of the present-day HI in galaxy clusters? A fraction clearly occurs in the spiral galaxies, some is in the HI-rich dwarfs, some will be in ultra-cluster space having been tidally or ram-pressure stripped from spiral galaxies. Is there any primordial gas present – the gas envisaged by the Steady State or evolutionary cosmological scenarios?

2.4. Primordial HI

One way of searching for the (remnants of the) primordial gas from which the cluster galaxies have formed at early epochs is to look in directions where high redshift intergalactic gas has already been found. The absorption features seen in the spectra of distant quasars provide good indicators of where primordial intergalactic gas may be situated. The bright radio-quiet quasar PHL 957 at a redshift of z = 2.69 has a strong broad Lyman-alpha absorption at z = 2.3099 with an HI column density of 1-2 $\times$ 10^{21} atoms cm^{-2}. In 1976 when we were making the radio H-line search for the emission associated with this feature, the protocluster model of Sunyaev and Zeldovich was much in vogue; this model predicted column densities of this order for a HI mass of 10^{14} $M_\odot$. Our search of this line of sight with the 250-ft telescope at the red-shifted frequency of 429 MHz showed that the associated HI mass was $<3 \times 10^{13}$ $M_\odot$ (Davies, Booth and Pedlar, 1977).

Our next step in the search for primordial HI was to use the 250-ft radio telescope to make a survey of a significant area of sky at z = 3–5 looking for the HI signature of cluster-sized masses. These were the pancakes predicted by Sunyaev and Zeldovich (1975) and also required in most scenarios of galaxy formation. This blind survey was made at two frequencies 328 and 240 MHz corresponding to redshifted HI at z = 3.33 and 4.9 (Davies, Pedlar and Mirabel, 1978). The observations showed no clouds in 40 fields at each frequency with hydrogen masses $>3 \times 10^{15}$ $M_\odot$. Limits in many fields were $<3 \times 10^{14}$ $M_\odot$, the expected Sunyaev-Zeldovich HI mass. 10 percent of our fields should have contained clusters with this mass. Although no positive detections were made, this appeared to be a fruitful approach.

Modern cosmology is driven by high redshift data at many wavelengths. The radio search for high redshift HI is a high priority aim. The Giant Metrewave

Radio Telescope recently completed near Pune, India, will make sensitive searches using its large collecting area and high resolution. The highest priority international project for radio astronomy is the Square Kilometre Array (SKA) being considered for completion in 10 years time; this should be capable of detection 10^{12} $M_\odot$ of HI at z = 5 for example.

3. Sir Fred Hoyle – Visiting Professor, Manchester University

I will now break the astronomical narrative and describe a new phase in Fred Hoyle's connection with Manchester when he was appointed a Visiting Professor. This part of my story has its origins in the influential committees of the UK Science Research Council (SRC) in the late 1950s and the 1960s whose membership included Fred Hoyle and Bernard Lovell. Both had experience in establishing and running major research institutions, Bernard Lovell the Jodrell Bank observatory in the University of Manchester and Fred Hoyle the Institute of Astronomy in the University of Cambridge. They had both had vigorous encounters with the establishment and brought to the SRC a strong commitment to get things done. There was an empathy between them in which their common love of cricket played a part – Fred as a loyal Yorkshireman and Bernard from Gloucestershire for which he always had a sympathy but latterly he espoused Lancashire where he was for many years president of the Lancashire County Cricket Club. Bernard liked to tell the story of how when Fred was describing the spread of life in the Universe he made the comment when lamenting a failed Test series 'there must be a star somewhere with a planet having a cricket team good enough to beat the Australians'.

In the late 1960s Bernard Lovell was chairman of the Astronomy Space and Radio Board of SRC and Fred Hoyle was chairman of the Astronomy Committee. During this period they were both heavily involved in the appointment of a successor to Richard Woolley, the Astronomer Royal, and in the establishment of the Anglo Australian Observatory (Lovell 1985a,b). Margaret Burbidge covers this aspect of Fred Hoyle's life in an accompanying article (Burbidge, 2003).

During one of the Board meetings Fred discovered that Bernard Lovell had been given the positions of the four pulsars newly discovered by the Cambridge radio astronomy group. He was excited to learn that this discovery was to be followed up immediately with the 250-ft telescope as there was much astrophysics that could be learned. This ultimately became an exceptionally fruitful field of study for the world's large telescopes. An early request from Fred was that a search should be made for pulsars in the Andromeda Nebula. None have yet been found. Pulsars at this distance are 10^4 times weaker than the faintest pulsars in the Galaxy, so their detection must await the construction of SKA.

In 1972, following disputes about funding and appointments, Fred Hoyle retired as Director of the Institute of Astronomy in Cambridge. Fred and Barbara, his wife, then made their home in Cockley Moor, overlooking Ullswater in the Lake

District. There he could spend his time fell-walking and working on his favourite projects. Bernard Lovell tells of his amazement that Fred could work in isolation in the Lakes and asked him 'How can you work without a library?' and received the answer 'I do not need one'. After Fred and Barbara moved to Bournemouth, I regretted that I did not meet up with Fred on the fells above Ullswater; we have a house on the southern slopes of the Lakeland Fells and a convenient meeting place would have been on High Street south of Ullswater where the Romans drove one of their roads straight over the mountains from the coast to Penrith. Would Fred's opening words have been 'Veni, vidi, vici?' when taking the long view?

Bernard Lovell persuaded the Vice-Chancellor of the University of Manchester to appoint Fred Hoyle to a Visiting Professorship. He was delighted to accept and was allocated an office in the Physics Department. His lectures to postgraduate students and his seminars were much appreciated. Most of all, staff benefited from lively discussions with Fred about their own research projects which over the years included the logN-logS controversy, the nature of quasars, gravitational lenses and the cosmic microwave background (CMB). In later years these visits became less frequent as Fred divided his time between Cockley Moor, California and Cardiff.

This is a convenient point to say something about what I know of Fred's involvement with the Royal Astronomical Society (RAS). Jodrell Bank astronomers have provided Officers or Presidents of the RAS almost continuously from the mid1960s to the mid-1990s and accordingly have had an unparalleled insight into the operations of the Society during this period of unprecedented expansion. I remember clearly a meeting which, as Secretary, I organised following our move from the Burlington House rooms to the much larger Scientific Societies Lecture Theatre where the members were even sitting in the aisles when Fred gave a lively talk on evidence for life molecules in space.

Fred held the highest office of the RAS, the Presidency, from 1971–73, immediately following Bernard Lovell. In 1968 he gave the George Darwin Lecture entitled 'Highly Condensed Objects', and was awarded the Gold Medal of the Society in that year. The President at the time, D.H. Sadler, included the following sentence in his perceptive Gold Medal Address to Fred Hoyle,

> There can be few, if any, astronomers whose interests that have not been touched by his catholicity of mind, his fecundity of ideas, his lack of inhibitions and his facility of expression, ..., there is hardly a significant development in astronomy to which he has not contributed something.

4. The Cosmic Microwave Background

The discovery of the CMB, a uniform microwave radiation field in all directions, by Penzias and Wilson (1965) was a pivotal development in cosmology. Evolutionary cosmologists saw it as an opportunity to explore the earliest epochs of the Universe. The Steady State theory required an explanation for this. Fred spoke about these matters on his visits to Manchester. The sum of all the undetected radio sources was clearly not sufficient to explain the CMB. Its remarkable constancy and uniformity over the sky presented a challenge. Fred, along with Chandra Wickramasinghe, came up with a prescription in terms of submillimetre-sized iron particles generated from the debris of supernova explosions in galaxies which then thermalized by the galactic starlight radiation to a uniform 2.7K. This explanation was set within a scenario of a slowly oscillating and weakly expanding Universe with continuous creation – thereby maintaining a quasi Steady State view. He and his colleagues were thus able to justify a CMB temperature of ~3K and could explain the observed high degree of isotropy (Hoyle, Burbidge, Narlikar, 1995; Hoyle, 1997).

4.1. CMB RESEARCH AT JODRELL BANK – THE DIPOLE

After the discovery of the CMB in 1965, one of the first objectives was to search for anisotropies. The largest, the dipole, was 10^{-3} of the CMB temperature. This value implied a velocity of the local group of galaxies of 6000 km s^{-1} towards l = 270°, b = 30° relative to the more distant Universe. It was realized that such a high velocity should be detectable in the velocity field of the nearby galaxies and moreover the mass concentration responsible should be identifiable. We developed a radio-optical method for estimating the distance of a galaxy by combining the HI velocity width with an optical indicator such as the apparent diameter or luminosity (Hart and Davies, 1982). This work was extended to include 291 spiral galaxies with velocities in the range 500 to 5000 km s^{-1} having a good sky coverage (Staveley-Smith and Davies, 1989). We found a flow consistent with the CMB result. Only 150 km s^{-1} of this flow was infall into the nearby Virgo cluster; the major component was infall into an attractor in the Centaurus direction but beyond the Centaurus cluster.

4.2. CMB RESEARCH AT JODRELL BANK – CMB FLUCTUATIONS

The search for fluctuations has been a major preoccupation of CMB research in the 1980s and 1990s. Fred was always keen to talk about progress in the CMB on his visits to Jodrell Bank. His main interest was in the goodness of fit of the data to a blackbody spectrum. The angular spectrum of the fluctuations was a secondary concern at the time, although he considered that such fluctuations would most likely be a function of the matter distribution in the local Universe.

Big Bang cosmology in the early 1980s suggested that the amplitude of the CMB fluctuations would be $\Delta T/T \sim 10^{-4}$ to 10^{-3}. At Jodrell Bank we used the Mk II telescope at a frequency of 5 GHz to search for fluctuations in an angular range of 8' to 60'. No evidence for fluctuations was found down to a level of 3×10^{-4}, the lowest level established at that time (Lasenby and Davies 1983). Our next step in 1982 was to move our 10 GHz beam-switching radiometer to Izana Observatory, Tenerife, at an altitude of 2400m where the masking effects of water vapour in the

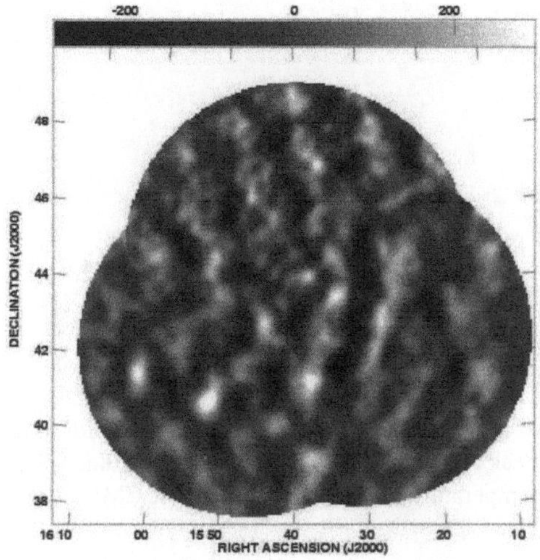

Figure 1. A mosaiced map from three pointings in a VSA field at RA = 15^h40^m, Dec = +42°. Individual CMB features can be seen. The strongest signals are 200 mJy per beam corresponding to 200 μK in a 20' beam.

atmosphere were substantially less than at Jodrell Bank. For nearly 10 years we successively pushed down the limit from 10^{-4} to 3×10^{-5} in various experiments at various frequencies before successfully detecting fluctuations at $\Delta T/T = 1.6 \times 10^{-5}$ on angular scales of ~5° (Hancock et al., 1994). Our experiments actually identified individual features in the CMB whereas the DMR on the COBE satellite made its detections via a statistical analysis of the CMB angular power spectrum.

Our next push forward in a strategy to improve sensitivity was to construct an interferometer at 30 GHz on the Tenerife site. This readily measured the fluctuation amplitudes on 1° and 2° scales and showed that the site was excellent for interferometric observations at frequencies around 30 GHz where Galactic foreground contamination was small and measurable. With this confidence-booster behind us, a collaboration was formed between Jodrell Bank Observatory, the Cavendish Laboratory, Cambridge and the Instituto de Astrofisica de Canarias to build a 14-element interferometric array at 30 GHz known as the Very Small Array. This system is capable of measuring the angular power spectrum of the CMB between ~2° and 10' corresponding to circular harmonics l = 50–1500.

The first results from the VSA are an encouraging addition to the many experiments world-wide covering a similar range of l but at many different frequencies with different techniques and where the Galactic foregrounds are due to different processes. Figure 1 shows a mosaic of 3 overlapping fields mapped with the VSA showing structure on a range of angular scales from 2° to 20' (Scott et al., 2002). The angular power spectrum of the region shown in Figure 1 combined with 2 similar-sized regions is plotted in Figure 2. The bright 'acoustic' peak at $l \sim 200$ is clearly evident as is the second peak at $l \sim 550$; later observations with the

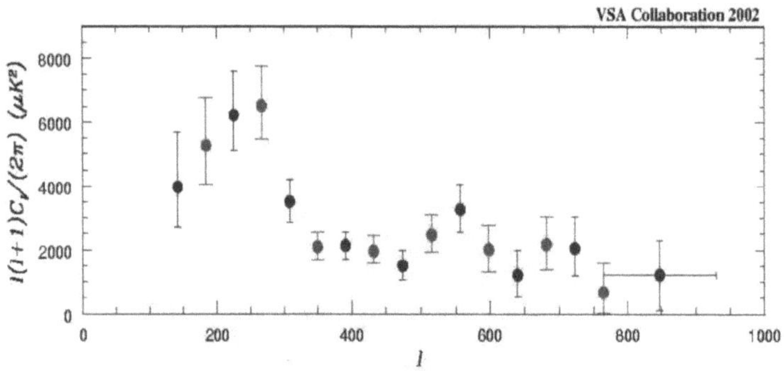

Figure 2. Combined power spectrum of three mosaiced VSA fields. The error bars represent 1σ limits. The first 'acoustic' peak at $l \sim 230°$ corresponds to an angular scale of 1°. The peak amplitude at $l \sim 230$ corresponds to temperature fluctuations in the 2.7K CMB of ~75 microKelvins. A second peak at $l \sim 550$ is evident.

VSA in its extended configuration show that the suggested third peak ~750 is in fact real. when data from all available experiments are combined, the first 3 peaks are clearly established. In 'conventional' Big Bang cosmology these 3 peaks are located near their expected values; their positions and amplitudes can be interpreted in terms of the cosmological parameters Ω_b, Ω_d, h, Λ etc. When combined with other cosmological data, firmer values of these parameters are obtained in the framework of the relevant cosmological theory. It is interesting to see that the Quasi Steady State theory finds an acceptable match with the current data set but where the fluctuations are due to the distribution of intergalactic matter (Narlikar et al., 2002). New measurements in the pipeline from the ground and from space (MAP and PLANCK) should further constrain the possible theoretical models.

5. Epilogue

The 24th General Assembly of the International Astronomical Union was held in Manchester on 7–18 August 2002. The IAU officers and the Local Organizing Committee had planned that Fred Hoyle should give an invited lecture during the proceedings. The University of Manchester was to confer an honorary doctorate at a special session. Fred's participation was widely anticipated. However at the last moment Fred's illness made this impossible although he was much looking forward to the event. This would have made a fitting tribute from the international astronomical committee.

References

Appleton, P.N., Davies, R.D. and Stephenson, R.J.: 1981, *MNRAS* **195**, 327.
Baade, W. and Minkowski, R.: 1954, *ApJ* **119**, 206.
Bondi, H.: 1952, *Cosmology*, Cambridge University Press.
Bondi, H. and Gold, T.: 1948, *MNRAS* **108**, 252.
Burbidge, E.M.: 2003 – this volume.
Davies, R.D.: 1956, PhD Thesis, University of Manchester.
Davies, R.D.: 1964, *MNRAS* **128**, 133.
Davies, R.D., Booth, R.S. and Pedlar, A.: 1977, *MNRAS* **181**, 1P.
Davies, R.D. and Jennison, R.C.: 1964, *MNRAS* **128**, 123.
Davies, R.D. and Lewis, B.M.: 1973, *MNRAS*.
Davies, R.D., Pedlar, A. and Mirabel, I.F.: 1978, *MNRAS* **182**, 727.
Field, G.: 1959, *ApJ* **129**, 536.
Hancock, S. et al.: 1994, *Nature* **367**, 333.
Hart, L. and Davies, R.D.: 1982, *Nature* **297**, 191.
Hoyle, F.: 1948, *MNRAS* **108**, 372.
Hoyle, F.: 1997, *Home is where the Wind blows*, Oxford University Press.
Hoyle, F., Burbidge, G. and Narlikar, J.V.: 1995, *Proc.Roy.Soc.A.* **448**, 191.
Humason, M.L., Mayall, N.U. and Sandage, A.R.: 1956, *AJ* **61**, 97.
Lasenby, A.N. and Davies, R.D.: 1983, *MNRAS* **203**, 1137.
Lovell, A.C.B.: 1985, *QJRAS* **26**, 393.
Lovell, A.C.B.: 1985, *QJRAS* **26**, 456.
Narlikar, J.V. et al.: 2002, astro-ph/0211036.
Penzias, A.A. and Wilson, R.W.: 1965, *ApJ* **142**, 419.
Sancisi, R., van Woerden, H., Davies, R.D. and Hart, L.: 1984, *MNRAS* **210**, 497.
Scott, P.F. et al.: 2003, *MNRAS* **341**, 1076.
Staveley-Smith, L. and Davies, R.D.: 1989, *MNRAS* **241**, 787.
Stebbins, J. and Whitford, A.E.: 1948, *ApJ* **108**, 413.
Sunyaev, R.A. and Zeldovich, Ya, B.: 1985, *MNRAS* **171**, 375.

MEETING WITH A REMARKABLE MAN: SIR FRED HOYLE MEMORIAL CONFERENCE

RANDALL MEYERS

Fred Hoyle's Universe 24–26 June 2002, Cardiff

The title of this conference – 'Fred Hoyle's Universe' – is very fitting considering the man we are here to commemorate. A man of whom it is fair to say – that most of what we think we know about the universe, in some significant way, was touched by his genius and creativity. Though the word creativity is not usually considered synonymous with scientific procedures, it is precisely this quality that stimulated and endeared Fred Hoyle to his peers and laymen alike. My brief two-day encounter with Fred revealed a man equally enthusiastic and inspired talking about mountain hiking and music, as he was about astrobiology and cosmology. A great scientist yes – but also a great human being, a real Renaissance man! It struck me during my talks with him that he was like a warrior, a warrior of scientific truth and fair play! And like all great warriors Fred exhibited courage and strength in the face of his many fierce scientific battles – defending his views and understanding against even the hardest of criticisms, yet ready to change or develop a problem if not solved to satisfaction. It is easy to see a great man's failings, because he has exhibited them with the same courage as his successes – to such a person, plus or minus are just parts of the same creative saga we call knowledge.

But, as everyone here well knows, Fred Hoyle was also the author of innumerable books – science fiction, popular astronomy, and books on groundbreaking research, essays, lectures and even an opera libretto! In fact I came across my first book of Hoyle's while in my early 20's – just beginning my career as a composer and performer in Europe. While browsing the myriad of used books stores around London I came across a volume of Fred Hoyle's entitled – 'Life Cloud – The Origin of Life In the Universe' 1978. This book was such an inspiration to my thinking that I attribute to it the awakening of my interest in astronomy and science in general. I have since read everything I could get my hands on with Hoyle's name on it – and just recently began to collect his works in earnest. I have been fortunate in my life to meet many outstanding people from various walks, such as music, art, literature and science; but few have left the distinct impression that I was in front of someone truly unique, great, someone remarkable. The two unforgettable days I spent with Fred Hoyle was just such a meeting – a meeting with a remarkable man!

Astrophysics and Space Science **285:** 321–323, 2003.

Universe – The Documentary Film

During the conference 'Fred Hoyle's Universe' a short 12 minute film was presented.

The film was edited from footage shot less than a year before Hoyle's passing, and therefore is to my knowledge the last filmed material of the great man before his death. The material is intended to be used in the larger, two part documentary film entitled 'Universe – The New Cosmology'. Though the project has had a rather long gestation period, due both to the complexity of the subject matter and the difficulty in financing this type of documentary, the project is now in full production. Both episodes have planned completion dates in 2003, with airing already in September-November 2003; at least in Scandinavia. Efforts are being taken to distribute the film worldwide, and several European countries have shown interest in being part of the project.

Though the film is not directly about the Quasi-Steady-State theory, many of Fred Hoyle's collaborators are included in the interviewed discussions. The main goal of the film is to give a broad voice to an otherwise generally lesser known and at times suppressed body of research, observations, and cosmological ideas. It is my belief that science is at an impasse, a kind of turning point, and like other social structures at similar evolutionary cross-roads, struggles for its identity and clarification. Even though history has frequently shown society examples of this, it is nevertheless difficult to accept, or even cognize, from within any one given historical moment. A crisis is generally perceived, but the direction it is taking may only be understood with perspective. It is my hope that this documentary will help bridge developments of the past four decades in cosmology (theory and observation), with the current state of discussion and empirical discoveries. There is therefore a strong historical bend to the filmed material, since Hoyle and several others interviewed go back to the very beginnings of the two predominant cosmologies. There is perforce a rather controversial flavour to the film as well, it being impossible to honestly discuss the theme without going against the orthodox presumptions of a big-bang beginning. Dogma is of course dangerous for society, but mostly for its direct opponents. The big-bang orthodoxy has demonstrated this aspect of its power control in virtually every possible way and on innumerable instances, and it is therefore my wish that these tendencies be adequately revealed to the general public.

Although focus is put on sociological struggles and controversies, the supporting science is of course also presented. Well known, and perhaps less well understood, subjects such as the cosmic background radiation, predictions of the age of the universe and the formation of the Large Scale Structures in the universe, as well as the light element abundances (Deuterium, He^3, He^4, and Li^7) as predicted by various models is discussed in depth. The two episodes touch on the nodal controversy of discordant redshifts, high redshift QSO's and their physical association to their respective low redshift 'parent galaxies', as presented by Halton

Arp and others since decades, and since decades ignored or slipped under the rug. A few QSO/galaxy examples are focused on in particular detail, such as: NGC 7603, Arp 220, NGC 1232, NGC 4319 + Mrk 205, and M82. High-end computer animations will be created to visualize the theorised mechanics of these objects for the viewer. The observations and spectrographic data will be discussed and presented. As mentioned before various sociological problems will be confronted. Systems like the 'Peer review system', and control of public resources, and the allocation of observation time on major telescopes for controversial, non-big-bang related programs etc.

Plasma astrophysics and plasma effects in the large scale structures of the universe are discussed. QSO's and non-Doppler related mechanisms (i.e. the Wolf effect), and proposed experiments related to the observed redshift anomalies are presented. A general history of the developments of this rather new approach to the questions in cosmology will be described in the second episode. Super-Computer simulations by Dr Peratt et al., and the theoretical developments by Alfven, Peratt, Falthammer and others are discussed, while extensive use of stock-shots from historical films will be incorporated. High-end computer animations as well as vector graphics and original composed symphonic music will be utilized to visualise the various concepts of both episodes for the viewer.

To my knowledge this will be the most comprehensive documentary film dealing with non-big-bang cosmology ever made. And it is my desire to make it as clear, expressive and honest as possible.

I would like to express my gratitude to Geoffrey and Margaret Burbidge, and Halton Arp for their sustaining help, wealth of knowledge, and warm friendship graciously given to me, without which this film would never have been made.

PART II: STELLAR STRUCTURES AND EVOLUTION

FRED'S CONTRIBUTIONS TO STELLAR EVOLUTION

LEON MESTEL
Astronomy Centre, University of Sussex, Falmer, Brighton BN1 9QJ, England

Abstract. Fred began work on stellar structure after Hans Bethe and Carl-Friedrich von Weizsäcker had independently established that the thermonuclear fusion of hydrogen into helium is the primary source of the energy radiated by the Sun and other main sequence stars. A joint paper with Ray Lyttleton included this temperature-sensitive process explicitly in the energy equation, effectively vindicating the essentials of the theory of homogeneous gaseous stars presented in Sir Arthur Eddington's celebrated monograph 'The Internal Constitution of the Stars'. Agreement with the solar luminosity can be obtained with two alternative values for the hydrogen content. In a subsequent paper, Fred argued convincingly in favour of the case with a very high rather than a moderate fraction of hydrogen. An epoch-making joint paper with Martin Schwarzschild followed the evolution of a low mass star through nuclear processing, from the main sequence into the giant domain in the Hertzsprung-Russell diagram. The slowly growing, burnt-out core becomes degenerate and nearly isothermal, while the photospheric boundary condition forces the expanding envelope to become largely convective. At the top of the giant branch, the degenerate core becomes hot enough for the fusion of helium into carbon; the consequent secular instability, noted first in studies of white dwarfs, brings the star down to the 'horizontal branch', the location of the short-period globular cluster Cepheids. Two subsequent papers with Brian Haselgrove studied in further detail the structure of both main sequence and giant stars.

1. The Background

The basic theoretical treatment of stellar structure had been presented by Sir Arthur Eddington in his celebrated monograph *The Internal Constitution of the Stars*. His self-gravitating, pressure-supported, homogeneous stellar models have a temperature gradient with a consequent energy outflow, presumed to be essentially radiative, and described by the diffusion approximation to the equation of transfer. H.A. Kramers's treatment of photoelectric absorption gave a tolerable approximation to the opacity coefficient κ. This outflow must be balanced by supply from sub-atomic sources, not as yet identified; however by some sleight of hand, Eddington was able to derive what he called his 'Mass-Luminosity' $(M-L)$ relation, but which is really a Mass-Luminosity-Radius $(M - L - R)$ relation in disguise. He then noted that his strong $L(M)$ dependence up the main sequence was accompanied by a quite modest, observationally inferred increase in the central temperature $T_c \propto GM/R$, implying that the actual energy sources must be strongly temperature-sensitive, as had been surmised also by Henry Norris Russell and Cecilia Payne-Gaposchkin. A remarkable prediction, though in fact not consistent with the hair-raising approximation that enabled him to reach his $M-L$ relation! And indeed in the late 30's, Hans Bethe, George Gamow and colleagues and Carl-Friedrich von Weizsäcker showed that at the predicted internal temperatures, the temperature-sensitive thermonuclear

Astrophysics and Space Science **285:** 327–337, 2003.

fusion process $4H^1 \rightarrow He^4$ liberates the right amount of energy, either through the CNO cycle or the p-p chain, so supplying the star with a built-in thermostat. It was this work, together with Tom Cowling's studies of pulsational stability, that seems finally to have convinced Sir James Jeans of the basic correctness of Eddington's gaseous rather than Jeans's preferred liquid stellar models (cf. Mestel, 1999 for a historical summary of these early controversies). However, a significant modification in the Eddington picture had been made in the early 30′s by Ludwig Biermann and Tom Cowling, who showed that the convective stability criterion enunciated by Karl Schwarzschild could be violated locally, yielding turbulent convection zones.

2. Early Work on Stellar Structure

Fred's paper (**1**) with Ray Lyttleton on stellar structure is probably the first to incorporate thermonuclear energy liberation at the approximate rate $E\rho T^{\eta}$, with $\eta \gg 1$, to calculate directly both the luminosity and the radius of a homogeneous, main sequence star of moderate mass, with chemical composition and so also mean molecular weight μ prescribed. The Kramers opacity law $\kappa \propto (1 - X^2)(\overline{Z^2/A})(\rho/T^{3.5})(1/g)$ is adopted, with X the fraction by mass of hydrogen, (Z, A) the atomic number and weight of the heavy elements responsible for most of the opacity, and g the 'guillotine factor'. If the variation of g with ρ and T is ignored, the derived homology relations are

$$L \propto E^{-1/(2\eta+5)} M^{(10\eta+31)/(2\eta+5)} \mu^{(14\eta+45)/(2\eta+5)} (1 - X^2)^{-(2\eta+6)/(2\eta+5)},$$
$$R \propto E^{2/(2\eta+5)} M^{(2\eta-7)/(2\eta+5)} \mu^{(2\eta-15)/(2\eta+5)} (1 - X^2)^{2/(2\eta+5)}, \tag{1}$$

showing the strong dependence of the luminosity L on M and μ and its weak dependence on the coefficient E, as is implicit in Eddington's M-L relation. Qualitatively similar formulae result when g is taken to vary according to Bengt Stromgren's inferred approximate relation $g \propto \rho^{1/2}T^{-3/4}$. For η large enough, as in the CNO cycle, the paper confirms the existence of a convective core, so effectively rediscovering Cowling's point-convective model.

Eddington had devoted a whole chapter of *I.C.S.* to the discrepancy by a factor 23 between the 'astronomical' coefficient of opacity, required to yield the observed luminosity, and the physical coefficient. In private and with reluctance, he ultimately accepted Stromgren's definitive demonstration that the discrepancy would disappear if his assumed near zero value for the H-content X were replaced by ≈ 0.35, yielding $\mu \approx 1$. He had himself seen this but rejected it because except for the most massive stars, the predicted radiation pressure p_{rad} would then be a small fraction of the total pressure, contrary to his earlier, rather mystical surmise that 'the stars happen when the two pressures are comparable' (*I.C.S*, p.16). Fred and Ray had adopted Stromgren's X-value, noting that p_{rad} is then important only in high masses, in particular extending the convective core through its effect on the adiabatic exponent γ. In fact, (1) shows that the opacity discrepancy disappears also when $\mu \simeq 0.5$, $(1 - X) \ll 1$, the much smaller opacity coefficient resulting from the low metal content $\propto (1 - X)$ being compensated by the small factor $\simeq \mu^{7.5}$ in the mass-luminosity-radius relation; but like others (e.g. Chandrasekhar, p.277) they had rejected such a high H-content as 'improbable'. However, in (**3**), Fred argued convincingly for $X \approx 1$, citing T. Dunham's recent observations of interstellar matter and the work by Henry Norris Russell and Stromgren on stellar atmospheres. (I would have included also, the seminal paper by Bill McCrea). He pointed out that the transition from electron scattering to photoelectric opacity would then occur at masses near $0.5M_{\odot}$ rather than near $5M_{\odot}$. He also claimed that his newly calculated $M - L$ curve gave a somewhat better fit with observation. Fred's conclusion – manifestly of such great cosmological importance – persists today, but with X replaced by $(X + Y)$ where Y is the 'primeval' He-content, in one sense or other of the term.

3. Stellar Evolution: The First Steps

Fred and Ray now turned their attention to the red giant stars. The paper (**2**) was the first suggestion in the UK and US that inhomogeneity in the mean molecular weight μ is responsible for the increase in stellar radii towards the red giant state. (Some earlier work by E.J. Öpik, published in Estonia, was then unknown.) Each model in this paper and in the later one (**5**) by the same authors, and in parallel work by C.M. and H. Bondi and by Li Hen and M. Schwarzschild, consists essentially of a homogeneous main sequence star of mass M_i, $\mu = \mu_i$, and containing an energy-generating Cowling-type convective core, surrounded by a homogeneous radiative envelope of mass M_e with $\mu_e < \mu_i$. The ratios μ_i/μ_e and M_i/M_e are effectively extra parameters, along with the total mass $(M_i + M_e)$ and the age. The computations by the Bondis were the most comprehensive, with some models having embarrassingly large radii.

Both the U.K. and the U.S. work assumed that meridian circulation currents within a rotating star would mix the processed, higher μ material from the core throughout all or part of the envelope. In particular, the U.K. papers assumed that an *isolated* star would be kept effectively homogeneous, with μ steadily increasing, but argued that a giant star would arise if there were subsequent very efficient *accretion* of interstellar gas of low μ (cf. the lecture by Hermann Bondi). Developments in the early 50's in both theory and observation led to a change in the direction of research. Detailed study of the meridian circulation process by Peter Sweet, Öpik, Schwarzschild and Mestel suggested very strongly that except possibly for very fast rotators, zero mixing between core and envelope was a far better approximation. Doubts also arose as to whether accretion of interstellar matter could be as efficient as required to yield giant models. Equally important, whereas in the Hertzsprung-Russell diagrams of field stars, the red giants form a blot to the right of the main sequence, the photoelectric studies by Allan Sandage and colleagues of globular and galactic clusters yielded much cleaner diagrams, implying that the evolution of stars of similar composition and forming coevally depends primarily on the one parameter M. Other parameters such as the rotation rate, or the amount and epoch of any accretion, should at most yield perturbations about one-parameter evolution.

When he was Hermann Bondi's Ph.D. student, my late colleague Roger Tayler did the work leading to his 1956 pioneering paper, on the evolution of unmixed stars, showing clearly the turn-off from the main sequence towards the giant domain, a highly important criterion for age estimation. When all the hydrogen in the convective core has been converted to helium, the burnt-out core attempts to become isothermal at a temperature high enough for energy generation to occur in a surrounding hydrogen shell; but such an equilibrium state is not possible if the core mass exceeds about 10% of the total mass (the Schönberg-Chandrasekhar limit). Sandage and Schwarzschild (1952) attempted to follow the evolution of a star with a burnt-out core and initially no other nuclear sources. Their models are explicitly time-dependent; the gravitationally contracting core acquires a modest temperature gradient, along which there is energy outflow at the expense of the released gravitational energy, fed into the expanding envelope by radiative transfer. Like some of the parametrically evolving models mentioned above, the evolution carries them beyond the observed giant branch.

4. The Evolution of Type II Stars

Paper (**6**) by Hoyle and Schwarzschild is without doubt outstanding among Fred's contributions to stellar evolution. The authors begin by modifying the initial Sandage-Schwarzschild picture by allowing for incipient degeneracy in the burnt-out core, which consequently remains virtually isothermal. They also revert to time-*in*dependent models, with the luminosity balanced by energy generation – through either the CNO-cycle or the p-p chain – in a thin shell surrounding the burnt-out core. (They accept that the thermal equilibrium assumption will not be appropriate in the initial phases near the main sequence, but their main concern is with the later evolution, for which thermal equilibrium has become valid before the star has reached the giant branch.) Energy transport through the envelope is taken to be radiative. Thus the evolution is initially again parametric: the models evolve through the shell's burning out and being added to the degenerate helium core, to be replaced by the next shell, while simultaneously the envelope expands. However, if their procedure is continued unmodified, they again find that the models would evolve through and beyond the giant branch, with only a moderate increase in luminosity.

The crucial modification introduced is the use of the correct physical surface boundary condition. In standard notation, the luminosity L, the photospheric radius R and the effective temperature T_e are related by

$$L = \pi ac R^2 T_e^4. \tag{2}$$

Below R, the photon mean-free-path is much less than the macroscopic scale-height, and the equation of radiative transfer is well described by the diffusion approximation. Above R, the mean-free-path $\to \infty$, and the integro-differential equation of radiative transfer must be used. In the simplest case of a 'grey' atmosphere, i.e. with the opacity κ independent of frequency, the Milne-Eddington treatment yields the condition on the photospheric optical depth τ:

$$\tau = \int_R^\infty \kappa\rho \mathrm{d}r \simeq 2/3. \tag{3}$$

Thus at R, the 'mathematical' boundary conditions $(\rho, T) \to 0$ are replaced by

$$T \to T_e, \qquad \rho \to \frac{\mu GM}{\kappa \Re T R^2}. \tag{4}$$

This is the condition that the radiation can escape, the definition of the photosphere. On the main sequence and in the early H-S models, the mathematical boundary conditions are adequate, yielding a radius little different from that given by (4). But as the envelope expands with $L \simeq$ constant and T_e systematically decreasing, it is found that ρ reaches the photospheric value when $T > T_e$; or equivalently, if the integration is continued out to $T = T_e$, it is found that ρ has fallen too low.

To resolve this dilemma, new models are constructed with envelopes that have deep convective zones, extending from below the photosphere to a great depth (well below that at which hydrogen ionization is complete). This suffices to keep ρ high enough to satisfy the surface condition. As the star approaches the giant region, in the low-temperature surface domains it is Rupert Wildt's negative hydrogen ion H^- which dominates the opacity and in fact sets a lower limit to T_e, so preventing the spurious evolution predicted by the earlier models. Instead, as the core mass increases, the star evolves up the giant branch, with a modest change in T_e but a large increase in R.

The main features of the H-S giant models are summarized as follows:

In the deep interior, the stellar luminosity L and the radius and temperature of the energy-generating shell are fixed by the prescribed fraction of the mass in the core, but are hardly affected by the photospheric radius R;

The surface condition fixes R and T_e so that the star radiates L supplied from below, requiring that the stellar envelope outside the energy-generating domain must be largely convective.

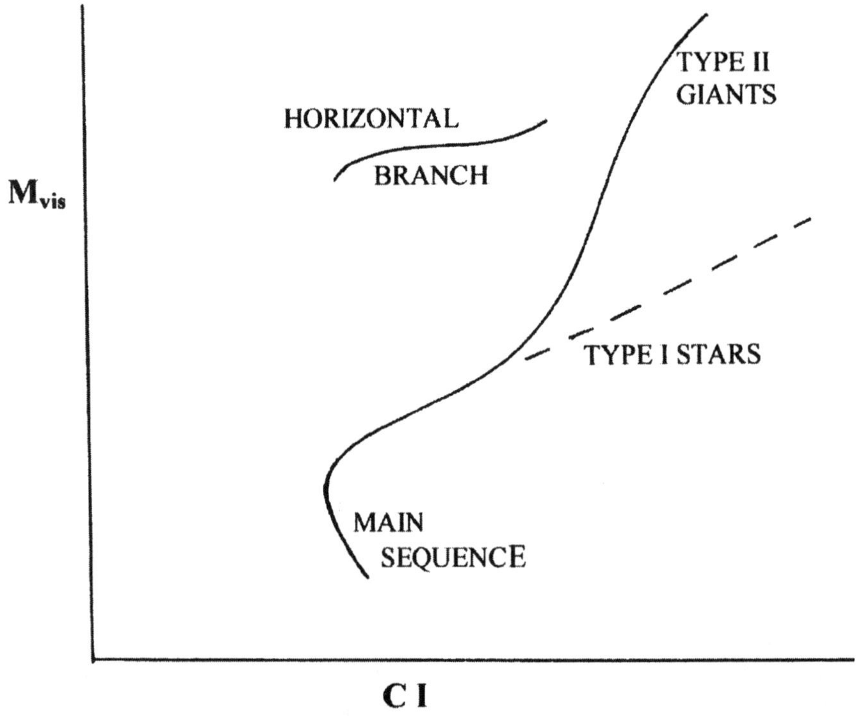

Figure 1. Summary of the Hoyle-Schwarzschild paper.

At the star approaches the top of the initial giant branch, the release of gravitational energy heats up the core centre to a temperature $\simeq 1.2 \times 10^8$ K at which helium fusion into carbon occurs through the Salpeter-Hoyle triple-alpha process. This onset of a temperature-sensitive reaction in the degenerate core leads to a thermal instability – the 'helium flash' (cf. below). It is conjectured that this moves the star down and to the left in the H-R diagram, to form the 'horizontal branch' models, each with a non-degenerate, He-burning core as well as the H-burning shell. The short-period variable stars are located on this branch.

Figure 1 summarizes the H-S paper. The authors argued that the higher metal content in the Type I stars would affect the opacity so as to change significantly the shape of the giant sequence. Later evolutionary computations, especially for higher mass stars, are summarized e.g. in Kippenhahn and Weigert 1990. The explicit time-dependence of Sandage and Schwarzschild is retained in the comparatively rapid evolution across the Hertzsprung gap, but otherwise the essentials of the Hoyle-Schwarzschild picture appear to have stood the test of time. John Faulkner will be elucidating some of the subtleties underlying the numerical results.

5. The H-S Work in Historical Context

The essence of Eddington's approach is that for homogeneous stars, it is the gross structure that fixes both L and R. Hydrostatic equilibrium requires a temperature gradient $\propto \mu M/R^2$, and the consequent outward transfer of radiation yields the luminosity $L(M, \mu, R)$. As noted, because of the modest variation through the star of the opacity $\kappa(\rho, T)$, L turns out to have a strong dependence on M and μ but a weak dependence on R. Equating L to the strongly temperature-sensitive energy generation in the centre then fixes $R(M, \mu)$, and $L(M, \mu)$. The stellar atmosphere adjusts itself to ensure that the photosphere achieves optical depth $\tau = 2/3$ at the temperature T_e given by (2), and at a radius close to that given by extrapolating the inner solution to zero T and p.

In current parlance, the inner and outer solutions are well-matched asymptotic approximations; in a favourite expression of Fred's (which he seems to have acquired from Eddington, and certainly passed on to me), the star as a whole is the 'dog', wagging the atmospheric 'tail'. This approach was however not accepted by E.A. Milne, who argued that the surface regions would react back strongly on the structure of the star as a whole. His claim that there existed alternative solutions to Eddington's, with homogeneous, main-sequence stars having cores degenerate in the Pauli-Fermi-Dirac sense, was contested successfully by Chandrasekhar, Cowling, Russell and Stromgren. However, in the H-S giant models, for which the expected physical evolution has led naturally to an inhomogeneous chemical composition with a *burnt-out* degenerate central core, surrounded by an energy-generating shell, the correct surface condition does indeed react on the structure of the star. The luminosity L continues to be fixed by an Eddington-type procedure, applied to the inner regions; but in order that L may emerge, the massive and as yet unprocessed envelope is forced to become largely convective, with a different $\rho - T$ distribution from that given by assuming pure radiative transport.

Fred was so impressed by this 'change of paradigm', that in the Cambridge lectures he was giving in the mid-50's, I understand he was saying that 'Milne was right after all'. However, during the rehearsal of a 1958 radio discussion, chaired by Harold Spencer-Jones, in which Fred, Tom, Hermann Bondi and I took part, Tom firmly rebutted this excessive tribute, emphasizing that Milne had been considering only homogeneous stars. Tom had been Milne's D.Phil. student, and had on occasion been given the unenviable task of defending Milne's idiosyncracies, even though his own studies, both earlier and later, had vindicated Eddington's approach against the criticisms of both Milne and Jeans. Fred had always expressed his respect for Tom's penetrating (if sometimes over-critical) contributions, and it is not surprising that he did not repeat his comment during the subsequent broadcast recording. (It is generally recognized that Milne's lasting contributions to stellar physics are in his studies of stellar atmospheres.)

As a postscript, I turn to the problem of a homogeneous pre-main sequence star, with central temperatures too low for any nuclear processing. Eddington's treatment of stellar structure – as brought out by Fred in Section 2 of his epoch-making pioneering paper (**4**) on nucleosynthesis – distinguishes between the two related but distinct questions: (a) what is the rate at which a given star radiates; and (b) what is the source of the energy that supplies the radiation? A pre-main sequence star lives on its gravitational capital, contracting according to the classical picture of Kelvin and Helmholtz, who of course took the solar luminosity from observation. A star that contracts homologously, transferring energy by radiative transport, has an effective energy source ε per gram $\propto T$. With a simple Kramers or electron scattering opacity, Eddington's theory again yields a luminosity $L \simeq L(M, \mu)$, i.e. with a weak or vanishing R-dependence, so predicting a nearly horizontal path in the H-R diagram. (Detailed computations by L. Henyey and colleagues found rapid convergence towards homologous contraction from a variety of initial $(\rho - T)$ distributions.) For a high mass star, this prescription suffices: well to the right of the main sequence, the predicted surface temperature T_e is high enough for the opacity to satisfy the condition (3). However, in his 1961 paper, C. Hayashi drew attention to the same lower limit to T_e, set essentially by the properties of the H^- ion. For a star of solar mass and of high radius, (2) yields T_e too low: an observer sees through to a hotter photosphere which emits a luminosity higher than Eddington's prediction. Again the star responds by becoming fully or largely convective, moving in a nearly vertical line in the H-R diagram, until it intersects the Henyey curve for its mass, at the appropriate radius (cf. Figure 2). Thus over the Hayashi phase, the surface condition fixes both the gross stellar structure and the luminosity.

6. Later Work

In 1959, Fred published two papers jointly with the late Brian Haselgrove. Paper (**7**) contains a comprehensive set of integrations of main sequence stars. The Type I sequence are given the solar chemical abundance, and have $M/M_\odot$ ranging from < 1 to 125. The Type II stars studied have $M/M_\odot$ varying by only a factor 2-3, and all have a lower metallic and CNO content than the Sun. One Type II sequence is given the same ('implausible') He content as Type I; the other a much lower He content. The resulting $M - L$ relationship is parametrized by $L \propto \bar{M}^q$, with $\bar{M} \equiv M/M_\odot$. When $\bar{M} \simeq 1$, $q \simeq 5$; $\bar{M} \simeq 2$, $q \simeq 4$; $\bar{M} \simeq 3 - 30$, $q \simeq 3$; and when $\bar{M} > 30$, $q \simeq 2$, the last result showing the increasing importance of radiation pressure. For low mass stars, with Kramers opacity and the p-p chain operating, R is nearly independent of M down to $T_e \simeq$ 7000K; but for lower masses, correct use of the surface boundary condition yields R decreasing with M, confirming the earlier conclusion of Donald Osterbrock. The authors conclude that the zero-age main sequences of the different populations show a small separation, suggesting that main-sequence fitting should yield reliable distance moduli.

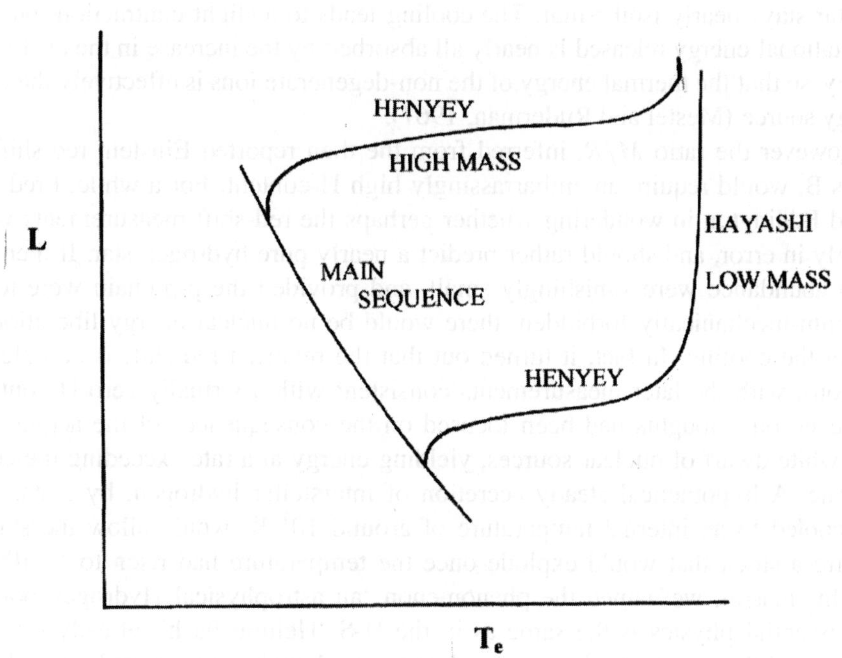

Figure 2. Schematic pre-main sequence evolution.

Paper (**8**) noted that the paper with Schwarzschild and a later paper had yielded $(6.2 - 6.5) \times 10^9$ yr as typical globular cluster ages. The monumental paper on nucleogenesis, known universally as 'B^2FH', had reduced the efficiency of the CNO-cycle by a factor 10-100. As a consequence, in Type II stars with $1 < \bar{M} < 1.5$ the p-p chain dominates, and the evolution time to the main sequence turn-off is increased to $> 10^{10}$yr, a result of some importance for cosmology.

7. White Dwarfs

I conclude by summarizing Fred's (largely, but not entirely, indirect) contributions to white dwarf theory, through his agreeing to accept me as his Ph.D. student (1948–51). Following Robert Marshak's (slightly emended) calculation of the thermal conductivity of an electron gas, the picture emerged of a white dwarf with a strongly degenerate, nearly isothermal bulk, surrounded by a thin but highly opaque, non-degenerate envelope, with a large local temperature gradient. In the absence of any remaining nuclear sources, the star cools towards the black dwarf state, but at a rate controlled by the envelope, which acts as a blanket. After some discussion, Fred agreed that this was not a tail-wagging-dog situation – the very high conductivity of a degenerate electron gas merely ensures that the bulk of the star stays nearly isothermal. The cooling leads to a slight contraction, but the gravitational energy released is nearly all absorbed by the increase in the exclusion energy, so that the thermal energy of the non-degenerate ions is effectively the only energy source (Mestel and Ruderman, 1967).

However the ratio M/R, inferred from the then reported Einstein red-shift in Sirius B, would require an embarrassingly high H-content. For a while, Fred followed Eddington in wondering whether perhaps the red-shift measurements were slightly in error, and should rather predict a nearly pure hydrogen star. If then the CNO abundance were vanishingly small, and provided the p-p chain were to be quantum-mechanically forbidden, there would be no nuclear energy liberation to off-set the cooling. In fact, it turned out that the reported red-shift is completely spurious, with the later measurements consistent with a virtually zero H-content. However, our thoughts had been focused on the consequences of the acquisition by a white dwarf of nuclear sources, yielding energy at a rate exceeding the cooling rate. A hypothetical steady accretion of interstellar hydrogen, by a star that had cooled to an internal temperature of around 10^7 K, would allow the star to acquire a stock that would explode once the temperature had risen to $\simeq 10^8$ K. For this reason, we named the phenomenon 'an astrophysical Hydrogen bomb'. The essential physics is the same as in the H-S 'Helium flash', already referred to; because the equation of state of a degenerate domain depends only weakly on the temperature, there is no thermostatic effect as in a classical gas – a degenerate stellar domain has a *positive* specific heat. We shall be hearing more of this from Don Clayton.

8. Valediction

Let me now express my thanks to Geoff, Chandra and Jayant for giving me this opportunity to honour Fred's memory and to say in public what I have said so often in private, offering my thanks to Fred for starting me on the road from which I never looked back.

References

Papers by F.H.

1. with R.A. Lyttleton: 1942, On the internal constitution of the stars, *MNRAS* **102**, 177.
2. with R.A. Lyttleton: 1942, On the nature of red giant stars, *MNRAS* **102**, 218.
3. 1946, The chemical composition of the stars, *MNRAS* **106**, 255.
4. 1946, The synthesis of the elements from Hydrogen, *MNRAS* **106**, 343.
5. with R.A. Lyttleton: 1949, The structure of stars of non-uniform composition, *MNRAS* **109**, 614.
6. with M. Schwarzschild: 1956, The evolution of Type II stars, *ApJ* Supplement Series, **2**, 1.
7. with C.B. Haselgrove: 1959, Main sequence stars, *MNRAS* **119**, 112.
8. with C.B. Haselgrove: 1959, The ages of Type I and Type II subgiants, *MNRAS* **119**, 124.

Other Authors Cited

Bondi, C.M. and Bondi, H.: 1950, Models for red giant stars, *MNRAS* **110**, 287; and 1951, **111**, 397.

Chandrasekhar, S.: 1939, *An Introduction to the Study of Stellar Structure*, University of Chicago Press (Dover Paperback edition 1957).

Eddington, A.S.: 1926, *The Internal Constitution of the Stars* (*I.C.S.*), Cambridge University Press (Dover paperback edition 1959).

Hayashi, C.: 1961, Stellar evolution in early phases of gravitational contraction, *PASJ* **13**, 450.

Hen, Li and Schwarzschild, M.: 1949, Red-giant models with chemical inhomogeneities, *MNRAS* **109**, 631.

Kippenhahn, R. and Weigert, A.S.: 1990, *Stellar Structure and Evolution*, Springer Verlag Berlin, New York.

Mestel, L., 1952, On the theory of white dwarf stars, *MNRAS* **112**, 583 and 598.

Mestel, L. and Ruderman, M.A.: 1967, The energy content of a white dwarf and its rate of cooling, *MNRAS* **136**, 27.

Mestel, L.: 1999, The early days of stellar structure theory, *Phys. Rep.* **311**, 295.

Sandage, A.R. and Schwarzschild, M.: 1952, Inhomogeneous stellar models, *ApJ* **116**, 463.

Tayler, R.J.: 1956, The evolution of unmixed stars, *MNRAS* **116**, 25.

FRED HOYLE, RED GIANTS AND BEYOND *

JOHN FAULKNER
UCO/Lick Observatory, Department of Astronomy and Astrophysics, UC Santa Cruz, CA 95064, USA

Abstract. The impact of Fred Hoyle's work on the structure and evolution of red giants, particularly his breakthrough contribution with Martin Schwarzschild (1955), is described and assessed. Working with his students in the early 1960s, Hoyle presented new physical ways of understanding some of the approximations used, and results obtained, in that seminal paper. His initial viewpoint on the critical role of the outer surface boundary condition was replaced by a more subtle, if related one, which emphasized the peculiar difficulty of storing much mass outside a dense stellar core. That viewpoint that – low-mass red giants are essentially white dwarfs with a serious mass-storage problem – is still extremely fruitful.

Recently, I have extended Hoyle's approach to explain not only many of the structural properties of red giants themselves, but also to link and unify the structures of low-mass stars from the main sequence through both the red giant and horizontal branch phases of evolution. Many aspects of these stars that had remained mysterious for decades have now fallen into place, and some questions have been answered that were not even posed before.

With red giants as the simplest example, this recent work emphasizes that stars, in general, may have at least *two* distinct but very important centres: (i) a *geometrical* centre, and (ii) a separate *nuclear* centre, residing in a shell outside a zero-luminosity dense core for example. This two-centre perspective leads to an explicit, analytic, asymptotic theory of low-mass red giant structure. In this theory, there arises a naturally important *in situ* measure of central compactness: the parameter $\rho_{sh}/\overline{\rho}_c$. That parameter, like others, is derived self-consistently and explicitly, and can be used to show how close a given model's properties are to ultimate asymptotic relationships.

The results obtained also imply that the problem of understanding why such stars become red giants is one of anticipating a remarkable yet natural structural bifurcation which occurs in them. In the resulting theory, both the ratio $\rho_{sh}/\overline{\rho}_c$ and products like $\rho_{sh} \cdot \overline{\rho}_c$ prove to be important, self-consistently derived quantities.

Two striking theorems involving such quantities express between them the very essence of red giant behaviour, proving analytically for the first time that stars with dense cores are necessarily (i) extremely luminous, and (ii) very large. Perhaps the most astonishingly unexpected single result is that for the very value Nature provides for the relevant nuclear energy-generating temperature exponent (CNO's $\eta = 15$), ρ_{sh} and $\overline{\rho}_c$ behave in a well-defined, precisely *inverse* manner. This emphasizes that the internal behaviour of such stars is definitely *anti-homologous* rather than *homologous*, thus showing how very unfortunate the term 'shell homology' is. Finally, I sketch a viewpoint which (i) links the structural and evolutionary behaviour of stars from the main-sequence through horizontal branch phases of evolution, and also (ii) has implications for post-main-sequence developments in more massive stars.

* The final text of this contribution was not received (The Editors).

Astrophysics and Space Science **285:** 339, 2003.

SOME REMARKS ON SOLAR NEUTRINOS

DOUGLAS GOUGH
Institute of Astronomy, Madingley Road, Cambridge, CB3 0HA, UK
Department of Applied Mathematics and Theoretical Physics, Silver Street, Cambridge, CB3 9EW, UK
Physics Department, Stanford University, California, USA

Abstract. In 1970 Fred Hoyle encouraged a study of solar neutrino production which led to a long-term investigation of the influence of what have become known as 'non-standard' processes (i.e. processes that are not accounted for in the relatively naively constructed so-called 'standard' theoretical solar models). The outcome is a very much sounder understanding of the structure and dynamics of the Sun, which has yielded a knowledge of conditions in the energy-generating core so precise that one can set quite tight reliable constraints on neutrino-producing nuclear reactions, and thereby provide an important contribution to the study of neutrino transitions.

Keywords: neutrinos, stellar stability, helioseismology

1. Preamble

I write here, somewhat unconventionally, about solar neutrinos partly because it was Fred who first set me working on the problem and partly because it taught me an extremely important lesson in science. It was in 1970, not long after I had returned to Cambridge as a member of graduate staff at the Institute of Theoretical Astronomy (IoTA). Only the ^{37}Cl Homestake neutrino data were available at that time, and their average value was only some 30 per cent of the current theoretical 'predictions'. According to the standard electroweak theory with massless neutrinos, they are dominated by the high-energy neutrinos emitted by the spontaneous decay of ^{8}B produced in the ppIII chain (e.g. Bahcall, 1989). The flux of those neutrinos is proportional to the abundance X_8 of ^{8}B which, through the apparently convectively stable core of the Sun varies as a rapidly increasing function of the temperature T: $X_8 \propto (1-X)(1+X)^{-1}X^2\rho T^{24.5}$, approximately, where ρ is density and X is the abundance of hydrogen. Notwithstanding the inward decline in X resulting from the fact that the consumption of hydrogen fuel increases inwards with the temperature, it follows that the ^{8}B neutrino production is strongly concentrated in a central region of the energy-producing core, even though T declines quite gently away from the centre of the Sun (Figure 1).

The discrepancy between theory and observation had posed what was called 'the solar neutrino problem'. Not unnaturally, the attempts to solve the problem involved incorporating into the models features that might lead to a *diminution* of the central temperature – such as imagining the presence of a magnetic field intense enough to relieve a significant fraction of the gas pressure in the solar core,

Astrophysics and Space Science **285:** 341–351, 2003.

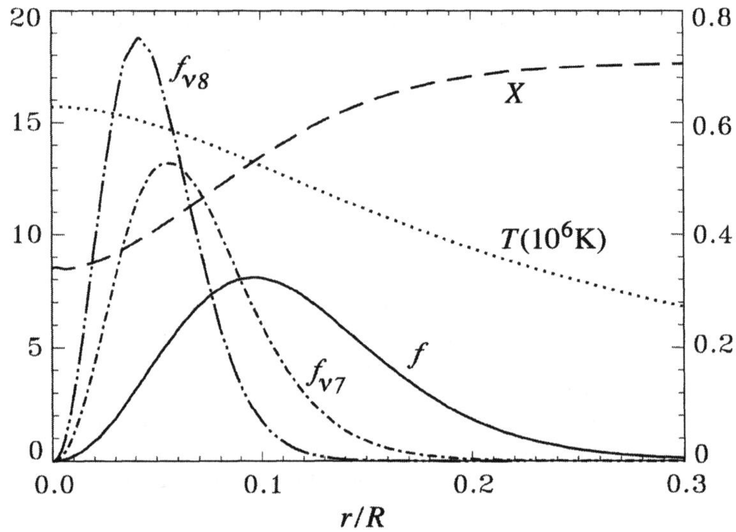

Figure 1. Neutrino production in a standard solar model. The ^{7}Be and ^{8}B neutrino luminosities, $L_{\nu i}$ with $i = 7$ and 8 respectively, are proportional to $\int_0^R f_{\nu i} dr$, where r is a radius variable and R is the radius of the Sun. The energy generation rate and the pp neutrino production rate are proportional to f; the pep production rate varies similarly (although not identically). The functions $f_{\nu i}$ and f are normalized to have unit integral. The temperature T and the hydrogen abundance X (whose ordinate scale is on the right of the diagram) are included for comparison.

or regarding the core to be rotating very rapidly, an idea which no doubt met, for other reasons, with Dicke's approval (Dicke and Goldenberg, 1967) – but all to no avail. However, Fred, as usual, was more imaginative. He asked me to pursue the idea that gravitational settling of helium (which was ignored in those days) might have displaced sufficient hydrogen from the innermost regions of the Sun's core to lower the ^{8}B production rate, despite the compensating *augmentation* of T which results from the readjustment that must be made to the initial helium abundance of the theoretical solar model to maintain the photon luminosity L at its observed value. In order to have a serious influence on the reaction rates, the degree of settling would need to be substantially greater than back-of-the-envelope estimates suggested, but perhaps a detailed calculation would reveal otherwise.

In those days I had too poor an understanding of the properties of stars to be able to estimate what the effect of introducing such a process might be. I therefore had to resort to a numerical computation of a full solar model. I had never carried out such a computation before (nor had I admitted that to Fred), but I was fortunate enough to have access to the computer code which Bohdan Paczyński had kindly left at IoTA for general use. That code was so clearly written that even someone like me who had never even seen such a code before could readily follow the logic and then modify the details to perform the task required. I was thus able to answer

Fred's question before Fred had become so impatient as to be moved to ask me what was going on.

Before discussing the answer, I wish to share the important lesson I learned whilst obtaining it. In order to check that I had programmed the neutrino production rate correctly, I naturally performed first what in those days was the so-called standard computation, namely the computation in which gravitational settling was ignored. I adopted for the uncertain parameters, such as nuclear-reaction cross sections and the total heavy-element abundance Z (which has an important influence on opacity) values which, after I had perused the astrophysical literature, seemed to me to be the most plausible. And with those I obtained a total solar neutrino flux of about 20 snu. That value was substantially greater than the 6–8 snu (e.g. Bahcall, Bahcall and Shaviv, 1968; Bahcall, Bahcall and Ulrich, 1969) that was the current theoretical 'prediction' of the day, obtained with what the experienced model-producers considered to be the most plausible parameters. So I needed next to learn how to improve my choice of plausible parameters. I went through an exercise of trial and error, monitoring the neutrino luminosity L_ν on the way. I could easily move it up and down by quite a bit, but oddly enough I could not reduce the flux below about 6–8 snu. Moreover, I noticed that the theoretical prediction that Ray Davis (1964) had adopted in the design of his experiment was, like my original value, much greater than that which was currently being promulgated. (It was actually twice my value, but that was before there had been a dramatic increase – by a factor 5 – in the experimental determination of the ^{3}He -^{3}He reaction cross section, which reduced the theoretical value of L_ν by some 30 per cent.) Evidently, before there were solar neutrino data to guide them, other theorists had been misled similarly. The lesson, at least for me, was clear: to obtain a real appreciation of the results of a complicated numerical calculation it is necessary to get one's hands dirty by performing a similar calculation oneself.

The outcome of introducing gravitational settling, in the crude manner in which I had, was dismal: the effect of the increased central condensation of the models on the nuclear reaction rates was generally to augment, not diminish, the ^{8}B production rate, and hence increase the ^{8}B contribution $L_{\nu 8}$ to L_ν, but by only about 10 per cent. It may have been the case that one could reduce $L_{\nu 8}$ if one imagined just the inner few per cent of the core to have been almost completely exhausted of hydrogen without substantially influencing the chemical composition in the rest of the core – I no longer remember – but such an occurrence could hardly have resulted from gravitational settling, because the rates of diffusive processes increase inwards faster than the rates of gravitational segregation, and therefore induce an abundance change which varies more and more gradually towards the centre of the star. Fred must have been very disappointed with my results, for he never again asked me to do calculation for him.

A second, lesser, lesson that I learned from this experience is that the relatively smooth modification that arises from a small perturbation such as gravitational settling causes the radiative interior of the solar model to respond almost homologously in global variables (such as pressure and temperature, whose derivatives appear in the equations of stellar structure) – the response of the outer convective envelope of any Sun-like star can also be estimated by appropriate homologous scaling (with respect to depth, rather than radius) adjusted to match onto the radiative interior, that scaling having negligible influence on the radiative interior because the weight of the convection zone is so small. Local deviations from homology (in more locally determined variables such as density) can be estimated by local power-law scaling (which is tantamount to linearizing the deviation in logarithmic variables). Therefore the outcome of introducing some change to the modelling can be estimated without recourse to a full numerical evolutionary calculation. Although the outcome may not always be in agreement with a full calculation (e.g. Gough, 1992), I suspect that it is generally a reliable guide. It certainly appears to work reasonably well for estimating the reaction to gravitational settling.

I conclude this story by pointing out that the more accurate modern calculations (e.g. Bahcall and Loeb, 1990; Christensen-Dalsgaard, Proffitt and Thompson, 1993), which, I might add, were not carried out until long after Fred had worried about the matter, confirm that gravitational settling has a rather minor impact on the structure of the solar core, and hence on neutrino production (and that it does indeed increase L_ν).

2. Solar Stability

That excursion into solar modelling prompted further thought into why theoretical neutrino fluxes could not be brought into line with observation. From the discussion in my preamble, it appears that the uncertainty in standard model calculations was perhaps a factor 2 or so, but that the lower limit was more robust. A serious constraint on standard models is that, because the models are in thermal balance, the rate of generation of energy by nuclear reactions in the core, $L_{\rm n}$, must be equal to the photon luminosity L observed at the surface. Admittedly, the generation of thermal energy and the generation of neutrinos, particularly the high-energy ^{8}B neutrinos, do not occur predominantly in the same location (Figure 1), but the gradients of both temperature T and pressure p are quite tightly constrained by the equations of stellar structure (and, by implication, ρ is too, because ρ is essentially determined by p, T and X, and X is determined by the nuclear reaction rates in terms of the initial, uniform, hydrogen abundance X_0): therefore one cannot easily envisage a dramatic change in $L_{\nu 8}$ if the value of $L_{\rm n}$ is to be preserved. As I discuss in the next section, only if Z is changed dramatically (which, through the model calibration, induces a relatively small change in X_0) can a standard model be found with $L_{\nu 8}$ in agreement with observation.

Of course we know now that an important reason for the difference between theoretical neutrino production rates and the neutrino flux measurements (and possibly the only significant reason) is that the two are not now expected to be equal, because electron neutrinos presumably decay into other flavours during their passage from the Sun to the Earth. With that possibility now firmly established – and I must emphasize that it is yet only a possibility, because the direct evidence for neutrino transitions (Kajita, 1999) does not involve the electron neutrinos which the Sun produces, although in most physicists' minds, including mine, the likelihood of it actually being the case is overwhelming – the general interest in the solar neutrino problem has naturally declined. But to the executors in the field, the matter is not yet resolved. It is still of considerable astrophysical interest to establish the reliability of solar models; and it is of importance to particle physics too, because it is not unlikely that for considerable time to come the Sun will remain the only source of electron neutrinos far enough away for us to be able to measure electron-neutrino decay. Therefore a knowledge of the physical conditions in the energy-generating core (and, indeed, outside the core too if the MSW effect is operating) must be central to determining with confidence the parameters (squared-mass differences and mixing angles) that characterize neutrino transitions.

It is common to many theoretical investigations of physical systems that the first step is to determine the steady states (if there are any) or the slowly evolving solutions of the governing differential equations. Subsequent steps involve dynamical studies, usually initiated as linearized perturbation analyses about the steady, or slowly evolving, solutions. Convection is a case in point, but I refrain from discussing that here because it is too difficult; almost invariably in stellar structure and evolution studies convection is parametrized in a very simple, local, time-independent fashion, and is not explicitly studied as a dynamical phenomenon. I shall do likewise. But what about other potential instabilities? Normally, in physics, if one finds a steady solution that is unstable, one concludes that it cannot persist in nature, and one no longer considers it as a possible representation of reality unless one has good reason to believe that the instability is ineffectual. But in the case of the Sun, history has been different. An instability has been known for three decades, yet it has been largely ignored, initially without justification. It now appears to be likely that its neglect has been fortunate, for without the niggling doubts that it would have engendered, solar modellers have felt free to investigate other, perhaps more pertinent, aspects of the physics, and so advance our knowledge of the properties of the models by more than they would have otherwise.

A further word about the instability and its ineffectuality is not out of place here, for the physics is interesting, not least because I, for one, first found the outcome to be counterintuitive. The mechanism of the instability itself is quite straightforward, and is not unlike an example of one of the two possible processes (the one that is now called the ϵ mechanism) that Eddington (1926) suggested as a possible driving mechanism for Cepheid pulsation. It is an instability of an almost adiabatic gravity wave acting as a thermodynamic heat engine (working almost in a cycle) in the core: compressed fluid is hotter and gains heat from accelerated nuclear reactions (and expanded, cooler fluid loses heat because the reduced reaction rates are insufficient to keep up with the general diffusive loss from the core); these processes were estimated to dominate over the normal diffusive losses of an oscillation mode which tend to cool warm fluid and warm cool fluid, and therefore overall the heat engine does positive work to augment the kinetic and potential energies of the oscillation. Eddington dubbed such motion 'overstable'. Of course, in order to be sure that the nuclear driving is actually sufficient to dominate the diffusive damping in the Sun, a proper stability calculation had to be performed, but that (Christensen-Dalsgaard et al., 1974), and all other subsequent calculations that do not contain an acknowledged error, indicated that the instability does indeed occur. What are the consequences? Either the motion saturates, or it triggers a different kind of (subcritical) flow which then takes over. In either case the mean stratification of the Sun must be changed (and the constraint $L_{\mathrm{n}} = L$ might even be violated), and with it so too is the neutrino flux changed. By how much?

Some years later Dziembowski (1983) studied an idealized model of the weakly nonlinear development of such an instability. He did so via what in dynamical-systems theory are called amplitude equations: the motion is presumed to be comprised of a superposition of components whose spatial structures are assumed to be identical to corresponding linear modes but whose amplitudes are determined by substituting the superposition into the full nonlinear dynamical equations and projecting the outcome onto each of the spatial structures of the modes. Dziembowski considered the coupling between the (parent) unstable gravity mode in question and a pair of similar stable daughter modes whose beat frequency resonates with the frequency of the parent. The daughters, acting in harmony, drain energy from the parent and thereby counteract the driving, limiting the amplitude of the parent. The rate at which they can do this is proportional to the product of their amplitudes, and because, in the Sun, diffusion is so weak, their amplitudes can grow sufficiently for the energy-draining process to be quite effective: the amplitude of the parent mode is therefore limited at a very low value, too low for the mode to have a significant effect on the mean structure of the Sun. It is interesting how the situation is different in the laboratory. In thermohaline convection, for example, a situation with stability characteristics similar to those of the solar core, growth of overstable gravity modes to large amplitudes does appear to occur, triggering subcritical thermohaline convective flow, which then modifies the stratification of the fluid dramatically. Indeed, the analogy between the thermohaline situation and the Sun had been invoked originally (Dilke and Gough, 1972) in discussing what the nonlinear outcome of the overstability in the Sun might have been, implicitly assuming, without justification, that the lower thermal damping in the Sun would have a lesser impact on limiting the amplitude of the gravity mode. But, as Dziembowski pointed out, the nonlinear regime is not like that: as the thermal diffusion coefficient decreases, the relative damping rate of the daughter modes decreases, and in the steady state (the only situation studied by Dziembowski) their amplitudes must therefore increase in order for them to dissipate the energy that the parent mode is extracting from the background stratification. Consequently their ability to extract energy nonlinearly from the parent mode increases yet more, and so decreases the limiting amplitude of the parent mode.

To achieve the steady state considered by Dziembowski, a precise resonance is required between daughter and parent modes: the frequencies must match to about a part in Q, where $Q \simeq 10^{11}$ is the quality factor of the parent mode. This is not impossible, because as order n and degree l of g modes becomes very large, the eigenfrequencies, which are proportional to $l/\sqrt{n^2+l^2}$, become very closely spaced, and with daughters having sufficiently high n and l an adequately close resonance can be found. But, of course, if n and l are very large, the damping rates are very high: that the daughter modes are effective in severely damping the parent is therefore not immediately assured. So Dziembowski used a probablistic argument to estimate the expectation of the amplitude of the parent: it was that value which was found to be too low to be of significance to the stratification of the Sun.

It was also of some concern that in order for the system to get close to the steady state, phase coherence between daughters and parent must be maintained (to within a part in Q) for several characteristic growth times of the parent, namely of order Q oscillation periods. That can hardly be realized in the Sun, because surely changes in the structure, associated with the solar cycle, perhaps, or even simply the general main-sequence evolution, must upset the delicate balance which is required for such precise resonance. Jordinson (2002) has recently investigated the dynamics of the oscillations under such changing circumstances, taking into account all parent-daughter interactions as they pass through resonance. The outcome was a parent amplitude precisely the same as Dziembowski's expectation. What really matters for appropriate phase maintenance is not the growth-time of the parent; it is the characteristic time spent near resonance, provided that that is long enough for the daughters to be excited by the parent essentially to their limiting amplitudes: more rapid evolution of the background state reduces that time, and therefore the energy extracted by a particular daughter pair, but it proportionately increases the rate of encountering other near-resonant daughter pairs, and the two effects cancel.

The discovery of the g-mode instability has spawned other conjectures concerning macroscopic motion in the solar core. Ghosal and Spiegel (1991) proposed that a shell of weak convection exists at the edge of the core, where the abundance of ^{3}He (which, when thrown out of balance with the pp chain of nuclear reactions, induces an increase in the temperature-sensitivity of the energy generation and thereby enhances the driving of both oscillatory and direct convective flow) is greatest. Subsequently, Cumming and Haxton (1996) proposed weak convective flow throughout the core. This hypothesis also implies that the nuclear reactions are thrown out of local balance, thereby modifying the neutrino fluxes. But helioseismological analysis, which I discuss in the next section, suggests that if any such motion were present, it must be so slow as to have only a minor effect on the distribution of hydrogen and helium, and by implication it would be unlikely to influence neutrino production significantly.

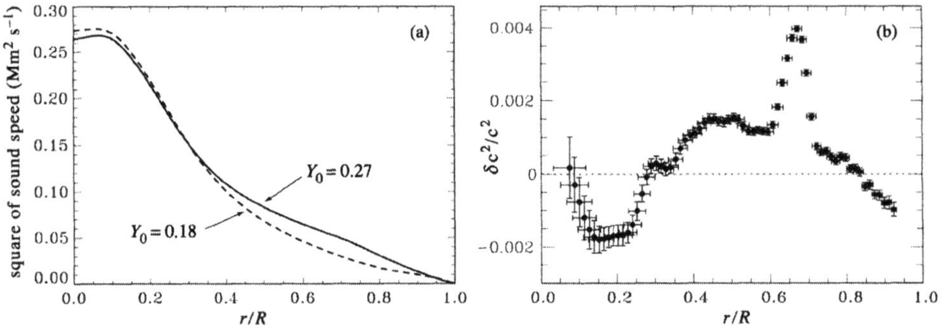

Figure 2. (a) Squares of the sound speed, c^2, in two solar models, obtained from the work of Christensen-Dalsgaard et al. (1979); the model with low Y_0 (and $Z = 0.001$) has been contaminated by infalling material to render the photospheric abundances consistent with observation. The ^{37}Cl neutrino fluxes, scaled to modern values, are 8.2 snu ($Y_0 = 0.27$, $Z = 0.02$) and 2.7 snu ($Y_0 = 0.18$); the observed value is 2.6 ± 0.2 snu. (b) Relative differences $\delta c^2/c^2$ between the spherically averaged squared sound speed in the Sun (inferred seismologically) and the squared sound speed in the standard model of Christensen-Dalsgaard et al. (1996). The seismic data are spatial averages, weighted by localized kernels whose characteristic widths (approximately full widths at half maxima) are denoted by the horizontal bars; the vertical bars represent formal standard errors (which are correlated). The theoretical model was computed with $Z_0/X_0 = 0.0245$, and $Y_0 \simeq 0.27$.

I close this section by mentioning that on several occasions since solar neutrinos were recognized as presenting a problem it has been suggested that gravity waves generated at the base of the convection zone (and even acoustic waves) might be an important agent in transporting heat and material. Although the issue is not closed, it seems at present to be unlikely that the waves are generated with sufficient amplitude to be of direct global importance to neutrino production.

3. Helioseismological Diagnosis

It had been known from quite early days (Iben, 1968) that solar models with very low heavy-element abundance Z, and consequently also low initial helium abundance Y_0, also have low neutrino luminosity, L_ν. Indeed, models can be constructed with L_ν compatible with the Homestake neutrino-flux data. Spectroscopic analysis is not compatible with the idea that the Sun is so deficient in Z (e.g. Anders and Grevesse, 1989; Grevesse and Noels, 1993). However, there has been discussion of the possibility that to account for the spectroscopic observations the directly observable surface layers have been contaminated by infalling cometary material, masking the interior composition (e.g. Joss, 1974). Although the contamination rate appears today to be too low to have the required effect, it may have been greater in the past, particularly at times when the Sun was passing through the spiral arms of the Galaxy. The contamination has little effect on the structure of the core. A model with $Z \simeq 10^{-3}$ (and $Y_0 \simeq 0.18$) can reproduce $L_{\nu 8}$, although, like all other standard models, it cannot reproduce the neutrino energy spectrum to within the standard error of modern measurements. Such a model has a relatively low central density, and a somewhat lower temperature than models with more favoured composition (because X is high and because the energy generation rate, approximately proportional to $\rho X^2 T^4$, is constrained in standard solar models by L_n); the central sound speed is higher. In the envelope, however, the sound speed is substantially lower (Figure 2a: one might naively have thought that this follows directly from considering how c^2 varies outwards from the core, after appropriate application of a theorem quoted by Faulkner and Swenson (1988), taking due account of the effect of the lower value of Z on the opacity, which controls the temperature gradient required to transport the luminosity, and the fact that X is greater in the low-Z model, leading to a lower mean molecular mass; but for a sound understanding, more thought is required). The sound-speed profiles of the two models in Figure 2a differ substantially, by greater than 30 per cent in some places. Therefore it is straightforward to distinguish between them seismologically.

I should point out that even though one might imagine the chemical composition of the radiative interior of a low-Z solar model to be masked by cometary debris, if such a model were actually to be representative of the Sun it would raise a serious issue in cosmology, because the initial helium abundance Y_0 of such a model is rather lower than the abundance resulting from the nuclear processing that is thought to have occurred immediately after the Big Bang.

It was evident from even the first inversion of helioseismic data that yielded the sound speed in the radiative interior (Duvall et al., 1985) that the low-Z model depicted in Figure 2a does not describe the Sun: the sound speed in the Sun differs from the model with $Y_0 = 0.27$ by barely the thickness of the solid curve. This result established directly that the neutrino problem was not resolvable simply by adjusting standard solar models. The relative difference between the Sun and a modern standard model (in which gravitational settling has been included) is illustrated in Figure 2b; although it is substantially greater than the standard errors in the inversion, it clearly confirms that the model reproduces the spherically averaged stratification within a few tenths per cent. Other standard models (e.g. Bahcall and Pinsonneault, 1992; Turck-Chièze et al., 2001) are similar. The present best estimate of the stratification of the Sun, inferred from more recent SOI data (Schou, 1999), is tabulated by Gough and Scherrer (2002).

It is evident from the close agreement of the model with the Sun that extensive mixing of chemical elements throughout the core of the Sun cannot have occurred: only a few tenths per cent smoothing of X appears to be permitted if all the helioseismic constraints (including those on the density) are taken into account. The conclusions of Dziembowski (1983) and Jordinson (2002) are therefore upheld. There is perhaps some room for weak macroscopic motion in the outer reaches of the core; that could have interesting dynamical implications, but it is unlikely to have a dramatic impact on neutrino production. The standard model appears to be a good first approximation to the Sun, and thus it will provide a useful tool for neutrino physics.

4. Concluding Remarks

The discovery of μ-neutrino transitions (Kajita, 1999) has offered a potential escape from the so-called neutrino puzzle: if μ neutrinos can undergo transition, then so, presumably, can the electron neutrinos that are produced in the Sun. Indeed, by combining results from SuperKamiokande and the Sudbury Neutrino Observatory, Ahmad et al. (2001; see also 2002) have shown that, with suitable transition rates, not only can standard solar models be reconciled with all the extant neutrino flux measurements, but that the measurements actually seem to imply a ^{8}B flux that is close to the prediction of the models.

Bahcall (2001) has declared that the close correspondence between observation and standard solar models is a triumph for the theory of stellar evolution. It is a triumph for helioseismological inference too. But, as is evident in Figure 2b, the difference between a typical model and the Sun is substantially greater than the standard measurement errors. For the benefit of neutrino physics, at least, those differences must be understood, and not merely explained away. They may be simply the product of minor errors in the microphysical parameters, such as opacity and the nuclear reaction rates, that are used to construct the standard models. But it is possible, although it seems less and less likely as time goes on, that they are symptoms of more serious deficiencies in standard modelling, and of that possibility we must continue to beware.

Acknowledgements

I am very grateful to Di Sword for preparing the LaTeX file, and to Richard Sword for helping with the diagrams.

References

Ahmad, Q.R. et al.: 2001, *Phys. Rev. Lett.* **87**, 071301–071306.
Ahmad, Q.R. et al.: 2002, *Phys. Rev. Lett.* **89**, 011301-1–5.
Anders, E. and Grevesse, N.: 1989, *Geochim. Cosmoch. Acta* **53**: 197–214.
Bahcall, J.N.: 1989, *Neutrino Astrophysics*, Cambridge Univ. Press.
Bahcall, J.N.: 2001, *Nature* **412**, 29–31.
Bahcall, J.N. and Loeb, A.: 1990, *ApJ* **360**, 267–274.
Bahcall, J.N. and Pinsonneault, M.H.: 1992, *Rev. Mod. Phys.* **64**, 885–926.
Bahcall, J.N., Bahcall, N.A. and Shaviv, G.: 1968, *Phys. Rev. Lett.* **20**, 1209–1212.
Bahcall, J.N., Bahcall, N.A. and Ulrich, R.K.: 1969, *ApJ* **156**, 559–568.
Christensen-Dalsgaard, J., Dilke, F.W.W. and Gough, D.O.: 1974, *MNRAS* **169**, 429–445.

Christensen-Dalsgaard, J., Gough, D.O. and Morgan, J.G.: 1979, *A&A* **73**, 121–128.
Christensen-Dalsgaard, J., Duvall, T.L. Jr, Gough, D.O., Harvey, J.W. and Rhodes E.J. Jr: 1985, *Nature* **315**, 378–382.
Christensen-Dalsgaard, J., Proffitt, C.R. and Thompson, M.J.: 1993, *Ap J. Lett.* **403**, L75–L78.
Christensen-Dalsgaard, J. et al.: 1996, *Science* **272**, 1286–1292.
Cumming, A. and Haxton, W.C.: 1996, *Phys. Rev. Lett.* **77**, 4286–4289.
Davis, R.: 1964, *Phys. Rev. Lett.* **12**, 303–305.
Dicke, R.H. and Goldenberg, H.M.: 1967, *Phys. Rev. Lett.* **18**, 313–316.
Dilke, F.W.W. and Gough, D.O.: 1972, *Nature* **240**, 262–264.
Dziembowski, W.A.: 1983, *Sol. Phys.* **82**, 259–266.
Eddington, A.S.: 1926, *The Internal Constitution of the Stars*, Cambridge Univ. Press.
Faulkner, J. and Swenson, F.J.: 1988, *Ap J. Lett.* **329**, L47–L50.
Ghosal, S. and Spiegel, E.A.: 1991, *GAFD* **61**, 161–178.
Gough, D.O.: 1992, in: G. Chardin, O. Fackler and J. Trân Thanh Vân (eds.), *Progress in Atomic Physics, Neutrinos and Gravitation*, Editions Frontières, Gif sur Yvette, Proc. XXVII Rencontres de Moriond, pp. 25–35.
Gough, D.O. and Scherrer, P.H., 2002, in: M. Huber, J. Geiss, J. Bleeker and A. Russo (eds.), *The Century of Space Science*, Kluwer, Dordrecht, in press.
Grevesse, N. and Noels, A.: 1993, in: N. Prantzos, E. Vangioni and M. Cassé (eds.), *Origin and Evolution of the Elements*, Cambridge Univ. Press.
Iben, I. Jr: 1968, *Phys. Rev. Lett.* **21**, 120–1212.
Iben, I. Jr and Mahaffy, J.: 1976, *Ap J. Lett.* **209**, L39–L43.
Jordinson, C., 2002, *MNRAS*, in press.
Joss, P.C.: 1974, *Ap J.* **191**, 771–774.
Kajita, T.: 1999, *Nucl. Phys. B (Proc. Suppl.)* **77**, 123–132.
Schou, J.: 1999, *Ap. J. Lett.* **523**, L181–L184.
Turck-Chièze, S., Nghiem, P., Couvidat, S. and Turcotte, S., 2001, *Sol. Phys.* **200**, 323–342.

NOVAE AS THERMONUCLEAR LABORATORIES

DONALD D. CLAYTON

Abstract. Fred Hoyle undertook a study of observational consequences of the thermonuclear paradigm for the nova event in the years following his 1972 resignation from Cambridge University. The most fruitful of these have been in the areas of gamma-ray astronomy, by which one attempts to measure the level of radioactivity in the nova envelope, and of presolar grain studies in laboratories, by which one measures anomalous isotopic ratios that fingerprint condensation in the thermonuclear event. This work summarizes progress with these two astronomical measures of the novae.

1. Introduction

Fred Hoyle took a pioneering view of everything to which he turned his attention. This includes the physical paradigm for the common nova phenomenon. I address that little-known theme of Hoyle's career. That theme went back far, at least to Hoyle's suggestion to Mestel of his 1952 study (Mestel, 1952) related to what would become nova theory. Following Hoyle's resignation from Cambridge University in 1972 his ability to travel for research increased markedly; and it was in that context that we met five times in the mid 1970s to seek new ideas for testing the nova as thermonuclear laboratory. Three working meetings were held at Rice University, which was able to pay his travel expenses and an honorarium for special public lectures; one meeting was at the Cardiff sponsored Workshop on Isotopic Anomalies, held at Gregynog in central Wales <1976 Hoyle with T. Gold>; and one meeting was at his home in Dockray, above Ullswater <1974 Hoyle and Clayton in Hoyle's Cockley Moor home>. Relevant photographs cited in this paper are available on the Clemson University web site for the history of nuclear astrophysics `http://photon.phys.clemson.edu/wwwpages/PhotoArchive`; and I guide the reader to them by <year title> in its photolist.

The guiding concept of the research program that we undertook was to discover experimental tests of the correctness of the modern thermonuclear paradigm for novae, which has seldom been stated better than by Hoyle (1955), whose creative description came far ahead of its time:

> If the outer skin of hydrogen should ever become degenerate an explosive condition must arise in a way already discussed in connection with the 'popping' of the (supernova) core. The important practical difference between the present case and the popping of the core is that the explosion now occurs on the outside of the star where it should be much easier to observe.
>
> The results of a precise calculation can be expressed in two ways. If the energy released were entirely converted into motion, the exploding material would

Astrophysics and Space Science **285:** 353–361, 2003.

> blow outward from the star with an average speed of about 1,000 kilometers per second. If on the other hand the energy were all converted into heat the star could keep shining for several weeks at a rate that exceeded the sun by 100,000 times. This suggests a relation with the exploding stars known as novae. That the observed ejection velocity of about 2,000 kilometers per second is higher than the average speed suggested by our calculations is probably important. It suggests that most of the exploding hydrogen is retained by the strong gravitational field of the star: that the heated hydrogen expands but for the most part does not possess enough speed to leave the star... which suggests that the amount of material actually expelled by the nova is very small, amounting to no more than about a hundredth of a percent of the whole mass of the star. After a time the main mass of the heated hydrogen must cool off and sink back to its former state. The process must then be repeated.

A timely stimulus to our research was the publication of new dynamic models based on this paradigm (Starrfield et al., 1972). Starrfield et al. (1972, 1974) presented 1-D hydrodynamical calculations that included the hot-hydrogen-burning reactions in conjunction with convective mixing. Their results enabled us to describe our theoretically devised tests without the need for constructing our own computer models. Three factors led us to believe that new astronomical tests might be possible: (1) time-dependent nuclear reactions of the hot CNO cycle produced unusual isotopic compositions and abundant radioactivity; (2) convective mixing of the hydrogen-burning zone at the base of the envelope to its surface; (3) new evidence for dust condensation in nova ejecta. Empowered by our mutual research visits we published three papers using this scenario during 1974–76. They were the first occasions on which gamma-ray lines from nova radioactivity and the isotopic signatures of presolar grains from novae were discussed, and they provided targets for empirical tests of nucleosynthesis and of nova theory. These are especially active new research areas today.

2. Radioactivity and Gamma Rays

Clayton and Hoyle (1974) described a nova atmosphere bathed in gamma rays as the result of intensely radioactive matter in the hydrogen envelope above the white dwarf core. Surface convection carries radioactive nuclei (primarily ^{13}N, ^{14}O, ^{15}O, ^{18}F, ^{22}Na and ^{26}Al) from the hot hydrogen burning at the base of the envelope to the surface from which the gamma rays can escape. Clayton and Hoyle argued that this radiation was detectable and that detection affords an outstanding opportunity to test the thermonuclear paradigm for the nova event. See <1973 Hoyle and Clayton at Rice> for a photograph of work on that paper and <1974 Hoyle and Clayton in Hoyle's Cockley Moor home> for a photograph at our second meeting following this paper's submission. In spirit this paper was similar to earlier arguments

TABLE I
Solar masses of radioactive nuclei ejected

Nova	M_{wd}	^{13}N	^{18}F	^{22}Na
CO	0.8	1.5×10^{-7}	1.8×10^{-9}	7.4×10^{-11}
ONe	1.15	2.9×10^{-8}	5.9×10^{-9}	6.4×10^{-9}

(Clayton and Craddock, 1965; Clayton, Colgate and Fishman, 1969) for diagnosing supernova physics in analogous ways.

Clayton and Hoyle did not treat the escape of continuum × rays and gamma ray lines with the modern Monte Carlo transport; that formulation had to wait in the nova case for Leising's 1986 Ph.D. thesis (Leising and Clayton, 1987), which was, to this writer's knowledge, the only treatment of any astronomical object by Monte-Carlo gamma-ray transport prior to the galvanizing appearance of supernova 1987A in February 1987, after which it became common. Instead they reasoned that the number of continuum gammas that escape is equal to the number emitted in the outer 8 gcm^{-2} of expanding envelope, producing a continuum flux of hard × rays near 0.001 $\mathrm{cm}^{-2}\ \mathrm{s}^{-1}$ for a nova at 1 kpc distance. That continuum has proven to be hard to detect because it exists for only a few minutes after the thermonuclear runaway. About 7 photons per cm^2 were predicted to reach the earth from such a nova, making the observational problems severe but not impossible. Leising and Clayton (1987) showed that inclusion of ^{18}F could increase the anticipated continuum fluence from positron annihilation by an order of magnitude. For a photograph see <1986 Leising and Clayton at Rice>. This difficult but worthy observational goal will require a large-angle continuous monitor of the sky in the hard-x-ray band, but it has the potential to demonstrate radioactive details of the thermonuclear event.

A less severe observational prospect was argued by Hoyle and Clayton to lie in ^{22}Na, with 2.6 yr lifetime, and which emits a nuclear deexcitation line at 1.275 MeV in addition to a positron. The lifetime allows the nova ejecta to thin prior to decay, so that all of the radioactive ^{22}Na nuclei can be seen; i.e. essentially all 1.275 MeV gamma rays will escape. The delay also affords years of planning time following the event. Clayton and Hoyle (1974) estimated a line flux as great as $10^{-3}\ \mathrm{cm}^{-2}\ \mathrm{s}^{-1}$ from a nova at 1 kpc.

Modern expectations for the masses of radioactive nuclei ejected by novae as calculated by the Barcelona group (Hernanz, Gomez-Gomar and Jose, 2001) are listed for two cases in Table I: namely, the low-mass CO nova and the high-mass ONe nova.

Quite evidently, the ONe white dwarf ejects 100 times more ^{22}Na owing to the much larger abundance of its seed nucleus, ^{20}Ne, in a stellar core that had previously been able to burn carbon before settling to a white dwarf. For this reason it had been anticipated that ONe novae were the more likely objects in which to detect the 1.275 MeV line.

3. OSSE, COMPTEL and INTEGRAL

Mission planning for the Oriented Scintillation Spectrometer Experiment (OSSE) and the Compton Telescope (COMPTEL) for Compton Gamma Ray Observatory used both the nova and supernova expectations for gamma-ray lines from radioactivity. Owing to their scientific excitement and uniqueness, those expectations were useful in developing scientific support for that NASA mission. But during the 1991–2000 mission only upper limits could be established for these fluxes by OSSE and COMPTEL. Three studies of novae during the CGRO mission, N Her 1991, N Cyg 1992 and N Vul 1984 #2, found that the ejected mass of ^{22}Na was less than 10^{-8} solar masses (Leising, 1997). By comparison with Table I it is evident that the observational limit is almost equal to the ^{22}Na mass expected from ONe novae. It is greatly to be hoped that a suitable nova for the SPI spectrometer aboard INTEGRAL will occur nearby (< 3 kpc) during its mission (scheduled launch October 2002).

By treating the gamma-ray background in a specific way the COMPTEL team (Iyudin et al., 2001) was able to construct a ^{22}Na light curve for slow nova N Cas 1995 by combining many lengthy observations of it over a three-year period. Their published light curve in the 1.275 MeV line is reproduced here in Figure 1. It is striking that the series of measurements (error bars) follows so well a theoretical light curve for a slow nova expanding at 150 km/s (sequence of small crosses). If this detection holds up it runs counter to the belief that the ONe novae, which make detectable levels of ^{22}Na, are fast novae rather than slow. Even though this exciting fit can be regarded only as a marginal detection at this time, it illustrates better than words the nature of the diagnostic test advanced by Clayton and Hoyle (1974). Future gamma-ray astronomy space missions will continue to rely on the ideas of their paper.

4. Nova Presolar Grains in the Solar System

It is fitting that when the Apollo 11 crew returned the first lunar samples in 1969, Fred Hoyle was invited to be banquet speaker for the first lunar science conference in Houston. Hoyle had a long history of studies of the origin of the solar system as well as of pioneering studies to explain the very existence of interstellar dust from which the solar system was assembled. The experimental techniques that were being supported by NASA for study of lunar samples in laboratories around the world were found doubly useful when a large carbonaceous meteorite fell in Mexico, also in 1969. Samples of it stimulated new studies of meteorites that

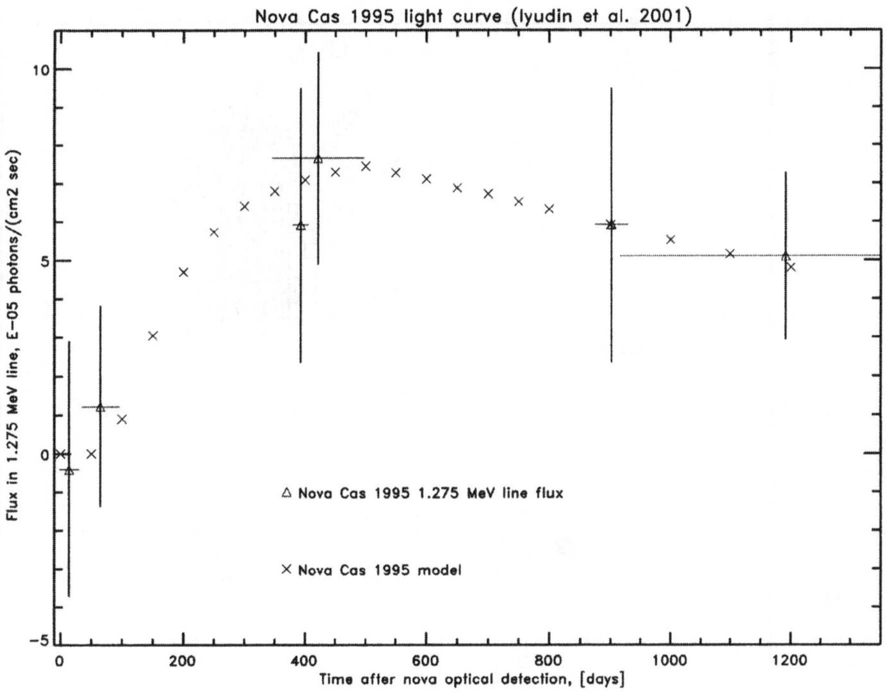

Figure 1. The 1.275 MeV light curve of nova N Cas 95 obtained from COMPTEL data (Iyudin et al., 2001). The wide field of view enabled many few-week observations over almost four years. The crosses show the expected transparency for nova matter expanding slowly near 150 km/s. This is a marginal detection of radioactive ^{22}Na (Clayton and Hoyle, 1974) in ejecta from a nova.

led to the first identifications of isotopic anomalies in solar system material and to pinning down the carriers of those isotopic anomalies until, by 1987, presolar grains themselves were being routinely extracted from the meteorites (Bernatowicz and Zinner, 1997).

In a second paper, Clayton and Hoyle (1976) presented arguments before the discovery of presolar grains that such grains from nova explosions should exist in the interstellar medium and that their presence in solar system samples might be recognized isotopically. They interpreted the observations of bright infrared excess after two months of Nova Serpentis 1970 (Geisel, Kleinmann and Low, 1970) as evidence of the thermal condensation of dust there. Clayton and Wickramasinghe (1976) modeled that dust condensation more explicitly. Clayton and Hoyle (1976) regarded the grains to be graphite, which Hoyle and Wickramasinghe (1962) had argued to condense in red giants and to provide a source of interstellar grain cores; but their considerations apply equally well to SiC grains, which have the measurement advantage of a larger concentration of silicon. Clayton and Hoyle set forth anticipated isotopic peculiarities that would be expected in presolar nova grains. Their predictions inspired searches for nova grains after the 1987 isolation of

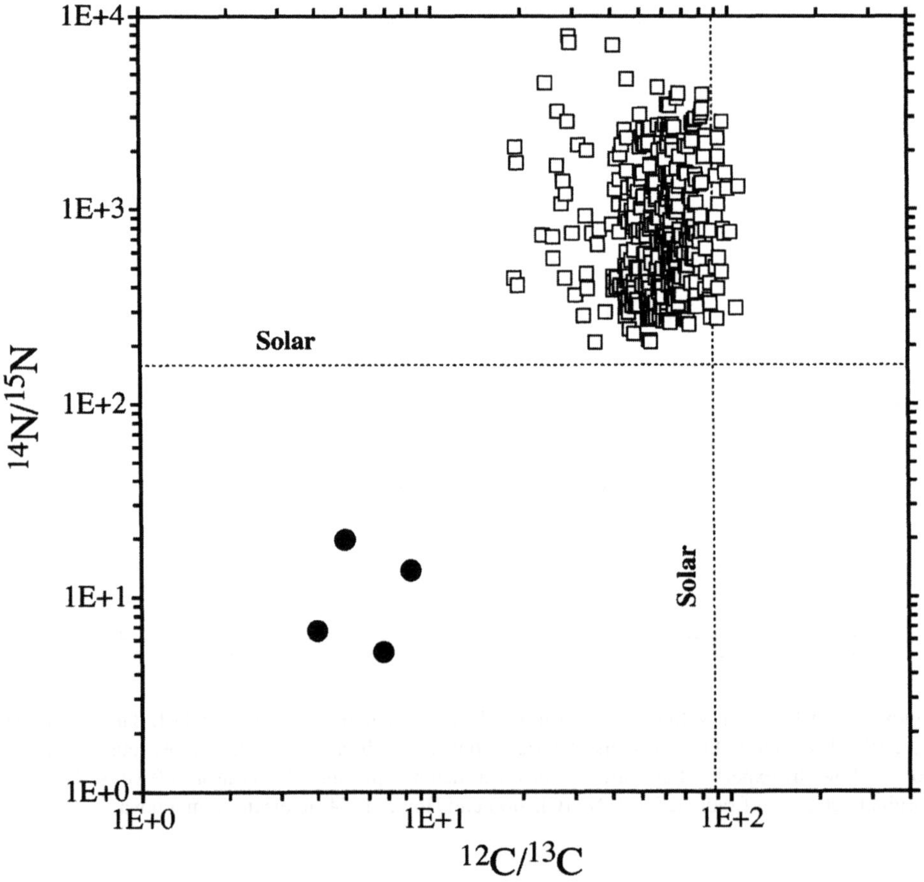

Figure 2. Two families of presolar SiC grains are located by their $^{14}N/^{15}N$ and $^{12}C/^{13}C$ isotopic ratios. Terrestrial materials plot at the intersection of the two dashed lines. Mainstream grains, which constitute about 93% of SiC grains in meteorites, are open squares; the four candidate nova SiC grains (Amari et al., 2001) are solid circles. Note the huge variations requiring the logarithmic scales. The ^{15}N and ^{13}C richness of the four nova grains were predicted markers of nova grains (Clayton and Hoyle, 1976). More abundant mainstream grains clearly differ from the nova grains. They arise from carbon star winds. Adapted from Amari et al. (2001).

presolar carbonaceous refractory grains. The discovery of presolar grains silenced doubters that man could ever hold in his hand a solid object that existed before the earth and planetary rocks came into existence. Laymen and scientists alike were stunned to learn that this had happened. Hoyle lived barely long enough to hear these predictions come true in laboratory studies by Amari et al. (2001) utilizing sputtering-ion mass spectrometry (SIMS). See <2000 Clayton and Hoyle in Bournemouth> for a photo of our final discussion of this discovery.

Figure 2, adapted from Amari et al. (2001), contrasts the location of the four nova SiC grains from those of AGB star grains in the $^{14}N/^{15}N$ vs. $^{12}C/^{13}C$ plot.

See <2001 Amari and others in Ringberg> for a photo of Amari just before her presentation on nova grains. Each of these grains is about one micrometer in size, presenting a statistically secure sample of 10^{12}-10^{14} atoms, more than enough for isotopic analysis. Sputtering from different spots on a grain using SIMS yields identical isotopic ratios, assuring that the grains are indeed grown entirely within a special stellar atmosphere. The precision of these SIMS isotopic ratios, better than 1%, far exceeds astronomical capability in any stellar atmosphere. They are high-resolution observations of stellar ejecta from stars; but the stars must be identified. The four nova grains at lower left of Figure 2 possess very low $^{14}N/^{15}N$ ratios (^{15}N richness) and very low $^{12}C/^{13}C$ ratios (^{13}C richness). Clayton and Hoyle (1976, p. 495) had stated, 'the nitrogen gas ejected by the nova ... is ^{15}N-rich, so positive $^{15}N^*$ is a possible result of nova grains.' They used the symbol $^{15}N^*$ to represent 'excess ^{15}N' in comparison with the solar ratio $^{14}N/^{15}N = 272$. They worried, however, that if N did not condense effectively within SiC, as it is now known to do, that the ^{15}N excess might not be evident; but in truth it is very evident owing to refractory substitution of AlN within SiC during condensation. And they also stated, 'the first feature is that most of the stable carbon is ^{13}C'. These distinguishing predictions are now confirmed in Figure 2 if the identification of nova grains by Amari et al. is correct. By contrast the so-called 'mainstream SiC grains' are mostly ^{14}N-rich, the natural result of dredgeup of CN-burning material in red-giant carbon stars, and they possess moderate $^{12}C/^{13}C$ ratios near 60, a value typical of galactic carbon stars.

Carbon and nitrogen are but two of the isotopic ratios that are useful for the characterization of presolar grain families in a multi-isotopic space. Additional elements are used to delineate the grain families more clearly by the volumes that the families occupy within the multi-dimensional isotopic space (Amari et al., 2001 and references therein). Two of the four nova grains have had their $^{26}Mg/^{24}Mg$ ratios measured, and they contain so much excess ^{26}Mg and such large elemental Al/Mg ratio that that excess can be attributed to the condensation of live ^{26}Al within the nova grains in amount $^{26}Al/^{27}Al = 0.011$ and >0.08. Such large Al ratios are characteristic of hot-hydrogen burning, as Clayton and Hoyle (1976) emphasized in their predictions. In similar vein they pointed out the expectation of ^{22}Ne-rich neon in nova grains as a result of the condensation of radioactive ^{22}Na produced in the nova burning; however, such sensitive measurements on individual SiC grains have not yet been made for the four nova SiC grains of Figure 2.

Clayton and Hoyle (1976) also pointed to excess ^{30}Si abundance, which should be very evident in those novae having high peak burning temperatures. This prediction has been quantified much more accurately by Jose and Hernanz (1998), who calculated especially large ^{30}Si excesses in the ONe novae. Because the SiC grains (Figure 2) contain so much silicon, this old expectation has been testable by Amari et al. (2001). They demonstrated the existence of excess ^{30}Si in each of the four grains in the lower left of Figure 2. Percentage excesses of the $^{30}Si/^{28}Si$ ratio above the solar value were 112%, 24%, 12% and 15%. The much smaller variations (positive in two grains and negative in two) of $^{29}Si/^{28}Si$ in the same grains demonstrate that the variations constitute a ^{30}Si excess.

It can therefore be claimed that isotopic ratios within four elements confirm the nova identification for these four SiC grains. It had taken 25 years since the 1976 predictions to find the presolar grains and to characterize them isotopically in the lab.

Immediate puzzles for these grains focus on the small sizes of their isotopic excesses in comparison with those generated in the burning zones of the novae and on the microphysics of the condensation process for SiC in novae. To the former it can be said (Amari et al.) that no more than a few percent of the atoms within the condensed grains can be those produced directly in the burning zone. It is not yet clear if that degree of dilution by mixing within the accreted envelope is to be expected, but it does not seem unreasonable. The second issue confronts the oxygen richness (O/C >1) in the nova, both in the accreted envelope and in the burning zone itself. This oxidizing environment requires that in thermal equilibrium SiC would not condense; however, Clayton, Liu and Dalgarno (1999) and Clayton, Deneault and Meyer (2001) have shown how radioactivity enables kinetic condensation of carbon to proceed by disrupting the relatively inert CO molecule, which would otherwise remove carbon from chemical availability. See <1998 Clayton with Weihong Liu>. If the radioactive environment during 20-80 days after nova outburst is not strong enough to dissociate CO, there may exist sufficient ultraviolet in the postexplosive ejecta to accomplish the same result.

5. Coda

The initial successes described in this paper ensure that future observations of thermonuclear novae will feature astronomical gamma-ray searches for both prompt hard $\times$ rays and for the radioactive ^{22}Na gamma-ray line during the first years following a nearby nova and, here on earth, isotopic and condensation studies of identified presolar grains from novae. Fred Hoyle's name will be forever associated with such studies. His advocacy of these scientific targets constitutes a very successful, if less-well-known, aspect of his multifaceted contributions to astrophysics.

Acknowledgements

I thank Sachiko Amari, Anatoly Iyudin and Mark Leising for helpful support of aspects of this presentation. My work with Fred Hoyle on these goals was supported over many years by NASA's High Energy Astronomy Program and by its Cosmochemistry Program.

References

Amari, S., Gao, X., Nittler, L.R., Zinner, E., Jose, J., Hernanz, M. and Lewis, R.S.: 2001, *ApJ* **551**, 1065.

Bernatowicz, T. and Zinner, E.: 1997, *Astrophysical Implications of the Laboratory Study of Presolar Materials*, AIP Conf. Proc. 402, AIP, Woodbury.

Clayton, D.D.: 2000, *Photo Archive for the History of Nuclear Astrophysics*, <http://photon.phys.clemson.edu/wwwpages/PhotoArchive>. Specific photos are referenced as <year title> of its PhotoList.

Clayton, D.D., Colgate, S.A. and Fishman, G.J.: 1969, *ApJ* **155**, 75.

Clayton, D.D. and Craddock, W.: 1965, *ApJ* **142**, 189.

Clayton, D.D., Deneault, E. and Meyer, B.S.: 2001, *ApJ* **562**, 480.

Clayton, D.D. and Hoyle, F.: 1974, *ApJ* **187**, L101.

Clayton, D.D. and Hoyle, F.: 1976, *ApJ* **203**, 490.

Clayton, D.D., Liu, W. and Dalgarno, A.: 1999, *Science* **283**, 1290.

Clayton, D.D. and Wickramasinghe, N.C.: 1976, *Ap. Sp. Sci.* **42**, 463.

Hernanz, M., Gomez-Gomar, J. and Jose, J.: 2001, in: S. Ritz, N. Gehrels and C. Schrader (eds.), *Gamma 2001*, AIP Conference Proceedings **587**, New York, pp. 498–507.

Hoyle, F.: 1955, *Frontiers of Astronomy*, Harper and Brothers, New York, pp. 161–162.

Hoyle, F. and Wickramasinghe, N.C.: 1962, *MNRAS* **124**, 417.

Iyudin, A., Schoenfelder, V., Strong, A.W., Bennett, K., Diehl, R., Hermsen, W., Lichti, G.G. and Ryan, J.: 2001, in: S. Ritz, N. Gehrels and C. Schrader (eds.), *Gamma 2001*, AIP Conference Proceedings **587**, New York, pp. 508–512.

Jose, J. and Hernanz, M.: 1998, *ApJ* **494**, 680.

Leising, M.D., and Clayton, D.D.: 1987, *ApJ* **323**, 159.

Leising, M.D.: 1997, in: C. Dermer, M. Strickman and J. Kurfess (eds.), *Proc. Fourth Compton Symposium*, AIP Conference Proceedings **410**, New York, pp. 163–170.

Mestel, L.: 1952, *MNRAS* **112**, 598.

Starrfield, S., Sparks, W.M. and Truran, J.W.: 1974, *ApJ Suppl* **261**, 247.

Starrfield, S., Truran, J.W., Sparks, W.M. and Kutter, G.S.: 1972, *ApJ* **176**, 169.

THE *EDDINGTON* MISSION

IAN ROXBURGH
Astronomy Unit, Queen Mary, University of London, London E1 4NS, UK
Observatoire de Paris, Place Jules Janssen, 92195 Meudon, France

Abstract. The *Eddington* mission was given full approval by the European Space Agency on the 23rd May 2002, with launch scheduled for 2007/8. Its science objectives are stellar evolution and asteroseismology, and planet finding. In its current design it consists of 4 × 60 cm folded Schmidt telescopes, each with $6^o \times 6^o$ field of view and its own CCD array camera. *Eddington* will spend 2 years primarily devoted to asteroseismology with 1–3 months on different target fields monitoring up to 50,000 stars per field, and 3 years continuously on a single field monitoring upwards of 100,000 stars for planet searching. The asteroseismic goal is to be able to detect oscillations frequencies of stars with a precision 0.1–0.3 μHz, to probe their interior structure and the study the physical processes that govern their evolution.

Keywords: space missions: *Eddington*, stars: asteroseismology, stellar evolution; planets: planet finding, earth like planets, habitable planets

1. Preface

Fred Hoyle's books – The Nature of the Universe and Frontiers of Astronomy – played a major role in encouraging me to seek to become a professional astronomer; his graduate lecture courses in Cambridge – on Stellar Evolution and on Relativity and Cosmology – laid the foundations of my graduate education. His technical papers were a constant source of inspiration. Amongst Fred's many outstanding contributions to astronomy, his work on stellar evolution was gigantic and has stood the test of time. I like to think that were Fred alive today he would strongly support the *Eddington* mission whose goals are to test and advance our understanding of stellar evolution using the new technique of *asteroseismology*; the use of the oscillation frequencies of stars to determine their interior structure and physical processes that govern their evolution. Fred also made major contributions to our efforts to understand the origin of our solar system and I am sure he would also strongly support the parallel goal of the *Eddington* mission; to detect planets and planetary systems around other stars and so to improve our understanding of the formation of planetary systems.

2. The *Eddington* Mission

The *Eddington* mission was proposed in early 2000 to ESA in response to the 'Call for mission proposals for two flexi-missions (F2 and F3)', released in October 1999 in the context of the Horizons 2000 programme. The proposal was submitted by an international scientific team led by Roxburgh, Christensen-Dalsgaard and Favata (2000). It built on the work done over two decades on previous proposals that had been studied within ESA and CNES (*EVRIS, PRISMA, PRISMA2, STARS*). The mission has two complementary primary scientific aims, to produce the data on stellar oscillations necessary for understanding the interior structure and evolution of stars, and to detect and characterize habitable planets around other stars. Both scientific goals will be achieved by performing highly accurate time-resolved photometry on a large number of stars, with a simple payload which is essentially a wide field broad-band photometer.

The accurate light curves will be used to determine stellar oscillation frequencies, and thus allow the use of asteroseismic tools to probe the interior structure of stars, quantitatively determining, e.g., the size of convective regions, the structure of regions with steep changes in chemical composition and internal rotation and to therefore determine very accurate stellar ages. Some asteroseismic tools are already available (see e.g. Roxburgh, 2002) but much work needs to be done to extract the maximum science from the measurements that will be made by *Eddington*.

At the same time, planets as small as the Earth orbiting the target stars in the habitable zone will be found through the temporary drops in the stellar light caused by their transits; very many other planets and planetary systems will be detected providing the information needed to advance our understanding of the formation of planetary systems and hence of our own solar system.

Following the favourable reviews by ESA's scientific advisory bodies (the Astronomy Working Group and the Space Science Advisory Committee) *Eddington* was selected in March 2000 as one of the 4 missions for which an assessment study carried out in the course of the spring and summer of 2000 (Favata et al., 2000). Following presentation of the studies in September 2000, *Eddington* was selected with a 'reserve status'. Study activities continued in the course of 2001 and 2002, and in May 2002 *Eddington* was approved by ESA's Science Programme Committee to be implemented in the framework of the *Herschel* and *Planck* projects in ESA's rescoped science programme **Cosmic Vision**.

3. Overview of the Mission

DATA RIGHTS AND OBSERVING PROGRAMME

Eddington is a facility-type mission, for which the observing plan will be the result of a combination of an open, competitive proposal cycle with a broad community consulting process, and for which the resulting data are proposed to be available to the scientific community, without proprietary rights for individual proposers.

Payload configuration and data product

The primary *Eddington* data product is a set of relatively calibrated photometric light curves for each star in the field of view down to a predefined magnitude limit. To maximize the field of view and collecting area the payload is baselined to consist of 4 parallel very similar (possibly identical) instruments, each of them independently producing a light curve, which will be merged a posteriori. Each of the 4 instruments consists of 3 basic components: (1) a wide field optical telescope, (2) a mosaic CCD camera (named 'EddiCam') and (3) a data processing unit (DPU).

Mission procurement

ESA will have overall responsibility for the mission, and will in particular be responsible for the procurement of the spacecraft and of of the ground segment, for the integration of the payload and spacecraft units, for the launch and operations, the acquisition of the data, their reduction, archiving and their distribution to the holders of the data rights. The payload is planned to be procured through a partnership between ESA and a consortium of scientific institutes.

ESA is planned to operate the Scientific Operations Centre (SOC) which will process and archive the data from the instrument and deliver the scientific data products (relatively calibrated light curves) to the holders of the data rights.

4. The Current Mission Concept

The final form of the mission will be decided following detailed industrial studies and advice from the *Eddington* Science Team. The current working concept is as follows:

4×0.6m folded Schmidt telescopes each with $6^o \times 6^o$ field of view.
Mosaic CCD 'panoramic' cameras 8 E2V 42-C0 chips.
Orbit at L2 – 95% duty cycle – 5 year mission.
3 years on 1 field (planet finding + asteroseismology).
1 – 3 months on asteroseismology fields (total 2 years).
Launch: 2007/8.

Asteroseismology fields

Max. no. of stars per field: 2000 ($m_v < 11$), 50,000 ($m_v < 15$).
Precision on frequencies: 0.1 μHz (goal), 0.3 μHz (bottom line).
Frequency range: 1 μHz – 100 mHz (goal), 1 μHz – 10 mHz (bottom line).
Magnitude range: 3 – 15 (goal), 5 – 15 (bottom line).
Time sampling: $\leq$ 30 sec (baseline), $\leq$ 5 sec (some targets).

Chromatic information: to be determined.

PLANET FINDING FIELD *(also available for asteroseismology)*

No. of stars: > 100,000 ($m_v < 17$).
Time sampling: 30 sec (goal), 600 sec (bottom line).

Acknowledgements

First I wish to acknowledge the debt I owe to Fred Hoyle whose books, lectures, and technical papers were a source of inspiration and education.

I also wish to record my thanks to the many scientists who have contributed over two decades to convincing ESA to fly an asteroseismology mission by working on the precursor proposals *PRISMA, PRISMA2* and *STARS*, and especially to those who worked for the selection of *Eddington*. In particular I record the contribution to *Eddington* made by the Assessment Study Team, the successor *Eddington* Science Team, and technical support from ESA.

Finally, and most importantly, I wish to record my thanks to Fabio Favata, the ESA mission scientist for *Eddington*, both for his major contribution to, and leadership of, the project and especially for his navigational skills in charting a course from reserve to fully approved status in a time of decreased budget within ESA.

References

Favata, F., Roxburgh, I.W. and Christensen-Dalsgaard, J. eds.: 2000, with Aerts, C., Antonello, E., Catala, C., Deeg, H., Gimenez, A., Grenon, M., Pace, O., Penny, A., Schneider, J. and Waltham, N., Eddington, *A Mission to Map Stellar Evolution through Oscillations and to Find Habitable Planets*, Report of Assessment Study, ESA-SCI(2000)8.

Roxburgh, I.W., Christensen Dalsgaard, J. and Favata, F. eds.: 2000, with Antonello, E., Baade, D., Badiali, M., Baglin, A., Bedding, T., Brown, T., Catala, C., Collier, A., Dziembowski, W., Gilmore, G., Gimenez, A., Gough, D., Horne, K., Kjelsden, H., Leger, A., Penny, A., Preite-Matinez, A., Rivinus, Th., Schneider, J., Stefl, Z., Sterken, C. and Weiss, W., *Eddington – A Stellar Physics and Planet Finder Explorer*, Proposal submitted to ESA in response to the ESA F2/F3 call for proposals.

Roxburgh, I.W.: 2002, in: F. Favata, I.W. Roxburgh and D. Galadi (eds.), *The Tools of Asteroseismology*, The First Eddington Workshop 'Stellar Structure and Habitable Planet Finding', ESA SP-485, p. 75.

BLACK HOLE BINARY DYNAMICS

SVERRE J. AARSETH
Institute of Astronomy, Madingley Road, Cambridge CB3 0HA
E-mail: sverre@ast.cam.ac.uk

Abstract. We discuss a new N-body simulation method for studying black hole binary dynamics. This method avoids previous numerical problems due to large mass ratios and trapped orbits with short periods. A treatment of relativistic effects is included when the associated time-scale becomes small. Preliminary results with up to $N = 2.4 \times 10^5$ particles are obtained showing systematic eccentricity growth until the relativistic regime is reached, with subsequent coalescence in some cases.

1. Introduction

The problem of the formation and dynamical evolution of a black hole (BH) binary with massive components is of considerable topical interest. Several previous efforts employed direct integration methods to elucidate the behaviour of such systems but applications to galactic nuclei pose severe limitations with regard to the particle number which can be investigated. The formation is usually envisaged as the end product of two separate galactic nuclei (or one being a dwarf galaxy) spiralling together by dynamical friction, but there are other scenarios.

Notwithstanding the particle number limitations imposed by the need for accurate numerical treatment, much can be learnt about the early evolution of BH binaries by studying smaller systems. It is well established that the presence of two massive bodies in a stellar system leads to their rapid inward spiralling and inevitable formation of a dominant binary. In fact, this development was already discussed 30 years ago (Aarseth, 1972) for a relatively small system of $N = 250$ particles. Here a massive binary absorbed about 90% of the total cluster energy after only 50 initial dynamical (or crossing) times. This calculation was also the earliest demonstration of two-body regularization in an N-body context.

Upon formation, the subsequent binary evolution is subject to a steady shrinkage of the semi-major axis and ejection of particles resulting from slingshot interactions. Although the rate of shrinkage decreases significantly, the corresponding energy increase is fairly constant (Quinlan and Hernquist, 1997; Milosavljević and Merritt, 2001). These investigations employed two-body regularization in order to reduce the systematic errors associated with direct integration of hard binaries. Alternative studies based on a small softening of the Newtonian potential also yielded similar results (Makino, 1997). The long-term evolution is characterized by significant depletion of the central region which ceases to be representative of a realistic system. However, even the use of chain regularization (Mikkola and Aar-

Astrophysics and Space Science **285:** 367–372, 2003.

seth 1993) for treating compact subsystems is not sufficient to prevent numerical problems.

2. Special Binary Treatment

In view of the numerical problems outlined above, it is highly desirable to develop a more suitable integration method. A critical appraisal of chain regularization with two massive members reveals that their contribution to the Hamiltonian energy dominates and hence solutions of the equations of motion for any other members are subject to the loss of precision. This recognition led to the construction of a new method which is based on kinematical considerations (Mikkola and Aarseth, 2002). Briefly, a special time transformation is combined with the standard leapfrog scheme, thereby avoiding a Hamiltonian formulation. This allows extremely close two-body encounters to be studied without significant loss of accuracy. The interested reader is referred to the published description for more details.

The new method is based on including all the $N(N-1)/2$ interaction terms and solving the equations of motion by the Bulirsch-Stoer (1966) integrator. However, all the solutions need to be advanced with the same time-step which limits the practical membership severely. Hence this formulation can only be used to describe the motions of a compact subsystem. The implementation of such a solution method into a large N-body simulation code is somewhat analogous to that for chain regularization and has been outlined elsewhere (Mikkola and Aarseth, 1993; Aarseth, 1999). In the following we comment on some special features of the BH scheme.

The vital question concerning binary BH evolution is whether a stage can be reached where the gravitational radiation time-scale is sufficiently short for coalescence or significant shrinkage to occur. Previous simulations did not address this issue, mainly because the calculations were terminated prematurely for technical reasons. In the present scheme we have included the 2.5 post-Newtonian approximation for the most critical two-body interaction (Soffel, 1989). An estimate of the smallest semi-major axis which can be reached in a system of N particles can readily be made for a given mass ratio. This size is several orders of magnitude outside the relevant range for reasonable system parameters. However, small two-body separations can also be achieved if the eccentricity becomes large enough.

Although large eccentricities were not reported by other investigators, more careful calculations do show significant eccentricity growth during the late stages. Hence this behaviour justifies the extra cost of including the relativistic terms, but only when the corresponding time-scale is less than the expected calculation time. We have $a/\dot{a} \propto a^4(1-e^2)^{7/2}$ for the decay time, where a is the semi-major axis and e is the eccentricity. Since the time-scale is quite long for circular orbits, we need large values of e before activating the relativistic treatment which is implemented at

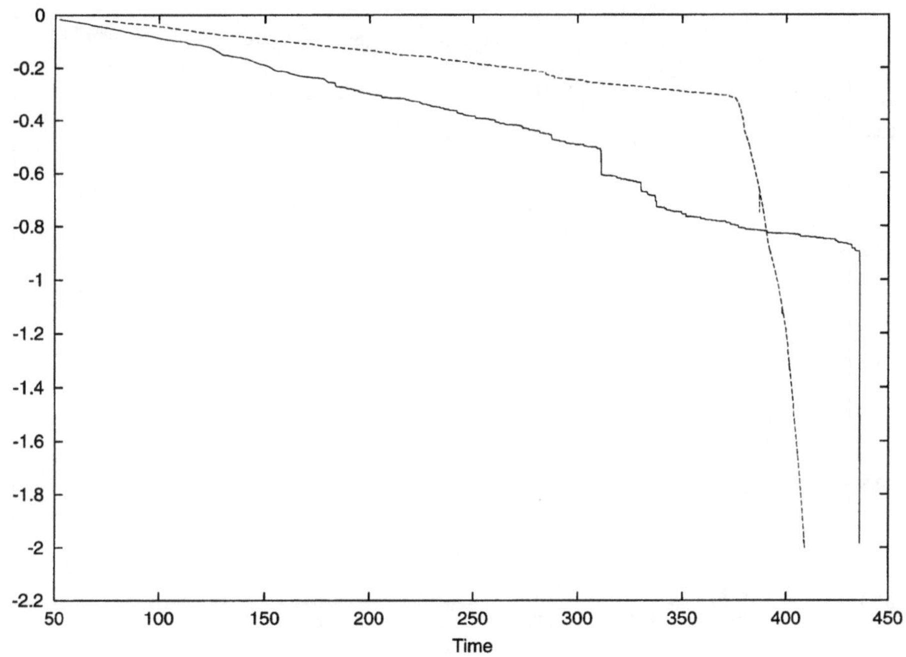

Figure 1. Binding energy of BH binaries. The upper curve (dashed line) is for $N = 2.4 \times 10^5$ and the lower curve (solid line) for $N = 1.2 \times 10^5$. Time is in scaled N-body units where one initial crossing time is 2.8 or about 1 Myr.

different levels of complexity. Thus we distinguish between the classical radiation term and two different expansion orders which describe the post-Newtonian acceleration and relativistic precession, respectively. Finally, coalescence is defined to take place if the BH separation becomes less than three Schwarzschild radii.

3. Numerical Results

The initial conditions consist or two cuspy dwarf galaxy models with N_0 equal-mass particles of mass $\bar{m}$ at the apocentre of an eccentric orbit ($e = 0.8$) having a separation of $8r_h$, where r_h is the local half-mass radius. A single BH of mass $m_{BH} = (2N_0)^{1/2}\bar{m}$ is placed at the centre of each system. The availability of the special-purpose GRAPE-6 supercomputer together with a fast workstation host allows quite large particle numbers to be studied. Here we report briefly on two recent simulations with $N_0 = 6 \times 10^4$ and 1.2×10^5 particles, making a total of 1.2×10^5 and 2.4×10^5 members, respectively. The two mass distributions soon combine into one slightly elongated system, with the dominant binary already formed at the centre after only about 20 crossing times in both models. Then follows a period of constant energy gain where the BH binary is advanced by standard two-body

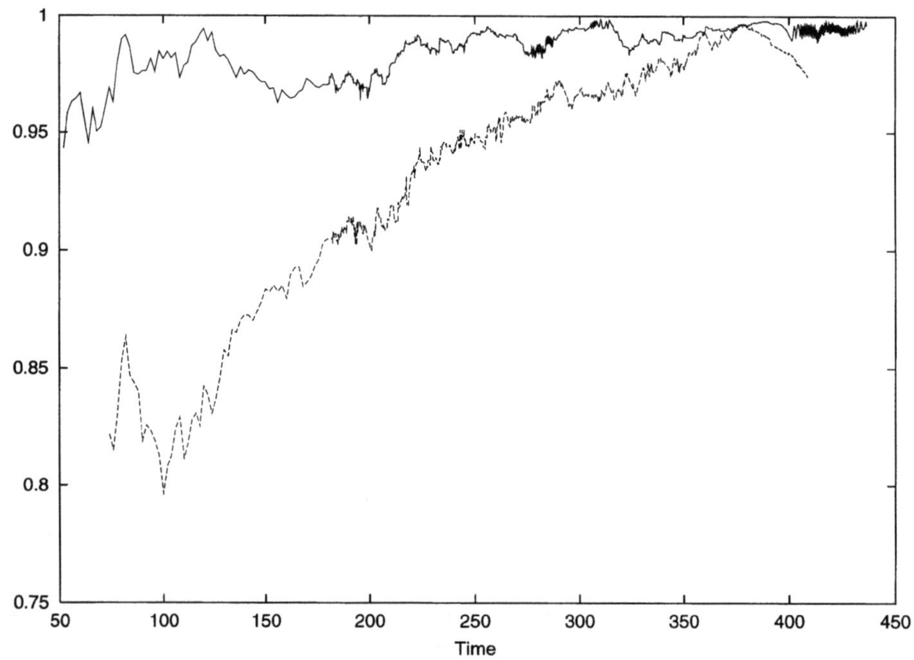

Figure 2. Eccentricity evolution of BH binaries. The upper curve is for $N = 1.2 \times 10^5$ and the lower curve for $N = 2.4 \times 10^5$.

regularization. A switch is made to the new method when the binary becomes super-hard; i.e. $a \leq 10^{-4} r_{\mathrm{h}}$. The subsequent slow evolution necessitates a large number of perturbed binary orbits to be studied.

The increase of the binding energy, $E_{\mathrm{BH}} = -m_{\mathrm{BH}}^2/2a$, is illustrated in Figure 1. As a result of the scaling procedure for two subsystems, the initial total energy is $E_{\mathrm{tot}} = -1.05$. Not shown on the plot is the final value $E_{\mathrm{BH}} = -61$ with the corresponding semi-major axis $a = 2.7 \times 10^{-7}$ for the smallest system, compared to $r_{\mathrm{h}} \simeq 1$ initially. Scale factors $r_{\mathrm{h}} = 4\,\mathrm{pc}$ and $\bar{m} = 1\,M_{\odot}$ were chosen. In other experiments we have demonstrated that a separation of three Schwarzschild radii can be reached without numerical problems. However, the end result of coalescence is ensured once the eccentricity starts to decline significantly, in which case the present purpose is achieved.

The strong binary evolution gives rise to the ejection of high-velocity particles by the slingshot mechanism. Although the effective mass ratio is about 1000 in the largest system, these ejections still result in significant recoil velocities acquired by the binary. Hence the typical velocity of the central object exceeds the standard equipartition value by a considerable factor, which has implications for the so-called loss-cone effect. However, it should be emphasized that the present results cannot be scaled directly to systems with much larger mass ratios.

The eccentricity evolution of the two systems is shown in Figure 2. Although the initial eccentricities are relatively high, the trend is for a gradual increase superimposed on fluctuations due to external perturbations. Two subsidiary maxima, $e_{\max} \simeq 0.998$ and 0.997, are first reached in the smaller system, followed by temporary declines before the final stage where coalescence sets in. The eccentricity growth is more pronounced in the second model. Note that $\dot{e} < 0$ during the final approach to coalescence.

The above examples should be considered as tests of the method rather than giving definite results. In this respect the outcome was highly successful, demonstrating that the numerical scheme is both efficient and accurate. Needless to say, the very large number of binary periods involved ($\sim 10^7$) represents a massive computational effort. However, the binary BH problem is a fundamental one and its study by the direct numerical approach is bound to be fruitful.

Postscript

Finally, it may be appropriate to record the story of how I got started on the *N*-body problem.

During my second year as a research student (spring 1961), Fred Hoyle returned from one of his regular trips to California.

He showed me a picture of the Hercules galaxy cluster given to him by the Burbidges and wondered how long the irregular shape could be maintained.

Faced by this intriguing question, he suggested looking at ways to calculate the dynamical interactions on a computer.

I had already acquired some experience of using the famous EDSAC-2 computer and had embarked on a project to produce synthetic rotation curves for disk galaxies by calculating the gravitational forces between concentric rings.

However, now I faced the formidable *N*-body problem in its full glory with nothing to lean on; only later did we discover the only paper on the subject, written in German.

As usual when a new project starts to take shape and further progress appears to be blocked, one tries find a good excuse to give up. I consulted Chandrasekhar's Stellar Dynamics and worked out that the relaxation time would be too long for anything significant to happen. However, this argument did not deter Fred, who made me go back to basics and attack the problem in a logical way.

Eventually we had a simple computer program which proved adequate to investigate the dynamics of galaxy clusters with up to 100 members. One of the models started as a V-shaped system with some halo objects to represent the Hercules cluster (MNRAS 126, 223, 1963). Although these beginnings were quite modest when judged by today's standards, the early computer program contained two basic ideas (individual time-steps and force polynomials) which are still key features in modern codes.

Consequently, the original thesis topic proposed by Fred Hoyle turned out to be very timely and further developments occupied my entire career as a research worker.

References

Aarseth, S.J.: 1972, Binary evolution in stellar systems, in: M. Lecar (ed.), *Gravitational N-Body Problem*, D. Reidel Publishing Co., Dordrecht, pp. 88–98.

Aarseth, S.J.: 1999, From NBODY1 to NBODY6: The growth of an industry, *PASP*, **111**, 1333–1346.

Aarseth, S.J. and Zare, K.: 1974, A regularization of the three-body problem, *Cel. Mech.* **10**, 185–205.

Bulirsch, R. and Stoer, J.: 1966, Numerical treatment of differential equations by extrapolation methods, *Num. Math.* **8**, 1–13.

Makino, J.: 1997, Merging of galaxies with central black holes, II. Evolution of the black hole binary and the structure of the core, *Astrophys. J.* **478**, 58–65.

Mikkola, S. and Aarseth, S.J.: 1993, An implementation of N-body chain regularization, *Cel. Mech. Dyn. Ast.* **57**, 439–459.

Mikkola, S. and Aarseth, S.J.: 2002, A time-transformed leapfrog scheme, *Cel. Mech. Dyn. Ast.* **84**, 343–354.

Milosavljević, M. and Merritt, D.: 2001, Formation of galactic nuclei, *Astrophys. J.* **563**, 34–62.

Quinlan, G.D. and Hernquist, L.: 1997, The dynamical evolution of massive black hole binaries, II. Self-consistent N-body integrations, *New Astron.* **2**, 533–554.

Soffel, M.H.: 1989, *Relativity in Astrometry, Celestial Mechanics and Geodesy*, Springer, Berlin, p. 141.

PART III: COSMOLOGY

NUMERICAL COINCIDENCES AND 'TUNING' IN COSMOLOGY

MARTIN J. REES
Institute of Astronomy, Madingley Road, Cambridge, CB3 0HA, UK

Abstract. Fred Hoyle famously drew attention to the significance of apparent coincidences in the energy levels of the carbon and oxygen nucleus. This paper addresses the possible implications of other coincidences in cosmology.

1. Introduction

Hermann Bondi's classic book 'Cosmology' was, for many of us, an inspiring introduction to the science of the cosmos. In a chapter entitled 'Microphysics and Cosmology', Bondi lists the famous dimensionless constants, and mentions the well-known coincidence, first highlighted by Dirac, between the ratio of the electrical and gravitational forces within a hydrogen atom and the ratio of the Hubble radius to the size of an electron. He says: 'These coincidences are very striking, and few would deny their possible deep significance, but the precise nature of the connexion they indicate is not understood and is very mysterious.'

I am not sure to what extent Fred Hoyle was influenced in this matter by Bondi, but he certainly took this problem seriously too. He also, through his famous realisation of the C^{12} resonance level's significance, made a celebrated addition to the list of cosmic coincidences. Moreover, in pondering their significance he was led to conjecture that the so called 'constants of nature' might not be truly universal. In 'Galaxies, Nuclei and Quasars' Hoyle writes that 'one must at least have a modicum of curiosity about the strange dimensionless numbers that appear in physics.' He goes on to outline two possible attitudes to them. One is that 'the dimensionless numbers are all entirely necessary to the logical consistency of physics'; the second possibility is that the numbers are not in the broadest sense universal, but that 'in other places their values would be different' Hoyle favoured this latter option because then 'the curious placing of the levels in C^{12} and O^{16} need no longer have the appearance of astonishing accidents. It could simply be that since creatures like ourselves depend on a balance between carbon and oxygen, we can exist only in the portions of the universe where these levels happen to be correctly placed.'

With these texts as my motivation, I'd like to summarise briefly how the issue looks today. I believe that Fred's conjecture is now even more attractive, though the 'portions of the universe' between which the variation occurs must now be interpreted as themselves vastly larger than the domain our telescopes can actually observe – perhaps even entire 'universes' within a multiverse.

Astrophysics and Space Science **285:** 375–388, 2003.

But before dwelling further on these coincidences, it might be worth noting that the 'coincidence' that Dirac and Bondi discussed does not in itself now cause puzzlement. There is really just one very large number in physics: it is e^2/Gm_p^2 (or, equivalently, the reciprocal of the 'gravitational fine structure constant' $\alpha_G^{-1} = \left(\frac{\hbar c}{Gm_p^2}\right)$ which is larger by 137. The Chandrasekhar mass exceeds the proton mass by $\alpha_G^{-3/2}$. Stars are so large because gravity is so weak: Dicke (1961) also realised that they are also long-lived for the same reason. To present Dicke's estimate for stellar lifetimes in a slightly different way, we can define a characteristic time (cf Salpeter 1964) equal to

$$t_s = \frac{M_* c^2}{L_{\rm Ed}} = \frac{2}{3}\left(\frac{e^2}{m_e c^3}\right)\left[\left(\frac{e^2}{Gm_p^2}\right)\left(\frac{m_p}{m_e}\right)\right]$$

This is the time it would take a body to radiate its rest mass energy if had the 'Eddington luminosity' where where radiation pressure balances gravity, and if electron scattering provided the main opacity. The lifetime of an actual star is obtained by multiplying t_s by by various factors: the efficiency of nuclear energy (~ 0.007); the ratio of total pressure to radiation pressures (> 1); and the ratio of actual opacity to electron-scattering opacity (> 1). However, the key point (evident from the second way I have written the expression for t_s above) is that stellar lifetimes are longer than the light travel time across the electron by the factor in square brackets which involves Dirac's large number. If we are observing the universe when its age is of order the age of a star (and there is such a time in a big bang model) then Dirac's 'coincidence' would naturally be satisfied.

2. Do the 'Special' Values of the Constants Need an Explanation?

If we ever established contact with intelligent aliens, how could we bridge the 'culture gap'? One common culture (in addition to mathematics) would be physics and astronomy. We and the aliens would all be made of atoms, and we'd all trace our origins back to the 'big bang' 13.7 billion years ago. We'd all share the potentialities of a (perhaps infinite) future. But our existence (and that of the aliens, if there are any) depends on our universe being rather special. Any universe hospitable to life – what we might call a *biophilic universe* – has to be 'adjusted' in a particular way. The prerequisites for any life of the kind we know about — long-lived stable stars, stable atoms such as carbon, oxygen and silicon, able to combine into complex molecules, etc — are sensitive to the physical laws and to the size, expansion rate and contents of the universe. Indeed, even for the most openminded science fiction writer, 'life' or 'intelligence' requires the emergence of some generic complex structures: it can't exist in a homogeneous universe, not in a universe containing only a few dozen particles. Many recipes would lead to stillborn universes with no atoms, no chemistry, and no planets; or to universes too short-lived or too empty to allow anything to evolve beyond sterile uniformity.

Consider, for example, the role of gravity. Stars and planets depend crucially on this force; however, we could not exist if gravity were much stronger than it actually is. A large, long-lived and stable universe depends quite essentially on α_G^{-1} being exceedingly large. Gravity also amplifies 'linear' density contrasts in an expanding universe; it then provides a negative specific heat so that dissipative bound systems heat up further as they radiate. There's no thermodynamic paradox in evolving from an almost structureless fireball to the present cosmos, with huge temperature differences between the 3 degrees of the night sky, and the blazing surfaces of stars. So gravity is crucial, but the weaker it is, the grander and more prolonged are its consequences.

Newton's constant G need not be fine-tuned – merely exceedingly weak so that α_G^{-1} is indeed very large. However, the natural world is much more sensitive to the balance between other basic forces. If nuclear forces were slightly stronger than they actually are relative to electric forces two protons could stick together so readily that ordinary hydrogen would not exist, and stars would evolve quite differently. Some of the details are still more sensitive, as Hoyle emphasised.

Even a universe as large as ours could be very boring: it could contain just black holes, or inert dark matter, and no atoms at all. Even if it had the same ingredients as ours, it could be expanding so fast that no stars or galaxies had time to form; or it could be so turbulent that all the material formed vast black holes rather than stars or galaxies. – an inclement environment for life. And our universe is also special in having three spatial dimensions. A four dimensional world would be unstable; in two dimensions, nothing complex could exist.

The distinctive and special-seeming recipe characterising our universe seems to me a fundamental mystery that should not be brushed aside merely as a brute fact. Rather than re-addressing the classic 'fine tuning' examples, I shall focus on the parameters of the big bang – the expansion rate, the curvature, the fluctuations, and the material content. Some of these parameters (perhaps even all) may be explicable in terms of a unified theory: or they may be somehow derivable from the microphysical constants. But, irrespective of how that may turn out, it is interesting to explore the extent to which the properties of a universe – envisaged here as the aftermath of a single big bang – are sensitive to the cosmological parameters.

3. The Cosmological Numbers

Traditionally, cosmology was the quest for a few numbers. The first were H, and q. Since 1965 we've had another : the baryon/photon ratio n_b/n_γ. This is believed to result from a small favouritism for matter over antimatter in the early universe – something that was addressed in the context of 'grand unified theories' in the 1970s. (Indeed, baryon non-conservation seems a prerequisite for any plausible inflationary model. Our entire observable universe, containing at least 10^{79} baryons, could not have inflated from something microscopic if baryon number were strictly conserved).

In the 1980s non-baryonic matter became almost a natural expectation, and Ω_b/Ω_{DM} is another fundamental number.

We now have the revival of the cosmological constant lambda (or some kind of 'dark energy', with negative associated pressure, which is generically equivalent to lambda).

Another specially important dimensionless number tells us how smooth the universe is. It's measured by

- The Sachs-Wolfe fluctuations in the microwave background
- the gravitational binding energy of clusters as a fraction of their rest mass
- or by the square of the typical length scale of mass- clustering as a fraction of the Hubble radius.

It's of course oversimplified to represent this by a single number, but insofar as one can, its value (let's call it Q) is pinned down to be 10^{-5}. (Detailed modelling of the fluctuations introduces further numbers: the ratio of scalar and tensor amplitudes, and quantities such as the 'tilt', which measure the deviation from a pure scale-independent Harrison-Zeldovich spectrum.)

4. Anthropic Requirements for a Universe

We can make a list of what would be required for a big bang to yield an 'anthropically allowed' universe – a universe where complexity, whether humanoid or more like a black cloud, could unfold. The list would include the following:

- Some inhomogeneities (i.e. a non-zero Q): clearly there is no potential for complexity if everything remains in a uniform ultra-dilute medium
- Some baryons: complexity would be precluded in a universe solely made of dark matter, with only gravitational interactions.
- At least one star (probably, though perhaps superfluous for black-cloud-style complexity)
- Some second-generation stars: only later-generation stars would be able to have orbiting planets, unless heavy elements were primordial.

It is interesting to engage in 'counterfactual history' and ask what constraints these various requirements would impose on hypothetical universes with different characteristics – in particular, with different values of:

- The fluctuation amplitude Q (which is 10^{-5} in our actual universe)
- The cosmological constant
- The baryon/photon ratio (about 10^{-9} in our universe)
- The baryon/dark matter ratio Ω_b/Ω_{DM}(about 0.2 in our universe).

4.1. THE FLUCTUATION AMPLITUDE

First, we might explore what a universe would be like which was initially smoother (Q smaller) or rougher (Q larger) than ours.

Were Q of order 10^{-6}, there would be no clusters of galaxies; moreover, the only galaxies would be small and anaemic. They would form much later than galaxies did in our actual universe. Because they are loosely bound, processed material would be expelled from shallow potential wells; there may therefore be no second-generation stars, and so no planetary systems. If Q were even smaller than 10^{-6}, there would be no star formation at all: very small structures of dark matter would turn around late, and their constituent gas would be too dilute to undergo the radiative cooling that is a prerequisite for star formation. (In a lambda-dominated universe, isolated clumps could survive for an infinite time without merging into a larger scale in the hierarchy. So eventually, for any $Q > 10^{-8}$, a 'star' could form – but by that time it might be the only bound object within the horizon).

Hypothetical astronomers in a universe with $Q = 10^{-4}$ might find their cosmic environment more varied and interesting than ours. Galaxies and clusters would span a wider range of masses. The biggest clusters would be 30 times more massive than any in our actual universe. There could be individual 'galaxies' – perhaps even disc galaxies – with masses up to that of the Coma cluster and internal velocity dispersions up to 2000 km/sec. These would have condensed when the universe was only 3.10^8 years old, and when Compton cooling on the microwave background was still effective

However a universe where Q were larger still – more than (say) 10^{-3} – would be a violent and inhospitable place. Huge gravitationally-bound systems would collapse, trapping their radiation and unable to fragment, soon after the epoch of recombination. (Collapse at, say, 10^7 years would lead to sufficient partial ionization (via strong shocks) to recouple the baryons and the primordial radiation.) Such structures, containing the bulk of the material, would turn into vast black holes. It is unlikely that galaxies of any kind would exist; nor is it obvious that much baryonic material would ever go into stars: even if so, they would be in very compact highly bound systems).

(Note that, irrespective of these anthropic constraints on its value, Q has to be substantially less than one in order to make cosmology a tractable subject, separate from astrophysics, This is because the ratio of the largest structures to the Hubble radius is of order $Q^{1/2}$. Numbers like Ω and H *are only well-defined* insofar as the universe possesses 'broad brush' homogeneity – so that our observational horizon encompasses many independent patches each big enough to be a fair sample. This wouldn't be so, and the simple Friedmann models wouldn't be useful approximations, if Q *weren't* much less than unity.)

According to most theories of the ultra-early universe, Q is imprinted by quantum effects: microscopic fluctuations, after exponential expansion, give rise to the large-scale irregularities observed in the microwave background sky, and which are the 'seeds' for galaxies and clusters. But as yet no theories pin down Q's value.

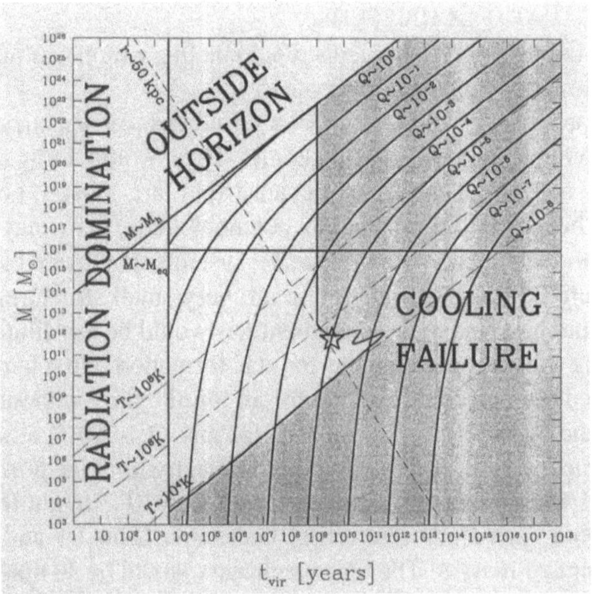

Figure 1. The domains in the which bound structures can form, for different values of Q (from Tegmark and Rees, 1998).

4.2. Λ OR DARK ENERGY

Theorists are even further from understanding Λ. Indeed, the naive guess is that Λ should be least 120 powers of 10 larger than it could be in our actual universe – unless there were some cancellation mechanism. (Indeed, inflation models postulate an effective vacuum density that was indeed as high as this for a brief initial interval.)

The interest has of course been hugely boosted recently, through the convergence of several lines of evidence on a model where the universe is close to being 'flat', but with 4 percent in baryons, about 25 percent in dark matter, and the remaining (dominant) component in Λ or some time-dependent 'dark energy'. (Incidentally, the full resurrection of Λ would be a great 'coup' for de Sitter. His model, dating for the 1920s, not only describes inflation, but would then also describes future aeons of our cosmos with increasing accuracy. Only for the 50-odd decades of logarithmic time between the end of inflation and the present would it need modification!). For a universe with the actual observed values of Q, it is readily shown that a value of Λ more than 5-10 times higher than the apparent 'dark energy' density would have the 'anti-anthropic' consequence of precluding galaxy formation. This happens because the cosmic repulsion would then be so fierce that it would take over before any galaxies had a chance to form via gravitational instability.

4.3. Ω_b AND Ω_{DM}: THE BARYON/DARK MATTER DENSITY

Baryons are anthropically essential; there are firm lower limits on their requisite abundances, but they need not be the dominant constituent. (Indeed they are far from dominant in our actual universe). Lower n_b/n_γ and lower Ω_{DM} reduce the 'efficient cooling' domain in the Rees/Tegmark (1998) curves. reproduced in Figure 1.

If the photons outnumbered the baryons and the dark matter particles by a still larger factor than in our actual universe, then the universe would remain radiation-dominated for so long that the gravitational growth of fluctuations would be inhibited (Rees, 1980).

On the other hand, a higher value of $n_b/n_\gamma\ (1 + \Omega_{DM}/\Omega_b)$ reduces t_{eq} and allows gravitational clustering to start earlier. This reduces the minimum Q required for emergence of non-linear structures (cf. Aguirre, 2001)

(Note also that the mechanism that gives rise to baryon favouritism may be linked to the strong interactions, and therefore correlate with key numbers in nuclear physics.)

4.4. DELINEATING THE ANTHROPICALLY-ALLOWED DOMAIN

In the above, I have envisaged changing just one parameter, leaving the others with their actual values. But of course there may be correlations between them. For example, suppose that there were big bangs with a whole range of Q-values. Structures form earlier (when the matter density is higher) in universes with larger Q, so obviously a higher Q is anthropically-compatible with a higher Λ.

If we consider a two-dimensional situation where Q and Λ vary, then we find that there is an anthropically allowed area. There are (rather vaguely defined) upper and lower limits to Q (as already discussed) but within the range, there is an upper limit to Q (see Figure 2).

We can carry out the exercise, in as many dimensions as we wish, of delineating the anthropically-allowed domain in parameter space. (even though to quantify this is more difficult). To delineate the allowed domains is procedurally uncontroversial, but what about the motivation? It obviously depends on believing that the laws of nature could have been otherwise – unless there is some scientific validity in imagining 'counterfactual universes' this exercise seems vacuous.

5. Is it 'Scientific' to enquire about other Universes?

If our existence – or, indeed, the existence of any 'interesting' universe – depends on a seemingly special cosmic recipe, how should we react? There seem two lines to take: we can dismiss it as happenstance, or we can conjecture that our universe is a specially favoured domain in a still vaster multiverse.

(a) *Happenstance (or coincidence)*

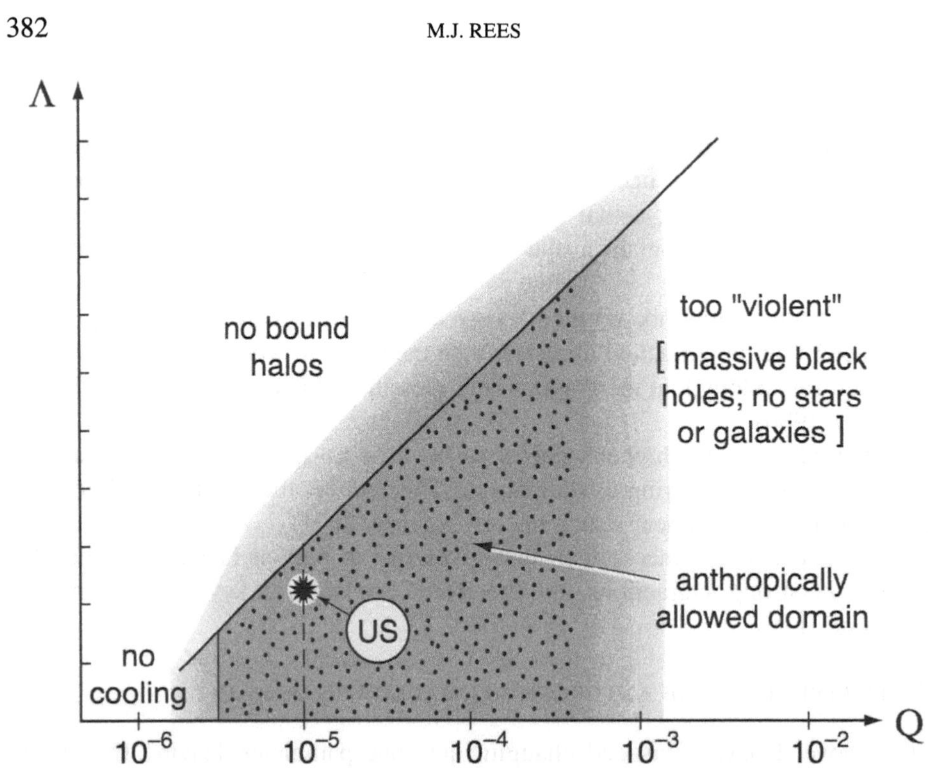

Figure 2. This shows in a two-dimensional parameter space Λ and Q. The upper and lower limits to Q are discussed by Tegmark and Rees (1998). The upper limit to Λ stems from the requirement that galactic-mass bound systems should form before the universe enters its accelerating de Sitter phase. Our universe (obviously) lies in the anthropically-allowed domain. But we cannot say whether it is at a typical location within that domain without a specific model for the probability distributions of Q and Λ in the ensemble.

Maybe a fundamental set of equations, which some day will be written on T-shirts, fixes all key properties of our universe uniquely. It would then be an unassailable fact that these equations permitted the immensely complex evolution that led to our emergence.

But I think there would still be something to wonder about. It's not guaranteed that simple equations permit complex consequences To take an analogy from mathematics, consider the Mandelbrot set. This pattern is encoded by a short algorithm, but has infinitely deep structure: tiny parts of it reveal novel intricacies however much they are magnified. In contrast, you can readily write down other algorithms, superficially similar, that yield very dull patterns. Why should the fundamental equations encode something with such potential complexity, rather than the boring or sterile universe that many recipes would lead to?

One hard-headed response is that we couldn't exist if the laws had boring consequences. We manifestly are here, so there's nothing to be surprised about. I think we would need to know why the unique recipe for the physical world should permit consequences as interesting as those we see around us (and which, as a byproduct, allowed us to exist)

(b) *A special universe drawn from an ensemble, or multiverse*

But there is another perspective – a highly speculative one, however. There may be many 'universes' of which ours is just one. In the others, some laws and physical constants would be different. But our universe wouldn't be just a random one. It would belong to the unusual subset that offered a habitat conducive to the emergence of complexity and consciousness. If our universe is selected from a multiverse, its seemingly designed or fine tuned features wouldn't be surprising. This might seem arcane stuff, disjoint from 'traditional' cosmology – or even from serious science. But my prejudice is to be openminded about ensembles of universe and suchlike, and even to suspect that we may not be able to account for some features of our own universe without invoking them.

First, a semantic digression: the word 'universe' traditionally denotes 'everything there is'. Therefore if we are to consider other domains of space time (originating in other big bangs) we should really define the whole ensemble as 'the universe', and introduce a new word – 'metagalaxy' for instance – to denote what observational cosmologists traditionally study. However, so long as this whole idea remains speculative, it is probably best to continue to denote what cosmologists observe as 'the universe', and to introduce a new term, 'multiverse', for the whole hypothetical ensemble.

Some might regard other universes — regions of space and time that we cannot observe (perhaps even in principle and not just in practice) – as being in the province of metaphysics rather than physics. Science is an experimental or observational enterprise, and it's natural to be troubled by invocations of something unobservable. But I think 'other universes' (in this sense) already lie within the proper purview of science. It is not absurd or meaningless to ask 'Do unobservable universes exist?', even though no quick answer is likely to be forthcoming. The question plainly can't be settled by *direct* observation, but relevant evidence *can* be sought, which could lead to an answer.

There is actually a blurred transition between the readily observable and the absolutely unobservable, with a very broad grey area in between (see Figure 3). To illustrate this, one can envisage a succession of horizons, each taking us further than the last from our direct experience:

(i) *Limit of present-day telescopes*

There is a limit to how far out into space our present-day instruments can probe. Obviously there is nothing fundamental about this limit: it is constrained by current technology. Many more galaxies will undoubtedly be revealed in the coming decades by bigger telescopes now being planned. We would obviously not demote such galaxies from the realm of proper scientific discourse simply because they haven't been seen yet.

(ii) *Limit in principle at present era*

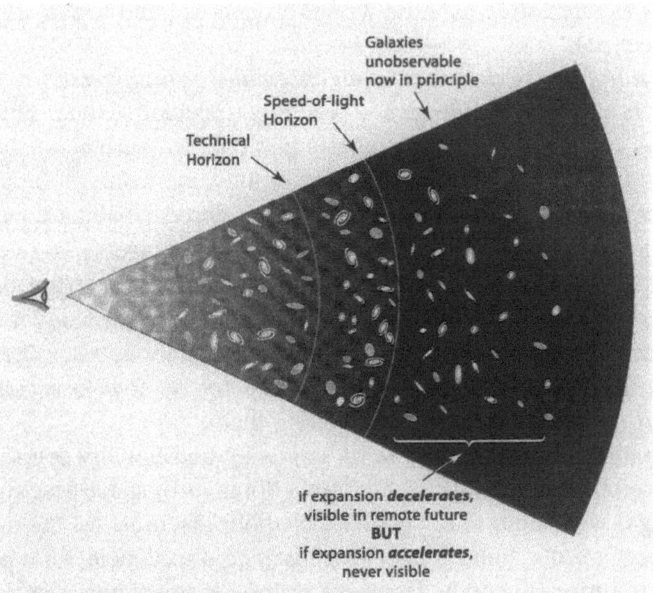

Figure 3. Extending horizons beyond the directly-observable.

Even if there were absolutely no technical limits to the power of telescopes, our observations are still bounded by the particle horizon, which demarcates the spherical shell around us at which the redshift would be infinite.

If our universe were decelerating, then the horizon of our remote descendants would encompass extra galaxies that are beyond our horizon today. It is, to be sure, a practical impediment if we have to await a cosmic change taking billions of years, rather than just a few decades (maybe) of technical advance, before a prediction about a particular distant galaxy can be put to the test. But does that introduce a difference of principle? Surely the longer waiting-time is a merely quantitative difference, not one that changes the epistemological status of these faraway galaxies?

(iii) *Never-observable galaxies from 'our' Big Bang*

But what about galaxies that we can *never* see, however long we wait? It's now believed that we inhabit an accelerating universe. As in a decelerating universe, there would be galaxies so far away that no signals from them have yet reached us; but if the cosmic expansion is accelerating, we are now receding from these remote galaxies at an ever-increasing rate, so if their light hasn't yet reached us, it never will. Such galaxies aren't merely *unobservable in principle now* – they will be beyond our horizon *forever*. But if a galaxy is now unobservable, it hardly seems to matter whether it remains unobservable for ever, or whether it would come into view if we waited a trillion years. (And I have argued, under (ii) above, that the latter category should certainly count as 'real'.)

(iv) *Galaxies in disjoint universes*

The never-observable galaxies in (iii) would have emerged from the same Big Bang as we did. But suppose that, instead of causally-disjoint regions emerging from a single Big Bang (via an episode of inflation) we imagine separate Big Bangs. Are space-times completely disjoint from ours any less real than regions that never come within our horizon in what we'd traditionally call our own universe? Surely not – so these other universes too should count as real parts of our cosmos, too

Whether other universes exist or not is a scientific question. Those who are prejudiced against the concept should regard the above step-by-step argument as an exercise in 'aversion therapy'. From a reluctance to deny that galaxies with redshift 10 are proper objects of scientific enquiry, you are led towards taking seriously quite separate space-times, perhaps governed by quite different 'laws'.

Linde, Vilenkin and others have performed computer simulations depicting an 'eternal' inflationary phase where many universes sprout from separate big bangs into disjoint regions of spacetimes – each such region itself vastly larger than our observational horizon. Guth, Harrison and Smolin have, from different viewpoints, suggested that a new universe could sprout inside a black hole, expanding into a new domain of space and time inaccessible to us. And Randall and Sundrum suggest that other universes could exist, separated from us in an extra spatial dimension; these disjoint universes may interact gravitationally, or they may have no effect whatsoever on each other.

None of these scenarios has been simply dreamed up out of the air: each has a serious, albeit speculative, theoretical motivation. However, one of them, at most, can be correct. Quite possibly none is: there are alternative theories that would lead just to one universe. Firming up any of these ideas will require a theory that consistently describes the extreme physics of ultra-high densities, how structures on extra dimensions are configured, etc. Perhaps, in the 21st-century theory, physicists will develop a theory that yields insight into (for instance) why there are three kinds of neutrinos, and the nature of the nuclear and electric forces. Such a theory would thereby acquire credibility. If the same theory, applied to the very beginning of our universe, were to predict many big bangs, then we would have as much reason to believe in separate universes as we now have for believing inferences from primordial nucleosynthesis about the first few minutes of cosmic history.

6. Universal Laws, or Mere Bylaws?

Some theorists, Frank Wilczek for instance, regard 'are the laws of physics unique?' as a key scientific challenge for the new century. The answer determines how much variety the other universes – if they exist – might display. If there were something uniquely self-consistent about the actual recipe for our universe, then the aftermath of any big bang would be a re-run of our own universe. But a far more interesting possibility (which is certainly tenable in our present state of ignorance of the

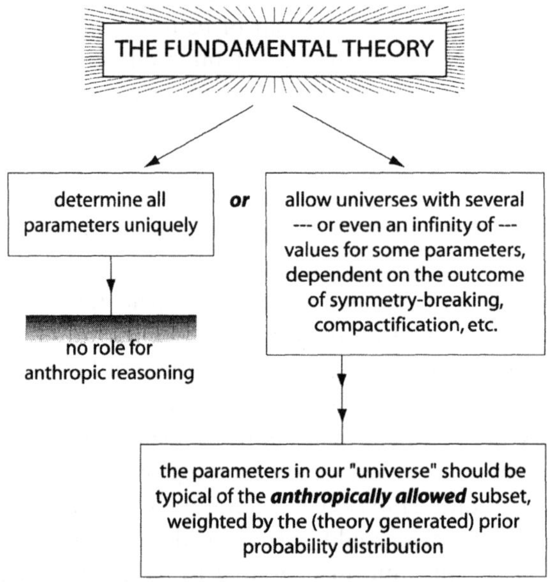

Figure 4. 'Decision tree'. Progress in 21-st century physics should allow us to decide whether anthropic explanations are irrelevant or, on the other hand, the best we can ever hope for.

underlying laws) is that *the underlying laws governing the entire multiverse may allow variety among the universes.* Some of what we call 'laws of nature' may in this grander perspective be *local bylaws*, consistent with some overarching theory governing the ensemble, but not uniquely fixed by that theory. Many things in our cosmic environment – for instance, the exact layout of the planets and asteroids in our Solar System – are accidents of history. Likewise, the recipe for an entire universe may be arbitrary.

More specifically, some aspects may be arbitrary and others not. As an analogy (which I owe to Paul Davies) consider the form of snowflakes. Their ubiquitous six-fold symmetry is a direct consequence of the properties and shape of water molecules. But snowflakes display an immense variety of patterns because each is moulded by its micro-environments: how each flake grows is sensitive to the fortuitous temperature and humidity changes during its growth.

If physicists achieved a fundamental theory, it would tell us which aspects of nature were direct consequences of the bedrock theory (just as the symmetrical template of snowflakes is due to the basic structure of a water molecule) and which are (like the distinctive pattern of a particular snowflake) the outcome of accidents.

The cosmological numbers in our universe, and perhaps some of the so-called constants of laboratory physics as well, could be 'environmental accidents', rather than uniquely fixed throughout the multiverse by some final theory. Some seemingly 'fine tuned' features of our universe could then only be explained by 'anthropic' arguments [see Figure 4]. Although this style of explanation raises hackles among some physicists it is analogous to what any observer or experimenter does when they allow for selection effects in their measurements: if there are many universes, most of which are not habitable, we should not be surprised to find ourselves in one of the habitable ones!

The entire history of our universe could just be an episode of the infinite multiverse; what we call the laws of nature (or some of them) may be just parochial bylaws in our cosmic patch. Such speculations dramatically enlarge our concept of reality. Putting them on a firm footing must await a successful fundamental theory that tells us whether there could have been many 'big bangs' rather than just one, and (if so) how much variety they might display. Until this fundamental issue is settled one way or the other, we won't know whether anthropic arguments are irrelevant or unavoidable.

7. Testing Multiverse Theories Here and Now

We may one day have a convincing theory that accounts for the very beginning of our universe, tells us whether a multiverse exists, and (if so) whether some so called laws of nature are just parochial by-laws in our cosmic patch. But while we're waiting for that theory – and it could be a long wait – we can check whether anthropic selection offers a tenable explanation for the apparent fine tuning. Such a hypothesis could even be refuted: this would happen if our universe turned out to be *even more specially* tuned than our presence requires.

We could apply this style of reasoning to the important numbers of physics (for instance, Λ) to test whether our universe is typical of the subset that that could harbour complex life. Most physicists would consider the 'natural' value of Λ to be large, because it is a consequence of a very complicated microstructure of space. Perhaps there is only a rare subset of universes where Λ is below the threshold that allows galaxies and stars to form. Λ in *our* universe obviously had to be below that threshold, But if our universe were drawn from an ensemble in which Λ was equally likely to take any value, we wouldn't expect it to be *too far below it*.

Current evidence suggests that if Λ constituted the 'dark energy', its actual value is 5-10 times below that threshold. That would put our universe between the 10th or 20th percentile of universes in which galaxies could form. In other words, our universe isn't significantly more special, with respect to Λ, than our emergence demanded. But suppose that (contrary to current indications) observations showed that Λ made no discernible contribution to the expansion rate, and was *thousands of times* below the threshold, not just 5–10 times. This 'overkill precision' would raise doubts about the hypothesis that Λ was equally likely to have any value, and suggest that it was zero for some fundamental reason (or that it had a discrete set of possible values, and all the others were well about the threshold).

I've taken Λ just as an example. We could analyse other important numbers of physics in the same way, to test whether our universe is typical of the habitable subset that that could harbour complex life. The methodology requires us to decide what values are compatible with our emergence. It also requires a specific theory that gives the probability of any particular value.

With this information, one can then ask if our actual universe is 'typical' of the subset in which we could have emerged. If it is an atypical member even of this subset (not merely of the entire multiverse) then our hypothesis would be disproved. Other parameters could be analysed similarly – testing in a multi-parameter space whether our universe is a typical member within the anthropically allowed domain.

As a two-dimensional example, consider the joint constraints on Λ and Q in Figure 2. We cannot decide whether our universe is typical without a theory that tells us what 'measure' to put on each part of the 2-dimensional parameter space. If high-Q universes were more probable, and the probability density of Λ were uniform, then we should be surprised not to find ourselves in a universe with higher Λ and higher Q.

These examples show that some claims about other universes may be refutable, as any good hypothesis in science should be. We cannot confidently assert that there were many big bangs – we just don't know enough about the ultra-early phases of our own universe. Nor do we know whether the underlying laws are 'permissive': settling this issue is a challenge to 21st century physicists. But if they are, then so-called anthropic explanations would become legitimate – indeed they'd be *the only type of explanation we'll ever have* for some important features of our universe.

Models with low omega, non-zero lambda two kinds of dark matter, and the rest may *seem* ugly. Some theorists are upset by these developments, because it frustrates their craving for maximal simplicity. I think we can learn a lesson from cosmological debates in the 17th century. Galileo and Kepler were upset that planets moved in elliptical orbits, not in perfect circles. Newton later showed, however, that all elliptical orbits could be understood by a single unified theory of gravity. Likewise our universe may be just one of an ensemble of all possible universes, constrained only by the requirement that it allows our emergence. But to regard this outcome as ugly may be as myopic as Kepler's infatuation with circles: Newton was perhaps the greatest scientific intellect of the second millennium. Perhaps his third-millennium counterpart will uncover a mathematical system that governs the entire multiverse.

References

Aguirre, A.: 2001, *Phys. Rev D* **64**, 3508.
Dicke, R.: 1961, *Nature* **192**, 446.
Rees, M.J.: 1980, *Physica Scripta* **21**, 614.
Salpeter, E.E.: 1964, *Astrophys. J.* **140**, 796.
Tegmark, M. and Rees, M.J.: 1998, *Astrophys. J.* **499**, 526.
Tegmark, M.: 2003, in: J.D. Barrow et al. (eds.), *Science and Ultimate Reality: from Quantum to Cosmos*, Cambridge University Press, in press.

WORKING WITH FRED ON ACTION AT A DISTANCE

J.V. NARLIKAR
Inter-University Centre for Astronomy and Astrophysics Post Bag 4, Ganeshkhind, Pune 411 007

Abstract. This paper reviews the work the author carried out with Fred Hoyle on the development of electrodynamics and gravitation as direct particle theories. In this account the author reviews how the work was started, and went through stages of increasing sophistication, e.g., extending the Wheeler-Feynman electrodynamics to curved spacetime, its consequences in different cosmologies, and the issues arising from its quantization. The resolution of ultraviolet divergences in quantum electrodynamics is also briefly discussed. The parallel development of a Machian theory of gravitation followed the lead from electrodynamics. In both theories one sees a strong link between the large scale structure of the universe and local physics, as might be expected from an action-at-a-distance framework. It is recalled why Fred considered this an important aspect of a physical theory.

1. Historical Background

I still vividly recall a wet afternoon in Varenna on Lake Como in northern Italy in the summer of 1961. I was a student participant in one of the annual summer schools held in this scenic resort. Our school was on various aspects of gravitational theories and observations and the lecturers included Bob Dicke, Alfred Schild, Joshua Goldberg, Bruno Bertotti and Fred Hoyle. We also had seminars from a few other scientists who passed by for a short duration. That day it was Hermann Bondi who lectured on in his inimitable style, though hampered by frequent sneezes brought about by hayfever. Nevertheless, his topic was extremely interesting.

What Bondi reported on that day concerned a very unusual aspect of electrodynamics, an aspect that linked it to cosmology. This was a follow up by a Canadian physicist named Jack Hogarth on the work done by John Wheeler and Richard Feynman in the 1940s on action at a distance electrodynamics (Wheeler and Feynman, 1945, 1949). Briefly, the history was as follows.

In fact the story begins a hundred years before the Wheeler-Feynman paper of 1945, with no less a person than Gauss. In a letter to Weber on March 19, 1845, Gauss wrote:

I would doubtless have published my researches long since were it not that at the time I gave them up I had failed to find what I regarded as the keystone, Nil actum reputans si quid superesset agendum, namely, the derivation of the additional forces – to be added to the interaction of electrical charges at rest, when they are both in motion – from an action which is propagated not instantaneously but in time as is the case with light.

Gauss's attempts came some two decades before the Maxwellian field theory and six decades before special relativity. The sucess of these two theories shifted

Astrophysics and Space Science **285:** 389–405, 2003.

the emphasis from action at a distance to fields and it was not until well into the present century that the problem posed by Gauss was solved.

A beginning was made by Schwarzschild (1903), Tetrode (1922), and Fokker (1929a,b, 1932), who independently formulated the concept of delayed action at a distance. The action principle as formulated by Fokker may be written in the following form:

$$J = -\sum_a \int m_a da - \sum_{a<b}\sum \int\int e_a e_b \delta(s_{AB}^2)\eta_{ik} da^i db^k. \tag{1}$$

In the above expression the charged particles are labeled, *a*,*b*,... with e_a and m_a the charge and mass of particle a. The worldline of a is given by the coordinate functions $a^i(a)$ of the proper time a. The spacetime is Minkowskian, so that

$$da^2 = \eta_{ik} da^i da^k, \tag{2}$$

with η_{ik}= diag $(-1, -1, -1, 1)$. The first term of J therefore discribes the inertial term. The second term describes the electromagnetic interaction between the worldlines of a typical pair of particles a, b. The delta function shows that the interaction is effective only when s_{AB}^2 , the invariant square of distance between typical world points A, B on the worldlines of a and b, vanishes. This implies delayed action: $s_{AB}^2= 0$ means that world points A and B are connected by a light ray.

Although this formulation met the requirement of relativistic invariance it gave rise to other difficulties. The major difficulty is as follows. For a typical point A on the worldline of a there are two points B_+ and B_- on the world line of b for which $s_{AB}^2 = 0$. The effect of A is felt at B_+ (at a later time) and at B_- (at an earlier time). Similarly, since the action principle guarantees the equality of action and reaction, the reaction from B_+ and B_- is felt at A. Thus there are influences propagating with the speed of light, not only into the future but also into the past. This led to a conflict with the principle of causality, which seems to hold in everday life. The other difficulties were of a less serious nature although not ignorable. For example, there was no 'self-action' ($a = b$ is avoided in the double sum) and so there did not appear to be any obvious way of accounting for rediation damping.

These difficulties were removed by Wheeler and Feynman (1945) by bringing into the discussion the important role of the absorber. In our above example, the reactions from B arrive at A instantaneously, whatever the spatial separation of a and b. So it becomes necessary to take into account the reaction from the entire universe to A. Although the remote particles are expected to contribute less, their total number is large enough to make the calculation nontrivial. The essence of the argument given by Wheeler and Feynman is described below.

To begin with, define the 4-potential at X due to particle b by

$$A_i^{(b)}(X) = e_b \int \delta(s_{XB}^2)\eta_{ik} db^k, \tag{3}$$

and the corresponding direct-particle field by

$$F_{ik}^{(b)} = A_{k;i}^{(b)} - A_{i;k}^{(b)}. \tag{4}$$

A direct particle field is not an ordinary field, because it does not have any independent degrees of freedom. The 4-potential *identically* satisfies the relations

$$A_{;k}^{(b)k} \equiv 0, \quad \Box A^{(b)k} \equiv \eta^{mn} A^{(b)k}{}_{;mn} = 4\pi j^{(b)k}, \tag{5}$$

where $j^{(b)k}(X)$ is the current density vector of the particle b at a typical point X, defined in the usual way. Thus although (5) resembles the Maxwell wave equation (and the guage condition) it represents identities.

The equation of motion of a typical charge a is obtained by varying its worldline and requiring $\delta J = 0$. We get the analogue of the Lorentz force equation in which the charge a is acted on by all other charges in the universe. It appears therefore that an alternative formulation of electrodynamics based on action at a distance has been found. However, this appearance is illusory.

We now turn to the difficulty introduced by the time symmetry of this formulation. Instead of being the retarded solution of (5), (3) is the time-symmetric half-advanced and half-retarded solution. The same applies to the direct-particle fields. Suppressing the indices i, k, we may write

$$F^{(b)} = \frac{1}{2}[F_{\text{ret}}^{(b)} + F_{\text{adv}}^{(b)}]. \tag{6}$$

This field is present in the past as well as the future light cone of B.

Wheeler and Feynman argued in the following way. If we move the charge b, it generates a disturbance that affects all other charges in the universe. Their reaction arrives back instantaneously. They then showed how to calculate such a reaction in a universe of static Minkowski type with a uniform distribution of electric charges. They found that the reaction to the motion of charge b can be calculated in a consistent fashion and comes out to be

$$R^{(b)} = \frac{1}{2}[F_{\text{ret}}^{(b)} - F_{\text{adv}}^{(b))}]. \tag{7}$$

Thus a test particle in the neighbourhood of charge b experiences a net total 'field'

$$F_{\text{tot}}^{(b)} = F^{(b)} + R^{(b)} = F_{\text{ret}}^{(b)}. \tag{8}$$

This is the pure retarded field observed in real life! The self- consistency of the argument follows from the fact that the reaction $R^{(b)}$ has been calculated by adding the $\frac{1}{2}F^{(a)}$ fields of all particles $a \neq b$ that have been excited by this total field $F^{(b)}$. Thus only the future light cone of B comes into play. The reaction from the future cancels the advanced component of $F^{(b)}$ and doubles its retarded component.

Also according to the Lorentz force equation, $R^{(b)}$ is the force arising from all *other* particles in the universe experienced by the particle B. This is nothing but the radiative reaction to the motion of b as obtained earlier by Dirac (1938) on empirical grounds. Earlier, Dirac's rule was difficult to understand within the context of the field theory, although it was known to give the right answer. Here the Dirac formula is understood as the consequence of a response of the universe to the local motion of the charge. Thus the theory not only gets round the problem of causality but it also accounts for the radiation damping formula.

Physically, what happens is the following. To the motion of b the future half of the universe acts as an absorber. It 'absorbs' all the 'energy' radiated by b, and in this process sends the reaction $R^{(b)}$, which does the trick! For this reason Wheeler and Feynman called this theory the *absorber theory of radiation*. The presence of the absorber is essential for the calculation to work. For example, it will not work in an empty universe surrounding the electric charge.

In the above self-consistent derivation there was still one defect: it was not unique. Another self-consistent picture was possible in which the net field near every particle was the pure advanced field and the radiative reaction was of opposite sign to that of (7). The two solutions are compared thus. In one we have the retarded solution while in the other we have the advanced solution. In the former, absorption in the future light cone is responsible while in the latter, it is the absorption in the past that plays the crucial role. The important role of the absorbers is that they convert the time-symmetric situation of an isolated charge, to a time-asymmetric one. It is, however, not possible to distinguish between the two without reference to some other independent time asymmetry.

Wheeler and Feynman realized this and linked the choice to thermodynamics. Given the usual thermodynamic time asymmetry, they argued that the latter situation would be highly unlikely (under the probability arguments of statistical mechanics) and that the usual asymmetry of initial conditions will favour the former (retarded) solution.

It was, however, pointed out by Hogarth (1962) that it is not necessary to bring thermodynamics into the picture at all. If one takes account of the fact that the universe is expanding, its past and future are naturally different. The reaction from the absorbing particles in the future light cone (designated by Hogarth collectively as the *future absorber*) does not automatically come out equal and opposite to the reaction from the past absorber. Thus the two pictures do not always follow in an expanding universe. Hogarth found that for the retarded solution to hold, the future absorber must be perfect and and past absorber imperfect; and vice versa for the advanced solution.

An absorber is perfect if it entirely absorbs the radiation emitted by a typical charge. In the static universe discussed by Wheeler and Feynman both the past and future absorbers are perfect; and this leads to the ambiguity mentioned earlier. However, Hogarth found that the ambiguity is resolved if the cosmological time asymmetry is taken into account. He found, for example, that in most big-bang models which expand forever, the advanced (and not the retarded) solution is valid. In the steady-state model (Bondi and Gold, 1948; Hoyle, 1948), only the retarded solution holds. In the big-bang models that expand and contract both the absorbers are perfect and the outcome is ambiguous. This was the work that Bondi reported on in Varenna.

2. Follow-up on the Absorber Theory with Fred Hoyle

Fred Hoyle and I were very impressed by this conclusion. Only a few months before we had argued against the claim by Martin Ryle and his group in Cambridge that their observations of radio source counts ruled out the steady state theory. Those arguments involved several details of observational errors and extrapolations of theory. Here, on the contrary was a clear cut conclusion that did not depend on such messy details. Because of its cosmological conclusions, we felt that this approach of action at a distance needed to be pushed further.

Though interesting, Hogarth's work was, however, incomplete in two aspects. First, he had not shown how to generalize the Fokker action to curved spacetimes needed to describe the expanding world models. Second, he had used collisional damping to decide upon the nature of absorbers, past and present; and this brought in thermodynamics by the back door! In fact, Feynman had criticised Hogarth's work on this very ground.

Later Fred and I (Hoyle and Narlikar, 1963) completed the work by first re-writing the Fokker action (1) in curve space as is necessary for any cosmological discussion. We too arrived at conclusions similar to Hogarth's but by using the *radiative damping* for producing absorption. This kept the asymmetry entirely within electrodynamics and cosmology and away from collective phenomena and thermodynamics.

However, a greater challenge lay ahead. If the action at a distance picture was to go further, on parity with the rival field theoretic picture, then it must be quantizable. For, quantum electrodynamics presents a far richer set of phenomena than its classical counterpart. Our next step was to demonstrate that it is indeed possible to extend the entire picture into the quantum domain and it is possible to describe the entire range of phenomena of quantum electrodynamics, such as spontaneous transition of the atomic electron, Compton scattering, pair creation, etc., even the Lamb shift and anomalous magnetic moment of the electron without recourse to field theory (Hoyle and Narlikar, 1969, 1971). This therefore removes any possible objection to the concept of action at a distance in so far as it is applicable to electrodynamics. In fact, the work two decades later, described briefly in Section 6 shows that even the infinities introduced by the so-called 'self interactions' can be eliminated in this formulation.

The crucial role played in the whole calculation is that of the *response of the universe*. In the classical calculation the steady-state universe generates the 'correct' response so that the local electric charges interact through retarded signals. The response from the big-bang models is of the wrong type. We are thus able to distinguish between the different cosmological models and decide on their validity or otherwise on the basis of the Wheeler-Feynman theory. We also see *why* charges interact through retarded signals: they do so because of the response of the universe. In the Maxwell field theory the choice of retarded solutions of Maxwell's equations is by an arbitrary fiat.

In the quantum calculation also it can be shown that an asymmetric phenomenon with respect to time, like the spontaneous downward transition of an atomic electron, is caused by the response of the universe. By contrast, the quantization of the Maxwell electromagnetic field ascribes such an asymmetry to the quantum vacuum and to the rather abstract rules of quantization.

The direct-particle approach therefore achieves for electrodynamics what Mach sought to achieve for inertia. By bringing in the response of the universe to a local experiment in electrodynamics we have essentially incorporated Mach's principle into electromagnetic theory. Given the correct response of the universe, we can almost decouple our local system from it, although strictly speaking the theory would not be possible without the universe.

Can the same prescription be applied to the original Mach's principle *per se*, to inertia and gravitation? We discuss this problem in the following section, for this issue also attracted Fred and me in the 1960s, parallel to our work with electrodynamics.

3. Inertia as a Direct-Particle Field

We now return to the problem of achieving a 'reconciliation' between general relativity and Mach's principle. To this end we shall look for a theory with the following properties:

(a) It has Mach's principle built into one of its postulates.

(b) It is conformally invariant.

(c) It does not have the conceptual difficulties associated with the case of a single particle in an otherwise empty universe.

(d) For a universe containing many particles the theory reduces to general relativity for most physical situations.

Some discussion is needed as to why the theory should be conformally invariant. The reasons are twofold. First, when we take note of the local Lorentz invariance of special relativity, the natural units to use are those in which the fundamental velocity c=1. The quantum theory, with which our theory should be consistent throws up another fundamental constant, the Planck constant related to the uncertainty principle. Thus it is natural to use units in which $\hbar$=1. This makes the classical action J, for example, dimensionless and the natural unit of mass is the Planck mass which we shall quantify by

$$m_P = \sqrt{\frac{3c\hbar}{4\pi G}}. \tag{9}$$

All masses are then expressible as numbers in the unit of this mass. With our choice of units, only one independent dimension out of the three length, mass and time, remains. Taking it as the dimension of mass, length goes as reciprocal of mass.

However, in a Machian theory, we expect the particle masses to be functions of space and time and as such not necessarily constant. Therefore, the standards of lengths and time intervals also may vary from one point to another. We therefore need laws of physics which are invariant with respect to this variation. Conformal invariance guarantees this.

Our second reason is based on the nature of action at a distance. Given that as in electrodynamics, the interaction propagates principally along null rays, we need an invariance that preserves the global structure of light cones. This again is guaranteed by conformal invariance. Just as Lorentz invariance identifies the light cone as an invariant structure *locally*, so does conformal invariance identify it *globally*.

A theory following these guidelines was developed by Hoyle and me (1964, 1966) and its broad features are described next. We begin by a second look at the Fokker action for electrodynamics, this time rewritten in a curved Riemannian space-time:

$$J = -\sum_a \int m_a da - \sum\sum_{a<b} 4\pi e_a e_b \int\int \bar{G}_{i_A k_B} da^{i_A} db^{k_B}. \tag{10}$$

Here, in going from (1) to (10) the first term of J needs a trivial modification: da is now computed with a Riemannian metric. The modification of the second term of J requires considerable thought. The $\delta(s_{AB}^2)\eta_{ik}$ is now replaced by $\bar{G}_{i_A k_B}$, a bivector propagator between A and B. It is the symmetric Green's function for the wave euqation

$$\Box \bar{G}_{ik_B} + R_i^l \bar{G}_{lk_B} = [-\bar{g}(X, B)]^{-1/2} \bar{g}_{ik_B} \delta_4(X, B). \tag{11}$$

Here $\bar{G}_{ik_B}$ behaves as a vector at X and B, respectively, with the indices i and k_B (the subscript X on i is suppressed for the convenience of writing). $\bar{g}_{ik_B}$ is the parallel propagator between X and B [see Synge (1960) for details] and $\bar{g}(X, B)$ its determinant. In the limit $g_{ik} \to \eta_{ik}$, $4\pi \bar{G}_{i_A k_B} \to \delta(s_{AB}^2)\eta_{ik}$. The detailed structure of this propagator has been studied by DeWitt and Brehme (1960).

The electromagnetic part of J is conformally invariant but the mechanical part (the first term) is not. We now compare (10) with the action for field theory of Maxwell and for general relativity. This action, denoted by $J^{(F)}$ is given by

$$J^{(F)} = \frac{1}{16\pi G}\int R(-g)^{1/2}d^4x - \sum_a \int m_a da -$$

$$\frac{1}{16\pi}\int F^{lm}F_{lm}(-g)^{1/2}d^4x - \sum_a \int A_i da^i. \quad (12)$$

The third and fourth term of $J^{(F)}$ represent, respectively, the free-field term and the field – particle interaction term. In the direct-particle theory the second term of J replaces these two terms of the field theory. The fields as such lose their independent status and are replaced by propagators connecting particle world lines. What can we do about the first two terms of (12)? The second term already exists in (10) and it is tempting to simply insert the first term into (10) as representing gravity.

This procedure, however, is contrary to the spirit of the direct-particle picture. The first term of (12), although containing geometrical information, has also the character of a field. Hence it is out of place.

The clue to the correct procedure that needs to be adopted is provided by a comparision of the last term of (12) with the second term of (10). If in the former we replace the potential A_i by a sum over the direct-particle potentials defined by a relation analogous to (3) for a curved space, we shall recover something that looks like the latter! In the same way we now replace the masses m_a by direct-particle fields defined in the following manner:

$$m^{(b)}(X) = \int \lambda_b G(X, B)db, \quad \lambda_b = \text{a coupling constant}. \quad (13)$$

$$m_a(A) = \lambda_a \sum_{b\neq a} m^{(b)}(A), \quad \lambda_a = \text{a coupling constant}. \quad (14)$$

The propagator $G(X, B)$ has to be biscalar since masses are scalars and we wish to preserve a symmetry between X and B. The action (10) is now changed to

$$J = -\sum\sum_{a<b} \int\int \lambda_a\lambda_b G(A, B)dadb$$

$$-\sum\sum_{a<b} 4\pi e_a e_b \int\int \bar{G}_{i_A k_B} da^{i_A} db^{k_B}. \quad (15)$$

What should be the exact form of $G(A, B)$? Taking a clue from electromagnetism, we expect it to be a symmetric Green's function of a scalar wave equation.

However, we also want the equation to be conformally invariant. These two requirements fix the form of the scalar propagator uniquely to within a multiplicative factor. We shall take $G(A, B)$ to satisfy the scalar wave equation

$$\Box G(X, B) + \frac{1}{6}R(X)G(X, B) = [-\bar{g}(X, B)]^{-1/2}\delta_4(X, B). \tag{16}$$

The wave operator is uniquely fixed by the requirement of conformal invariance.

Turning from these purely formal aspects to those of interpretation we note that (13) and (14) are essentially Machian ideas on inertia expressed mathematically. The mass of a particle a at its world point A is the sum of the contributions of all other particles in the universe. Thus requirement (a) has been met. Requirement (c) is also met, because for a single particle in an otherwise empty universe there is no action! The minimum number of particles required to define J is two. Thus for each of the two particles the other provides the 'background' in the Machian sense. The requirement of conformal invariance is also met by our choice of the propagator. It therefore remains to examine requirement (d).

So far we have concentrated on inertia and ignored gravity. The action (15) does not contain the gravitational term

$$\frac{1}{16\pi G}\int R(-g)^{1/2}d^4x$$

explicitly. Yet, as we shall see in the following section, the theory is fully capable of describing gravitational phenomena.

4. Conformal Gravity

Returning to the action (12) we note that when we try to derive the Einstein field equations by the Hilbert action principle, we get the Einstein tensor from the first term. This term does not exist any more in the direct particle action (15). Shall we get any gravitational term at all from (15) if we sought to perform the metric veriation $g_{ik} \rightarrow g_{ik} + \delta g_{ik}$? A look at the electromagnetic part of (12) does not inspire confidence that the answer to this question should be in the affirmative. There it is the third rather than the fourth term that contributes the energy tensor of electrodynamics, and it is the fourth term that was used in going over to (15). Nevertheless, a closer examination shows that the terms in (15) do give nontrivial answers when the metric variation is performed.

The reason for this is understood as follows. Consider the electromagnetic propagator $\bar{G}_{i_A k_B}$ connecting A and B respectively, on the world lines of a and b. Suppose we perform a variation in the space-time metric of a compact region Ω. Since the propagator is a global property of space-time strcture, it will change because of this change in structure of Ω. The change in the propagator is therefore expressible, in a first order calculation, as a functional of δg_{ik} over the volume Ω.

In the electromagnetic case the answer may be expressed in the following form

$$-\delta \sum_{a<b}\sum 4\pi e_a e_b \int\int \bar{G}_{i_A k_B} da^{i_A} db^{k_B} = -\frac{1}{2}\int T^{ik}\delta g_{ik}(-g)^{1/2} d^4x, \qquad (17)$$

where

$$T^{ik} = \frac{1}{8\pi}\sum_{a<b}\sum [\frac{1}{2} g^{ik} F^{(a)mn}_{\text{ret}} F^{(b)}_{mn\ \text{adv}} - F^{(a)i}_{l\ \text{ret}} F^{(b)kl}_{\text{adv}} - F^{(b)i}_{l\ \text{ret}} F^{(a)kl}_{\text{adv}}]. \qquad (18)$$

The details of this derivation are given by Narlikar (1974).

It is interesting to note in passing that this derivation resolves an ambiguity about the energy tensors of direct-particle electrodynamics. Wheeler and Feynman (1949) had discussed two tensors for this theory. Of these one was the canonical tensor given above by (18) and the other was the Frenkel tensor whose form differs from that given in (18) in having all the direct particle fields $F^{(a)lm}$ as the symmetric half-advanced-plus -half-retarded fields. Wheeler and Feynman had concluded:

From the standpoint of pure electrodynamics it is not possible to choose between the two tensors. The difference is of course significant for the general theory of relativity, where energy has associated with it a gravitational mass. So far we have not attempted to discriminate between the two possibilities by way of this higher standard.

As mentioned above, the usual prescription of metric variation uniquely yields the canonical tensor. The fact that we could get a nontrivial answer to the variational problem and that this resolves a long-standing ambiguity, reinforces our belief that we are proceeding along the correct path toward a theory of gravitation.

We now consider the variation of the first term of (15) as $g_{ik} \to g_{ik} + \delta g_{ik}$. We shall ignore the second term and concentrate on gravitation alone. Also for simplicity we begin by putting $\lambda_a = 1$ for all a. This does not alter the essential features of the theory.

The method is similar to that adopted for electromagnetism. We compute the change in the propagator $G(A, B)$ as the geometry changes in any compact region Ω. The details of this somewhat lengthy calculation are given elsewhere [see Hoyle and Narlikar (1974)]. We simply quote the result. The field equations turn out to be

$$\frac{1}{2}\phi(R_{ik} - \frac{1}{2}g_{ik}R) =$$

$$-T_{ik} + \frac{1}{6}[g_{ik}\Box\phi - \phi_{;ik}] + \frac{1}{2}[m_i^{\text{ret}} m_k^{\text{adv}} + m_k^{\text{ret}} m_i^{\text{adv}} - g_{ik} m^{l\ \text{ret}} m_l^{\text{adv}}], \qquad (19)$$

where,

$$m(X) = \sum_a \int G(X, A) da, \quad m_i = \partial m/\partial x^i, \qquad (20)$$

$$\phi(X) = m^{\text{adv}}(X)m^{\text{ret}}(X), \tag{21}$$

and m^{ret} and m^{adv} denote twice the retarded and advanced parts of $m(X)$, respectively. The energy tensor T_{ik} is the familar energy tensor for a system of particles $a, b, \ldots$ with masses as defined by the Machian presciption (13) and (14). Note that the masses are time symmetric. The function $m(X)$ satisfies the conformally invariant wave equation

$$\Box m + \frac{1}{6}Rm = N, \tag{22}$$

where

$$N(X) = \sum_a \int \delta_4(X, A)[-\bar{g}(X, A)]^{-1/2} da \tag{23}$$

is the invariant particle number density.

There are 10 equations in (19) and one equation (22) for the 11 unknowns g_{ik} and m. However, the divergence and trace of (19) identically vanish, showing that there are in fact five fewer independent equations. This is hardly surprising since four of these five are due to the general coordinate invariance (as in general relativity) while the fifth identity (the vanishing of trace) is due to conformal invariance. It is easy to verify that if $[g_{ik}, m]$ is a solution of these equations then so is $[\zeta^2 g_{ik}, \zeta^{-1} m]$ for an arbitrary well-behaved (i.e. of type C^2) nonvanishing finite function ζ of space and time. This arbitrary function is nothing but the expression of the arbitrariness of mass-dependent units discussed in Section 3.

Suppose now that it is possible to choose ζ such that

$$m^{ret}\zeta^{-1} = m_0 = \text{constant}. \tag{24}$$

Suppose also that the response of the universe is such as to cancel all advanced components and double the retarded ones so that the effective mass function is m^{ret}. Then the field equations are simplified to

$$R_{ik} - \frac{1}{2}g_{ik}R = -\kappa T_{ik}, \quad \kappa \equiv \frac{8\pi G}{c^4} = 6/m_0^2. \tag{25}$$

We shall later identify m_0 with the mass of the Planck particle. However, as seen above,we have arrived at the familiar equations of general relativity! The conformal frame for which (24) and (25) hold will be called the *Einstein frame*. We have thus completed the remaining part of the programme outlined at the beginning of Section 3.

The following points are worth emphasizing in the above derivation of Einstein's equations, which is so radically different from the standard ones (used for example, by Einstein in 1915 and by Hilbert later the same year).

(I) The approach to Einstein's equations is via the wider framework of a conformally invariant gravitation theory. Only in the limit of many particles in a suitably responding universe do we arrive at Einstein's equations. In the other limit of zero or no particles there is no theory! Thus it brings out the reason why the Machian paradox of one particle in an empty universe is not valid in the context of Einstein's equations. This reason does not emerge in the standard derivations of Einstein's equations.

(II) It is significant that the coupling constant κ is positive in this approach. This conclusion is unaffected by the change of sign of the coupling constants λ_a, λ_b, etc. (taken here as unity); nor is it affected by the choice of signature (i.e $---+$ instead of $+++-$) of the spacetime metric. The choice of the conformally invariant scalar propagator leads to the coupling constant being positive, i.e. to gravity being 'attractive'. In the standard derivation the coupling constant is fixed (in sign as well as magnitude) by a comparison with Newtonian gravity.

(III) A considerable discussion has gone on regarding the admissibility of the so-called λ-term in Einstein's equations. This is because this term could be accmmodated in Einstein's heuristic derivation or in Hilbert's action principle. It is worth emphasizing that the direct-particle approach to gravity given so far does not permit the λ-term. As we shall see later, this term does arise in a Machian way in the direct particle theory, *provided* we allow the wave equation (22) for inertia to be nonlinear. The present cosmological observations generally seem to require the cosmological constant (Bagla et al., 1996).

(IV) The condition (24) that leads to Einstein's equations needs to be reexamined carefully under two special circumstances. Near a typical particle a, we expect the mass function $m^{(a)}(X)$ to 'blow up' so that $m(x) \to \infty$, as $X \to A$, on the worldline of a. In order to make $\zeta^{-1}m(X)$ finite at A, we therefore require $\zeta \to \infty$,as $X \to A$. However, we have already ruled out such conformal functions by restricting ζ to finite values. Thus the transition to Einstein's equations is not valid as we tend to any typical source particle. The nature of the equations and their solutions near a particle in this theory have been discussed by Hoyle and Narlikar (1966) and by Islam (1968). The other aspect arises if there exist $m = 0$ hypersurfaces. Clearly we cannot choose a finite $\zeta (\neq 0)$ to make $m =$ constant > 0 on these hypersurfaces. If we insist on driving such solutions into the Einstein frame, we end up having spacetime singularity there. This has been pointed out by Kembhavi (1978).

5. Cosmological Constant and the Creation of Matter

In recent years, this theory has been further generalized and applied to cosmology to include the cosmological constant, as well as explicit description of creation of matter (see Hoyle et al., 1995).

Taking the cosmological constant into account first, we may ask whether (22) is the most general conformally invariant wave equation satisfied by a scalar function $m(X)$. The answer is *No!* The most general such equation is

$$\Box m + \frac{1}{6}Rm + \Lambda m^3 = N, \tag{26}$$

where Λ is a constant. This of course makes the scalar Machian interaction non-linear and more difficult to handle. However, it very naturally leads to the cosmological constant of the right magnitude at the present epoch. For, if we assume that for a single particle, the value of this constant is unity, then in the sum (20) leading to $m(X)$, because of the presence of a large number $\mathcal{N}$ of particles within the cosmological horizon, the cube term is less effective by the factor $\mathcal{N}^{-2}$. Thus, the factor Λ in (26) is of this order. With the identification of m_0, with the only possible fundamental quantity with the dimension of mass, viz, the Planck mass

$$m_0 = \sqrt{\frac{3\hbar c}{4\pi G}}, \tag{27}$$

one can write the familiar cosmological constant in the Einstein equations as

$$\lambda = -3\Lambda m_0^2. \tag{28}$$

With around 2.10^{60} Planck particles in the horizon, we get the value of $\lambda \sim -2.10^{-56}\text{cm}^{-2}$. It is of the right order, but *negative*. However, it leads to interesting and physically relevant cosmological models.

Let us consider matter creation next. The standard relativity theory starts with the assumption that matter can neither be created nor destroyed, that is, the worldlines of particles are endless. However, in the big bang singularity, all worldlines are incomplete, thus leading to a contradiction with the basic assumption. Indeed the presence of the singularity signifies the breakdown of the basic rules such as the action principle from which the equations of general relativity are obtained.

To account for the creation of matter in a nonsingular fashion, we introduce the additional input into the theory that the particle worldlines are with ends. The same action as before then describes this theory but the endpoints generate extra terms in the field equations.

For details of this work see the paper Hoyle et al(1995). The field equations in the 'constant mass' conformal frame then take the form:

$$R_{ik} - \frac{1}{2}g_{ik}R + \lambda g_{ik} = -\kappa[T_{ik} - \frac{2}{3}(c_i c_k - \frac{1}{4}g_{ik}c^l c_l)]. \tag{29}$$

The scalar c-field arises from the contribution to inertia from ends of particle worldlines. These are the contributions of the Planck particles created but which last a very short time scale$\sim 10^{-43}$ second.

Sachs et al. (1996) have solved these field equations and obtained a series of cosmological models which are a combination of two kinds:(i) models with creation of matter and (ii) models without creation of matter. The generic solution is known as the *Quasi Steady State Cosmology*. It is a nonsingular model which has a de Sitter type expansion with short term oscillations superposed on it. The former represents the creative and the latter the noncreative mode. The QSSC is being proposed as an alternative to the standard hot big bang cosmology (Hoyle et al., 2000).

6. Finite resolution of the Self-Energy Problem

After this excursion into gravitation, let me return to electrodynamics. In a recent paper Hoyle and Narlikar (1993) had shown that with suitable cosmological boundary conditions, like the de Sitter horizon, there is a cut off on high frequencies that otherwise lead to divergent integrals in the standard electromagnetic field theory. Thus the electron self-energy problem and the various radiative corrections of quantum electrodynamics can be handled without subtraction of one infinity from another. In this sense the direct particle theory fares better than the field theory. This effect comes about in the following way.

How are the high-frequency contributions to various integrals of quantum electrodynamics which diverge in normal field theory calculations, get 'damped' in the absorber theory? The de Sitter line element which is used in the steady state theory, (or its asymptotic form in the quasi-steady state cosmology),

$$ds^2 = c^2dt^2 - \exp(2Ht)\left[dr^2 + r^2(d\theta^2 + \sin^2\theta d\phi^2)\right] \tag{30}$$

can be written in a manifestly conformally flat form

$$ds^2 = (1 - H\tau)^{-2}\left[c^2d\tau^2 - dr^2 + r^2(d\theta^2 + \sin^2\theta d\phi^2)\right] \tag{31}$$

by a time-transformation

$$H\tau = 1 - \exp(-Ht) \tag{32}$$

where H is the Hubble constant. The fact that $\tau < H^{-1}$, suggests that there is an event horizon in the future absorber. This property in turn tells us that the response of the future absorber is cut off at a frequency

$$k_{\max} = \omega_{\text{eff}}/HT, \tag{33}$$

where, ω_{eff} is the effective frequency of future absorber and T is the time duration of the local process. For a medium with N as the number density of charged particles in the future absorber, ν the typical plasma frequency and m the electron mass, one has

$$\omega_{\text{eff}} = \left[2\nu N e^2/mH\right]^{1/2} \tag{34}$$

where ν is defined in terms of the physical condition of the plasma. For a free electron, one may take $T = \hbar/mc^2$.

A cutoff at $k_{\max}$ results in a finite value for the observed mass of an electron in terms of its bare mass:

$$m_{\text{obs}} = m + \Delta m, \quad \Delta m = m \times \left[1 + (3e^2/2\pi)\ln(\hbar k_{\max}/mc^2)\right]. \tag{35}$$

For a free electron, the choice of $T \sim 10^{-21}$s, $H \sim 3.10^{-18}\text{s}^{-1}$ and $\omega_{\text{eff}} \sim 80\ \text{s}^{-1}$, yields Δm ~$0.15m$. *The important result is that radiative corrections are finite* and no subtraction of one infinity from another is required. This cutoff necessarily results from the 'local-distant' interaction that is characteristic of action at a distance.

7. Concluding Remarks

We thus have both in the case of electrodynamics and gravity, a close connection between the local and the distant parts of the universe. The connection is established through the Green's functions G, that is, the propagators of the interaction. These Green's functions establish the connection along the light cone and as such the action at a distance is not instantaneous. However, when one examines the details of the absorber theory of radiation in the electromagnetic case, it tells us that a signal from a local source at $r = 0$, sent out at $t = 0$, reaches a distant absorber particle at distance $r = R$ at $t = R/c$. The response of the absorber travels *backwards* in time and so reaches the original source at $t = R/c - R/c = 0$, i.e., instantaneously ! This mixture of advanced and retarded signals serves to create an instantaneous effect without violating relativistic invariance.

Does this, however, violate causality ? The answer is 'yes', but at a very minute level. As was discussed by Dirac (1938), the radiation reaction formula has an advanced component, and this may be interpreted as generating 'pre-acceleration'. Since in this framework, one is looking at a self-consistent solution, this acausal effect is not disturbing. At the quantum level, it may be more interesting as was pointed out by Hoyle and Narlikar (1995).

The quantum version of the Wheeler-Feynman theory involves an influence functional through which the local system interacts with the large scale cosmological boundary conditions. This local+cosmological interaction appears as transition probability for a local system, wherein all cosmological variables are integrated out. Phenomena like spontaneous transition or a collapse of the wavefunction are seen to arise from this interaction. This suggests that, the attempts to explain some of these pheneomena through local hidden variables having failed, the real clue to the mystery may lie in the response of the universe in the above fashion.

Experiments by Aspect et al. (1982a,b) inspired by Bell's inequality (1966) have generated considerable discussion on nonlocality of hidden variables, and apparent acausal effects across spacelike separations. The response of the universe provides an additional factor which has been so far ignored in such discussions.

To summarize, the ideas which go under the name Mach's Principle are capable of wider applications than thought earlier by Ernst Mach. One can use the Machian concept in electrodynamics where the response of the universe can play a key role in both classical and quantum electrodynamics. The action at a distance framework used here is consistant with special relativity as well as with causality. The formalism can be used to give an expression to inertia as a direct long range effect from the distant parts of the universe. From inertia, one can arrive at a theory of gravity which is wider in its applications than general relativity. The theory can be extended to incorporate the cosmological constant and the concept of creation of matter without spacetime singularity. It leads to a viable cosmological model, known as the quasi-steady state cosmological model.

Finally, the as yet unchartered territory for this framework lies in the direction of epistemological aspects of quantum mechanics, such as understanding the rationale behind the collapse of a wavefunction and the correlations found across spacelike separations by Aspect-type experiments.

These local $\leftrightarrow$ cosmological interactions occupied Fred Hoyle's interest right to the end. Although the bulk of our work on action at a distance was done in the 1960s, Fred liked to return to the topic from time to time, vide our work on electron self energy problem in 1992–93 and the genesis of the quasi-steady state cosmology in 1993–95.

References

Aspect, A. Dalibard, J. and Roger, G.: 1982a, *Phys. Rev. Lett.* **49**, 91.
Aspect, A. Dalibard, J. and Roger, G.: 1982b, *Phys. Rev. Lett.* **49**, 1804.
Bagla, J.S., Padmanabhan, T. and Narlikar, J.V.: 1996, *Comm. Astrophys.* **18**, 275.
Bell, J.S.: 1966, *Rev. Mod. Phys.* **38**, 447.
Bondi, H. and Gold, T.: 1948, *M.N.R.A.S.* **108**, 252.
DeWitt, B.S. and Brehme, R.W.: 1960, *Ann. Phys.* (New York) **9**, 220.
Dirac, P.A.M.: 1938, *Proc. Roy. Soc.* **A167**, 148.
Fokker, A.D.: 1929a, *Z. Phys.* **58**, 386.
Fokker, A.D.: 1929b, *Physica* **9**, 33.
Fokker, A.D.: 1932, *Physica* **12**, 145.
Hogarth, J.E.: 1962, Proc. Roy. Soc. **A267**, 365.
Hoyle, F.: 1948, *M.N.R.A.S.* **108**, 372.
Hoyle, F. and Narlikar, J.V.: 1963, *Proc. Roy. Soc.* **A277**, 1.
Hoyle, F. and Narlikar, J.V.: 1964, *Proc. Roy. Soc.* **A282**, 191.
Hoyle, F. and Narlikar, J.V.: 1966, *Proc. Roy. Soc.* **A294**, 138.
Hoyle, F. and Narlikar, J.V.: 1969, *Ann. Phys.* New York **54**, 207.

Hoyle, F. and Narlikar, J.V.: 1971, *Ann. Phys.* New York **62**, 44.
Hoyle, F. and Narlikar, J.V.: 1974, *Action at a Distance in Physics and Cosmology*, Freeman, San Francisco.
Hoyle, F. and Narlikar, J.V.: 1993, *Proc. Roy. Soc.* **A442**, 469.
Hoyle, F. and Narlikar, J.V.: 1995, *Rev. Mod. Phys.* **67**, 113.
Hoyle, F., Burbidge, G. and Narlikar, J.V.: 1995, *Proc. Roy. Soc.* **A448**, 191.
Hoyle, F., Burbidge, G. and Narlikar, J.V.: 2000, *A Different Approach to Cosmology*, Cambridge University Press.
Islam, J.N.: 1968, *Proc. Roy. Soc.* **A306**, 487.
Kembhavi, A.K.: 1978, *M.N.R.A.S.* **185**, 807.
Narlikar, J.V.: 1974, *J. Phys.* **A7**, 1274.
Sachs, R., Narlikar, J.V. and Hoyle, F., *A&A* **313**, 703.
Schwarzschild, K.: 1903, *Nachr. Ges. Wis. Gottingen* **128**, 132.
Synge, J.L.: 1960, *Relativity, the General Theory*, North Holland, Amsterdam.
Tetrode, H.: 1922, *Z. Phys.* **10**, 317.
Wheeler, J.A. and Feynman, R.P.: 1945, *Rev. Mod. Phys.* **17**, 156.
Wheeler, J.A. and Feynman, R.P.: 1949, *Rev. Mod. Phys.* **21**, 424.

GRAVITY FROM SPACETIME THERMODYNAMICS

T. PADMANABHAN
IUCAA, Pune University Campus, Pune 411 007
E-mail: nabhan@iucaa.ernet.in

Abstract. The Einstein-Hilbert action (and thus the dynamics of gravity) can be obtained by: (i) combining the principle of equivalence, special relativity and quantum theory in the Rindler frame and (ii) postulating that the horizon area must be proportional to the entropy. This approach uses the local Rindler frame as a natural extension of the local inertial frame, and leads to the interpretation that the gravitational action represents the free energy of the spacetime geometry. As an aside, one obtains an insight into the peculiar structure of Einstein-Hilbert action and a natural explanation to the questions: (i) Why does the covariant action for gravity contain second derivatives of the metric tensor? (ii) Why is the gravitational coupling constant positive? Some geometrical features of gravitational action are clarified.

Keywords: horizon, rindler, entropy, principle of equivalence, holography

1. Introduction and summary

The (i) existence of the principle of equivalence and (ii) the connection between gravity and thermodynamics are the two most surprising features of gravity. Among these two, the principle of equivalence finds its natural expression when gravity is described as a manifestation of curved spacetime. This – in turn – makes gravity the only interaction which is capable of wrapping up regions of spacetime so that information from one region is not accessible to observers at another region. Given the fact that entropy of a system is closely related to accessibility of information, it is inevitable that there will be some connection between gravity and thermodynamics (for a review, see Birrell and Davies, 1982 and Padmanabhan, 2002a). But, in contrast to the principle of equivalence, years of research in this field (see, for a sample Bekenstein, 1973; Hawking, 1975; Gerlach, 1976; 't Hooft, 1985; York, 1985; Zurek and Thorne, 1985; Bombelli et al., 1986; For an earlier attempt similar in spirit to the current paper, see Jacobson, 1995), has not led to something more profound or fundamental arising out of this feature.

This suggests that we should learn a lesson from the way Einstein handled the principle of equivalence and apply it in the context of the connection between thermodynamics and gravity. Einstein did not attempt to 'derive' principle of equivalence in the conventional sense of the word. Rather, he accepted it as a key feature which must find expression in the way gravity is described – thereby obtaining a geometrical description of gravity. Once the geometrical interpretation of gravity is accepted, it follows that there *will* arise surfaces which act as one-way-membranes for information and will thus lead to some connection with thermodynamics. It is,

Astrophysics and Space Science **285:** 407–417, 2003.

therefore, more in tune with the spirit of Einstein's analysis to *accept* an inevitable connection between gravity and thermodynamics and ask what such a connection would imply. I will now elaborate this idea further in order to show how powerful it is (Padmanabhan, 2002).

The first step in the logic, the principle of equivalence, allows one to define a coordinate system around any event $\mathcal{P}$ in a region of size L (with $L^2(\partial^2 g/g) \ll 1$ but $L(\partial g/g)$ being arbitrary) in which the spacetime is locally inertial. As the second step, we want to give expression to the fact that there is a deep connection between one-way-membranes arising in a spacetime and thermodynamical entropy. This, of course, is not possible in the local inertial frame since the quantum field theory in that frame, say, does not recognize any non trivial geometry of spacetime. But it is possible to achieve our aim by using a uniformly accelerated frame around $\mathcal{P}$. In fact, around any event $\mathcal{P}$ we have fiducial observers anchored firmly in space with $\mathbf{x}$ = constant and the four-velocity $u^i = g_{00}^{-1/2}(1, 0, 0, 0)$ and acceleration $a^i = u^j \nabla_j u^i$. This allows us to define a second natural coordinate system around any event by using the Fermi-Walker transported coordinates corresponding to these accelerated observers. I shall call this the local Rindler frame. [Operationally, this coordinate system is most easily constructed by first transforming to the locally inertial frame and then using the standard transformations between the inertial coordinates and the Rindler coordinates.] This local Rindler frame will lead to a natural notion of horizon and associated temperature. The key new idea will be to postulate that the horizon in the local Rindler frame also has an entropy per unit transverse area and demand that any description of gravity must have this feature incorporated in it.

What will such a postulate lead to? Incredibly enough, it leads to the *correct Einstein-Hilbert action principle* for gravity. Note that the original approach of Einstein making use of the principle of equivalence lead only up to the *kinematics* of gravity – viz., that gravity is described by a curved spacetime with a non trivial metric g_{ab} – and cannot tell us how the *dynamics* of the spacetime is determined. Taking the next step, using the local Rindler frame and demanding that gravity must incorporate the thermodynamical aspects lead to the action functional itself.

This approach also throws light on (what has been usually considered) a completely different issue: Why does the Einstein-Hilbert action contain second derivatives of the metric tensor? The new approach 'builds up' the Einstein-Hilbert action from its surface behaviour and, in this sense, shows that gravity is intrinsically holographic ('t Hooft, 1993; Susskind, 1995; for a recent review see Bousso, 2002). I use this term with the specific meaning that given the form of the action on a two dimensional surface, there is a way of obtaining the full bulk action. In the $(3 + 1)$ formalism, this leads to the interpretation of the gravitational action as the free energy of spacetime. Einstein's equations are equivalent to the principle of minimization of free energy in thermodynamics.

2. Gravitational Dynamics from Spacetime Thermodynamics

The principle of equivalence leads to a geometrical description of gravity in which g_{ab} are the fundamental variables. So we expect the dynamics of gravity to be described by some *unknown* action functional

$$A = \int d^4x\sqrt{-g}L(g, \partial g) \equiv \int d^4x\sqrt{-g}L(g, \Gamma) \tag{1}$$

involving g_{ab}s and their first derivatives $\partial_c g_{ab}$ or, equivalently, the set $[g_{ab}, \Gamma^i_{jk}]$ where Γs are the standard Christoffel symbols.

Given any Lagrangian $L(\partial q, q)$ involving only up to the first derivatives of the dynamical variables, it is *always* possible to construct another Lagrangian $L'(\partial^2 q, \partial q, q)$, involving second derivatives such that it describes the same dynamics (Lynden-Bell and Padmanabhan, 1994; Padmanabhan, 1996). This idea works for any number of variables (so that q can be a multicomponent entity) dependent on space and time. But I shall illustrate it in the context of point mechanics. The prescription is:

$$L' = L - \frac{d}{dt}\left(q\frac{\partial L}{\partial \dot{q}}\right) \tag{2}$$

While varying the L', one keeps the *momenta* $(\partial L/\partial\dot{q})$ fixed at the endpoints rather than q's. This is most easily seen by explicit variation; we have

$$\begin{aligned}\delta A' &= \int_{\mathcal{P}_1}^{\mathcal{P}_2} dt\left[\frac{\partial L}{\partial q}\delta q + \frac{\partial L}{\partial \dot{q}}\delta\dot{q}\right] - \delta\left(q\frac{\partial L}{\partial \dot{q}}\right)\Bigg|_{\mathcal{P}_1}^{\mathcal{P}_2} \\ &= \int_{\mathcal{P}_1}^{\mathcal{P}_2} dt\left[\frac{\partial L}{\partial q} - \frac{d}{dt}\left(\frac{\partial L}{\partial \dot{q}}\right)\right]\delta q - q\delta p\Bigg|_{\mathcal{P}_1}^{\mathcal{P}_2}\end{aligned} \tag{3}$$

If we keep $\delta p = 0$ at the end points while varying L', then we get back the same Euler-Lagrange equations as obtained by varying L and keeping $\delta q = 0$ at end points. Since $L = L(\dot{q}, q)$, the quantity $q(\partial L/\partial\dot{q})$ will also depend on $\dot{q}$ and the term $d(q\partial L/\partial\dot{q})/dt$ will involve $\ddot{q}$. Thus L' contains second derivatives of q while L contains only up to first derivatives. In spite of the fact that L' contains second derivatives of q, the equations of motion arising from L' are only second order for variation with $\delta p = 0$ at end points. It can be shown that, in the path integral formulation of quantum theory, the modified Lagrangian L' correctly describes the transition amplitude between states with given momenta (see p. 170 of Padmanabhan, 1996).

Thus, in the case of gravity, the *same* equations of motion can be obtained from another (as yet unknown) action:

$$A' = \int d^4x\sqrt{-g}L - \int d^4x\partial_c\left[g_{ab}\frac{\partial\sqrt{-g}L}{\partial(\partial_c g_{ab})}\right]$$

$$\equiv A - \int d^4x \partial_c(\sqrt{-g}V^c) \equiv A - \int d^4x \partial_c P^c \tag{4}$$

where V^c is made of g_{ab} and Γ^i_{jk}. Further, V^c must be linear in the Γ's since the original Lagrangian L was quadratic in the first derivatives of the metric. Since Γs vanish in the local inertial frame and the metric reduces to the Lorentzian form, the action A cannot be generally covariant. However, the action A' involves the second derivatives of the metric and we shall see later that that the action A' is indeed generally covariant.

To obtain a quantity V^c, which is linear in Γs and having a single index c, from g_{ab} and Γ^i_{jk}, we must contract on two of the indices on Γ using the metric tensor. (Note that we require A, A' etc. to be Lorentz scalars and P^c, V^c etc. to be vectors under Lorentz transformation.) Hence the most general choice for V^c is the linear combination

$$V^c = \left(a_1 g^{ck}\Gamma^m_{km} + a_2 g^{ik}\Gamma^c_{ik}\right) \tag{5}$$

where $a_1(g)$ and $a_2(g)$ are unknown functions of the determinant g of the metric (which is the only (pseudo) scalar entity which can be constructed from g_{ab}s and Γ^i_{jk}s). Using the identities $\Gamma^m_{km} = \partial_k(\ln\sqrt{-g})$, $\sqrt{-g}g^{ik}\Gamma^c_{ik} = -\partial_b(\sqrt{-g}g^{bc})$, we can rewrite the expression for $P^c \equiv \sqrt{-g}V^c$ as

$$P^c = \sqrt{-g}V^c = c_1(g)g^{cb}\partial_b\sqrt{-g} + c_2(g)\sqrt{-g}\partial_b g^{bc} \tag{6}$$

where $c_1 \equiv a_1 - a_2$, $c_2 \equiv -a_2$ are two other unknown functions of g. If we can fix these coefficients by using a physically well motivated prescription, then we can determine the surface term and by integrating, the Lagrangian L. I will now show how this can be done.

Let us consider a static spacetime in which all g_{ab}s are independent of x^0 and $g_{0\alpha} = 0$. Around any given event $\mathcal{P}$ one can construct a local Rindler frame with an acceleration of the observers with $\mathbf{x}$ = constant, given by $a^i = (0, \mathbf{a})$ and $\mathbf{a} = \nabla(\ln\sqrt{g_{00}})$. This Rindler frame will have a horizon which is a plane surface normal to the direction of acceleration and a temperature $T = |\mathbf{a}|/2\pi$ associated with this horizon. I shall postulate that the entropy associated with this horizon is proportional to its area or, more precisely,

$$\frac{dS}{dA_\perp} = \frac{1}{\mathcal{A}_P} \tag{7}$$

where $\mathcal{A}_P$ is a fundamental constant with the dimensions of area. It represents the minimum area required to hold unit amount of information and our postulate demands that this number be finite. Given the temperature of the horizon, one can construct a canonical ensemble with this temperature and relate the Euclidean action to the thermodynamic entropy (see, e.g. Padmanabhan, 2002b, 2002c)). Since the Euclidean action can be interpreted as the entropy in the canonical ensemble,

I will demand that the surface term in equation (4) should be related to the entropy S by $S = -A^{\rm E}_{\rm surface}$ (with the minus sign arising from standard Euclidean continuation Padmanabhan, 2002), when evaluated in the local Rindler frame with the temperature T. In particular, this result must hold in flat spacetime in Rindler coordinates. [We will see later that the action A' is generally covariant and will vanish in the flat spacetime, in the absence of the cosmological constant. It follows that the numerical value of the action A in the Rindler frame is the same as the surface term in equation (4).] In the static Rindler frame, the surface term is

$$A_{\rm surface} = \int d^4x \partial_c P^c = \int_0^\beta dt \int_{\mathcal{V}} d^3x \nabla \cdot \mathbf{P} = \beta \int_{\partial \mathcal{V}} d^2x_\perp \hat{\mathbf{n}} \cdot \mathbf{P} \tag{8}$$

I have restricted the time integration to an interval $(0, \beta)$ where $\beta = (2\pi/|\mathbf{a}|)$ is the inverse temperature in the Rindler frame. This is *needed* since the Euclidean action will be periodic in the imaginary time with the period β. We shall choose the Rindler frame such that the acceleration is along the $x^1 = x$ axis. The most general form of the metric representing the Rindler frame can be expressed in the form

$$\begin{aligned} ds^2 &= (1+2al)dt^2 - \frac{dl^2}{(1+2al)} - (dy^2+dz^2) \\ &= [1+2al(x)]\,dt^2 - \frac{l'^2}{[1+2al(x)]}dx^2 - (dy^2+dz^2) \end{aligned} \tag{9}$$

where $l(x)$ is an arbitrary function and $l' \equiv (dl/dx)$. [Since the acceleration is along the x-axis, the metric in the transverse direction is unaffected. The first form of the metric is the standard Rindler frame in the (t, l, y, z) coordinates. We can, however, make any coordinate transformation from l to some other variable x without affecting the planar symmetry or the static nature of the metric. This leads to the general form of the metric given in the second line, in terms of the (t, x, y, z) coordinates.] Evaluating the surface term P^c in (6) for this metric, we get the only non zero component to be

$$P^x = -2ac_2(g) - [1+2al(x)]\frac{l''}{l'^2}[c_1(g) - 2c_2(g)] \tag{10}$$

so that the action in (8) becomes

$$A = \beta P^x \int d^2x_\perp = \beta P^x A_\perp = S \tag{11}$$

where $A_\perp$ is the transverse area of the $(y-z)$ plane. The last equality identifies the entropy S, which is equal the Euclidean action, with the minus sign arising from standard Euclidean continuation. From our postulate (7) it follows that

$$-\frac{dS}{dA_\perp} = 2a\beta c_2(g) + \beta[c_1 - 2c_2](1+2al)\frac{l''}{l'^2} = -\frac{1}{\mathcal{A}_P} \tag{12}$$

For this quantity to be a constant independent of x for any choice of $l(x)$, the second term must vanish requiring $c_1(g) = 2c_2(g)$. An explicit way of obtaining this result is to consider a class of functions $l(x)$ which satisfy the relation $l' = (1 + 2al)^n$ with $0 \leq n \leq 1$. Then

$$\beta[c_1(l') - 2c_2(l')](1 + 2al)\frac{l''}{l'^2} = 2a\beta[c_1(l') - 2c_2(l')]n \tag{13}$$

which can be independent of n and x only if $c_1(g) = 2c_2(g)$. Further, using $a\beta = 2\pi$, we find that $c_2(g) = (4\pi \mathcal{A}_p)^{-1}$ which is a constant independent of g. Hence P^c has the form

$$\begin{aligned} -P^c &= \frac{1}{4\pi \mathcal{A}_P}\left(2g^{cb}\partial_b\sqrt{-g} + \sqrt{-g}\partial_b g^{bc}\right) = \frac{\sqrt{-g}}{4\pi \mathcal{A}_P}\left(g^{ck}\Gamma^m_{km} - g^{ik}\Gamma^c_{ik}\right) \\ &= -\frac{1}{4\pi \mathcal{A}_P}\frac{1}{\sqrt{-g}}\partial_b(gg^{bc}) \end{aligned} \tag{14}$$

The second equality is obtained by using the standard identities mentioned after equation (5) while the third equality follows directly by combining the two terms in the first expression. This result is remarkable and let me discuss it before proceeding further.

The general form of P^c which we obtained in (6) is not of any use unless we can fix (c_1, c_2). For static configurations, we can convert the extra term to an integral over time and a two-dimensional spatial surface. This is true for any system, but in general, the result will not have any simple form and will involve an undetermined range of integration over time coordinate. But in the case of gravity, two natural features conspire together to give an elegant form to this surface term. First is the fact that Rindler frame has a periodicity in Euclidean time and the range of integration over the time coordinate is naturally restricted to the interval $(0, \beta) = (0, 2\pi/a)$. The second is the fact that the surviving term in the integrand P^c is linear in the acceleration a thereby neatly canceling with the $(1/a)$ factor arising from time integration. [I will discuss these features more in section (3).]

Given the form of P^c we need to solve the equation

$$\left(\frac{\partial\sqrt{-g}L}{\partial g_{ab,c}}g_{ab}\right) = P^c = -\frac{1}{4\pi \mathcal{A}_P}\left(2g^{cb}\partial_b\sqrt{-g} + \sqrt{-g}\partial_b g^{cb}\right) \tag{15}$$

to obtain the first order Lagrangian density. It is straightforward to show (?) that this equation is satisfied by the Lagrangian

$$\sqrt{-g}L = \frac{1}{4\pi \mathcal{A}_P}\left(\sqrt{-g}\, g^{ik}\left(\Gamma^m_{i\ell}\Gamma^\ell_{km} - \Gamma^\ell_{ik}\Gamma^m_{\ell m}\right)\right). \tag{16}$$

This is the second surprise. The Lagrangian which we have obtained is precisely the first order Dirac-Schrodinger Lagrangian for gravity (usually called the Γ^2 Lagrangian). Note that we have obtained it without introducing the curvature tensor anywhere in the picture. Once again, this is unlikely to be a mere accident.

Given the two pieces, the final second order Lagrangian follows from our equation (4) and is, of course, the standard Einstein-Hilbert Lagrangian.

$$\sqrt{-g}L_{grav} = \sqrt{-g}L - \frac{\partial P^c}{\partial x^c} = -\left(\frac{1}{4\pi \mathcal{A}_P}\right) R\sqrt{-g}. \tag{17}$$

Thus our full second order Lagrangian *turns out* to be the standard Einstein-Hilbert Lagrangian. We have obtained this result by just postulating that the surface term in the action should be proportional to the entropy per unit area. This postulate uniquely determines the gravitational action principle and gives rise to a generally covariant action. The surface terms dictate the form of the Einstein Lagrangian in the bulk. The idea that surface areas encode bits of information per quantum of area allows one to determine the nature of gravitational interaction on the bulk, which is an interesting realization of the holographic principle.

I stress the fact that there is a very peculiar identity connecting the Γ^2 Lagrangian L and the Einstein-Hilbert Lagrangian L_{grav}, encoded in equation (17). This relation, which is purely a differential geometric identity, can be stated through the equations:

$$L_{grav} = L - \nabla_c \left[g_{ab} \frac{\partial L}{\partial(\partial_c g_{ab})} \right]; \quad L = L_{grav} - \nabla_c \left[\Gamma^j_{ab} \frac{\partial L_{grav}}{\partial(\partial_c \Gamma^j_{ab})} \right] \tag{18}$$

This relationship between the three terms defies any simple explanation in conventional approaches to gravity but arises very naturally in the approach presented here.

The solution to (15) obtained in (16) is not unique. However, self consistency requires that the final equations of motion for gravity must admit the line element in (9) as a solution. It can be shown. by fairly detailed algebra, that this condition makes the Lagrangian in (15) to be the only solution. In particular, since we are demanding the flat spacetime to be a solution to the field equations, the cosmological constant in the pure gravity sector must be zero. [This, of course, does not prevent a cosmological constant arising from the matter sector of the theory.]

3. Structure of Einstein-Hilbert action

The approach leads to new insights regarding the peculiar structure of Einstein-Hilbert and the $(3 + 1)$ formalism of gravity. To discuss these features, it is convenient to temporarily switch to the signature $(- + ++)$ so that the spatial metric is positive definite. We will foliate the spacetime by a series of space like hypersurfaces Σ with u^i as normal; then $g^{ik} = h^{ik} - u^i u^k$ where h^{ik} is the induced metric on Σ. From the covariant derivative $\nabla_i u_j$ of the normals to Σ, one can construct only three vectors $(u^j \nabla_j u^i, u^j \nabla^i u_j, u^i \nabla^j u_j)$ which are linear in covariant derivative operator. The first one is the acceleration $a^i = u^j \nabla_j u^i$; the second identically vanishes since u^j has unit norm; the third, $u^i K$, is proportional to the trace of the extrinsic curvature $K = -\nabla^j u_j$ of Σ. Thus V^i in the surface term in equation (4) must be a linear combination of $u^i K$ and a^i. In fact, one can show that (see, e.g. equation (21.88) Miser et al., 1973)

$$R = {}^3\mathcal{R} + K_{ab}K^{ab} - K^a_a K^b_b - 2\nabla_i(Ku^i + a^i) \equiv \mathcal{L} - 2\nabla_i(Ku^i + a^i) \qquad (19)$$

where $\mathcal{L}$ is the ADM Lagrangian. To prove this, we begin with the relation

$$R = -Rg_{ab}u^a u^b = 2(G_{ab} - R_{ab})u^a u^b \qquad (20)$$

and rewrite the first term using the identity:

$$2G_{ab}u^a u^b = {}^3\mathcal{R} - K_{ab}K^{ab} + K^a_a K^b_b \qquad (21)$$

As for the second term in (20), we note that $R_{abcd}u^d = (\nabla_a \nabla_b u_c - \nabla_b \nabla_a u_c)$ giving

$$\begin{aligned} R_{bd}u^b u^d &= g^{ac}u^b u^d R_{abcd} = (u^b \nabla_a \nabla_b u^a - u^b \nabla_b \nabla_a u^a) \\ &= \nabla_a(u^b \nabla_b u^a) - (\nabla_a u^b)(\nabla_b u^a) - \nabla_b(u^b \nabla_a u^a) + (\nabla_b u^b)^2 \\ &= \nabla_i(Ku^i + a^i) - K_{ab}K^{ab} + K^a_a K^b_b \end{aligned} \qquad (22)$$

Using (22) and (21) we can rewrite (20) in the form of (19).

Let us now use (19) to integrate $(R/16\pi)$ over a four volume $\mathcal{V}$ bounded by two space-like surfaces Σ_1 and Σ_2 (with normals u^i) and two time-like surfaces $\mathcal{S}_1$ and $\mathcal{S}_2$ (with normals n^i). The induced metric on the space-like surface Σ is $h_{ab} = g_{ab} + u_a u_b$ while the metric on the time-like surface $\mathcal{S}$ is $\gamma_{ab} = g_{ab} - n_a n_b$. These two surfaces will intersect on a two-dimensional surface $\mathcal{Q}$ on which the metric is $\sigma_{ab} = h_{ab} - n_a n_b = g_{ab} + u_a u_b - n_a n_b$. Integrating both sides of (19) over $\mathcal{V}$ we now get

$$\begin{aligned} A_{\mathrm{EH}} &= \frac{1}{16\pi}\int_{\mathcal{V}} R\sqrt{-g}\,d^4x = \frac{1}{16\pi}\int_{\mathcal{V}} \mathcal{L}\sqrt{-g}\,d^4x - \frac{1}{8\pi}\int_{\Sigma_1}^{\Sigma_2} K\sqrt{h}\,d^3x \\ &\quad -\frac{1}{8\pi}\int_{\mathcal{S}_1}^{\mathcal{S}_2}(a_i n^i)\sqrt{\sigma}\,d^2x\,N\,dt \end{aligned} \qquad (23)$$

where $g_{00} = -N^2$. In a static spacetime with a horizon: (i) $K = 0$ making the second term on the right hand side vanish. (ii) The integration over t becomes multiplication by β. (iii) Further, as the surface $\mathcal{S}_1$ approaches the horizon, the quantity $N(a_i n^i)$ tends to $(-\kappa)$ where κ is the surface gravity of the horizon, which is constant over the horizon. Using $\beta\kappa = 2\pi$, the last term gives, on the horizon, the contribution

$$\frac{\kappa}{8\pi}\int_0^{\beta} dt \int d^2x\,\sqrt{\sigma} = \frac{1}{4}\mathcal{A}_H \qquad (24)$$

where $\mathcal{A}_H$ is the area of the horizon. In the Euclidean sector the first term gives βE where E is the integral of the ADM Hamiltonian over the spatial volume. We thus get the result

$$A_{\rm EH}^{\rm Euclidian} = \frac{1}{4}\mathcal{A}_H - \beta E = (S - \beta E) \tag{25}$$

which is the free energy. For any static spacetime geometry, having a periodicity β in the Euclidean time, the Euclidean gravitational action represents the free energy of the spacetime; the first order term gives the Hamiltonian and the surface term gives the entropy.

More generally, the analysis suggests a remarkably simple, thermodynamical, interpretation of semiclassical gravity. In any static spacetime with a metric

$$ds^2 = -N^2(\mathbf{x})dt^2 + \gamma_{\alpha\beta}(\mathbf{x})dx^\alpha dx^\beta \tag{26}$$

we have $R = {}^3R - 2\nabla_i a^i$ where $a_i = (0, \partial_\alpha N/N)$ is the acceleration of $\mathbf{x} =$ constant world lines. Then, limiting the time integration to $(0, \beta)$, say, the Einstein-Hilbert action becomes

$$A = \frac{\beta}{16\pi}\int_{\mathcal{V}} d^3x N\sqrt{\gamma}\,{}^3R - \frac{\beta}{8\pi}\int_{\partial\mathcal{V}} (a^\alpha n_\alpha) d^2\mathcal{S} \equiv \beta E - S \tag{27}$$

where the first term is proportional to energy (in the sense of spatial integral of ADM Hamiltonian) and the second term is proportional to entropy in the presence of horizon. The variation of this action – which leads to Einstein's equation – is equivalent to the thermodynamic identity. (This result is explored in detail for spherically symmetric spacetimes in Padmanabhan, 2002c).

The surfaces Σ, $\mathcal{S}$ as well as the two surface $\mathcal{Q}$ on which they intersect will have corresponding extrinsic curvatures K_{ab}, Θ_{ab} and q_{ab}. In the literature, it is conventional to write the Einstein-Hilbert action as a term having only the first derivatives, plus an integral over the trace of the extrinsic curvature of the bounding surfaces. It is easy to obtain this form using the foliation condition $n^i u_i = 0$ between the surfaces and noting:

$$n_i a^i = n_i u^j \nabla_j u^i = -u^j u^i \nabla_j n_i = (g^{ij} - h^{ij})\nabla_j n_i = -(\Theta - q) \tag{28}$$

where $\Theta \equiv \Theta^a_a$ and $q \equiv q^a_a$ are the traces of the extrinsic curvature of the 2-surface when treated as embedded in the 4-dimensional or 3-dimensional enveloping manifolds. Using (28) to replace $(a_i n^i)$ in the last term of (23), we get the result

$$\begin{aligned} A_{\rm EH} &+ \frac{1}{8\pi}\int_{\Sigma_1}^{\Sigma_2} K\sqrt{h}\, d^3x - \frac{1}{8\pi}\int_{\mathcal{S}_1}^{\mathcal{S}_2} \Theta\sqrt{\sigma}\, d^2x\, N dt \\ &= \frac{1}{16\pi}\int_{\mathcal{V}} \mathcal{L}\sqrt{-g}\, d^4x - \frac{1}{8\pi}\int_{\mathcal{S}_1}^{\mathcal{S}_2} q\sqrt{\sigma}\, d^2x\, N dt \end{aligned} \tag{29}$$

In the first term on the right hand side, the ADM Lagrangian $\mathcal{L}$ contains ${}^3\mathcal{R}$ which in turn involves the second derivatives of the metric tensor. The second term on the right hand side removes these second derivatives making the right hand side equal to the quadratic Γ^2 action for gravity. On the left hand side, the second and third terms are the integrals of the extrinsic curvatures over the boundary surfaces which, when added to the Einstein-Hilbert action gives the quadratic action without second derivatives. (This is the standard result often used in the literature). Unfortunately, this form replaces the normal component of the acceleration $a^i n_i$ in (23) by $(\Theta - q)$ and combines q with ${}^3\mathcal{R}$ to get the first order Lagrangian. In the process, the normal component of the acceleration disappears and we miss the nice interpretation of Einstein-Hilbert action as the free energy of spacetime.

4. Conclusions

The approach adopted here is a natural extension of the original philosophy of Einstein; viz., to use non inertial frames judiciously to understand the behaviour of gravity. In the original approach, Einstein used the principle of equivalence which leads naturally to the description of gravity in terms of the metric tensor. Unfortunately, *classical* principle of equivalence cannot take us any further since it does not encode information about the curvature of spacetime. However, the true world is quantum mechanical and one would like to pursue the analogy between non inertial frames and gravitational field into the quantum domain. Here the local Rindler frame arises as the natural extension of the local inertial frame and the study of the thermodynamics of the horizon shows a way of combining special relativity, quantum theory and physics in the non inertial frame. I have shown that these components are adequate to determine the action functional for gravity and, in fact, leads to the Einstein-Hilbert action. This is remarkable because we did not introduce the curvature of spacetime explicitly into the discussion and – in fact – the analysis was done in a Rindler frame which is just flat spacetime. The idea works because the action for gravity splits up into two natural parts *neither* of which is generally covariant but are related to each other by the remarkable identity (18) which – as far as I know – was not noticed before. The sum of the two parts is generally covariant but the expression for individual parts can be ascertained in the local Rindler frame specifically because these parts are *not* generally covariant.

The fundamental postulate we use is in equation (7) and it does *not* refer to any horizon. To see how this comes about, consider any spatial plane, say the $y - z$ plane, in flat spacetime. It is always possible to find a Rindler frame in the flat spacetime such that the chosen surface acts as the horizon for some Rindler observer. In this sense, any plane in flat spacetime must have an entropy per unit area. Microscopically, I would expect this to arise because of the entanglement over length scales of the order of $\sqrt{\mathcal{A}_P}$. We have defined in (7) the entropy per unit area rather than the total entropy in order to avoid having to deal with global nature of the surfaces (whether the surface is compact, non compact etc.). This approach also provides a natural explanation as to why the gravitational coupling constant is positive. It is positive because entropy and area are positive quantities.

The result emphasizes the role of two dimensional surfaces in fundamental physics. A two dimensional surface is the basic minimum one needs to produce region of inaccessibility and thus entropy from lack of information. When one connects up gravity with spacetime entropy it is is inevitable that the coupling constant for gravity has the dimensions of area in natural units. The next step in such an approach will be to find the fundamental units by which spacetime areas are made of and provide a theoretical, quantum mechanical description for the same. This will lead to the proper quantum description of spacetime with Einstein action playing the role of the free energy in the thermodynamic limit of the spacetime.

Acknowledgement

I thank Apoorva Patel and K. Subramanian for several useful discussions. I thank the organizers of the 'Fred Hoyle's Universe' conference (Cardiff, June, 2002) for inviting me to give the talk and providing local hospitality.

References

Bekenstein, J.D.: 1973, *Phys. Rev. D* **7**, 2333.

Birrell, N.D. and Davies P.C.W.: 1982, *Quantum Fields in Curved Space*, Cambridge University Press, Cambridge.

Bombelli. L et al.: 1986, *Phys. Rev. D* **34**, 3, 73.

Bousso, R.: 2000, *The Holographic Principle*, hep-th/0203101.

Gerlach, U.H.: 1976, *Phys. Rev. D* **15**, 1479.

Hawking, S.W.: 1975, *Comm. Math. Phys.* **43**, 199.

Jacobson, T.: 1995, *Phys. Rev. Letts.* **75**, 1260.

Lynden-Bell, D. and Padmanabhan, T.: 1994, unpublished.

Miser, C.W., Thorne K.S. and Wheeler, J.A.: 1973, *Gravitation*, Freeman and co., p. 520.

Padmanabhan, T.: 1996, *Cosmology and Astrophysics – Through Problems*, Cambridge university press, p. 170; p. 325.

Padmanabhan, T.: 2002a, *Mod. Phys. Letts. A* **17**, 923. [gr-qc/0202078].

Padmanabhan, T.: 2002b, *The Holography of Gravity encoded in a Relation between Entropy, Horizon Area and the Action for gravity* [Second Prize essay; Gravity Research Foundation Essay Contest, 2002] and elaborated in *Mod. Phys. Letts. A* **17**, 1147 [hep-th/0205278].

Padmanabhan, T.: 2002c, *Class. Quan. Grav.* **19**, 5387, [gr-qc/0204019].

Susskind, L.: 1995, *J. Math. Phys.* **36**, 6377.

't Hooft, G.: 1993, *Dimensional Reduction in quantum gravity*, gr-qc/9310026; *The holographic principle*, hep-th/0003004.

't Hooft, G.: 1985, *Nucl. Phys.* **B256**, 727.

York, J.: 1985, *Phys. Rev. D* **15**, 2929.

York, J.W.: 1988, in: W.H. Zurek et al. (eds.), *Between Quantum and Cosmos*, Princeton University Press, Princeton, p. 246.

Zurek, W.H. and Thorne, K.S.: 1985, *Phys. Rev. lett.* **54**, 2171.

A STATISTICAL EVALUATION OF ANOMALOUS REDSHIFT CLAIMS

W.M. NAPIER
Armagh Observatory, College Hill, Armagh BT61 9DG, Northern Ireland
University of Cardiff

Abstract. Claims that ordinary spiral galaxies and some classes of QSO show periodicity in their redshift distributions are investigated using recent high-precision data and rigorous statistical procedures. The claims are broadly upheld. The periodicites are strong and easily seen by eye in the datasets. Observational, reduction or statistical artefacts do not seem capable of accounting for them.

1. Introduction

Claims that there exists a class of extragalactic redshifts best described as 'anomalous' have appeared in the literature for about 30 years. Amongst these claims are: QSOs and galaxies of widely different redshifts are connected by bridges of material; faint galaxies in clusters are systematically redshifted in relation to the dominant galaxy; there is a periodicity of $\sim$72 km s^{-1} in the Coma cluster; there is a global redshift periodicity of 36.2 km s^{-1} for wide-profile field galaxies in the galactocentric frame of reference; and a periodicity of 0.089 in $\log_{10}(1 + z)$ exists in the redshifts z of QSOs close to bright, active nearby galaxies.

In spite of their persistence, these claims are almost universally ignored. The phenomena do not fit current cosmologies and the statistics have been criticised (not always fairly). In recent years however, new high-precision data have become available, and with the advent of fast computers, these claims may now be put to the test using rigorous statistical procedures and extensive Monte Carlo simulations. Probably, also, examination of such claims is best carried out by individuals not involved in their creation. Consequently, the author with colleagues has been engaged on a long-term project to examine the status of the 'anomalous redshift' claims from a strictly objective standpoint. In the present paper I review the state of play of this project, updating previously published results with somewhat more comprehensive statistical analyses.

Three 'anomalous redshift' claims have so far been tested, namely the 72 km s^{-1} periodicity of the Coma cluster, the 36 km s^{-1} galactocentric periodicity of wide-profile field spirals, and the periodicity of QSOs close to nearby bright, active spiral galaxies. All three claims are upheld: the null hypothesis that these periodicities do not exist in the data examined can be rejected with high confidence.

Astrophysics and Space Science **285:** 419–427, 2003.

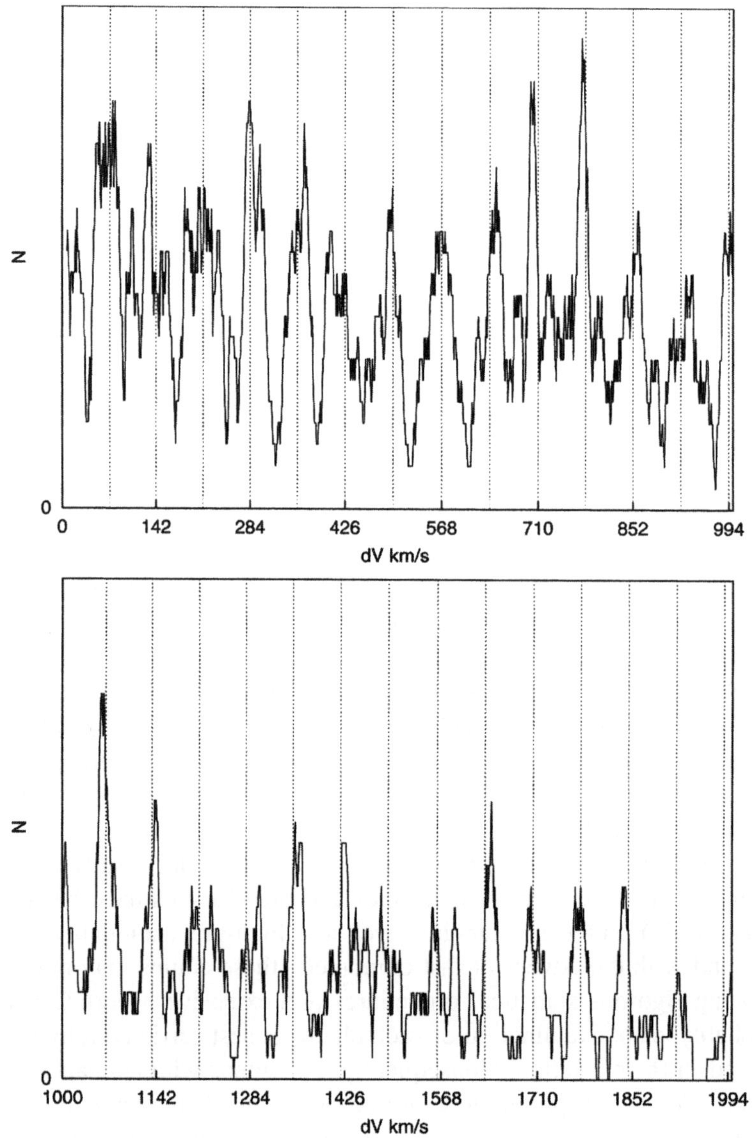

Figure 1. Upper: Differential redshifts of 48 spiral galaxies in the Virgo cluster. Data are plotted in the galactocentric frame of reference as described in the text, and a rectangular window of width 13 km s^{-1} has been applied corresponding to the rms sum of the formal redshift uncertainties. The vertical dotted lines are the best-fit periodicity of 71 km s^{-1}. Upper: 0–1000 km s^{-1}. Lower: 1000–2000 km s^{-1}.

2. The 72 km s^{-1} Claim

The Virgo cluster is the nearest rich cluster of galaxies, and had not previously been used to test for redshift periodicity claims. It therefore constitutes an unbiased sample with which to test the Coma cluster claim. Figure 1 shows the diferential redshifts of 48 spirals in the Virgo cluster, plotted in the galactocentric frame of reference. For this plot the latter was taken to be the IAU-approved

$$V_{\odot} = 220\text{km s}^{-1}, l_{\odot} = 90.0°, b_{\odot} = 0.0°$$

where ($V_{\odot}$, $l_{\odot}$, $b_{\odot}$) are respectively the speed, galactic longitude and galactic latitude of the Sun's velocity vector around the nucleus of the Galaxy. The selection details are described in Guthrie and Napier (1990); broadly, these are spiral galaxies avoiding the core of the Virgo cluster and with redshifts determined to formal accuracies $\sigma \leq 10$ km s^{-1}. A periodicity is clearly present, which power spectrum analysis shows to be 71 km s^{-1}.

A simple procedure to test the significance of this periodicity is to construct synthetic Virgo clusters and examine them for periodicity in identical fashion to the procedure adopted for the real Virgo cluster, that is, by searching for the strongest signal in some range of 'solar velocity space'. The synthetic clusters must be constructed so as to reproduce all the features of the real Virgo cluster in every respect except that of the periodicity under test. Thus the positions of the 48 Virgo cluster galaxies in the sky are preserved, their redshifts are sampled with replenishment, and for each redshift chosen, a small random displacement is added, sufficient to smear out the periodicity under test but not enough to alter the broad redshift distribution. A synthetic Virgo cluster so constructed is identical to the real one in all respects except that any fine structure up to this range has been smeared out.

For each model cluster so constructed one then corrects for a range of solar motions, applying power spectrum analysis to the corrected redshift distribution and recording the highest peak occurring in, say, the range 70–75 km s^{-1}. Since the solar motion is subject to uncertainty one in effect searches for the highest peak in a volume of velocity space prescribed by its three components. Each such 3-dimensional search constitutes a single trial, yielding a single number I_{max}, namely the highest power found anywhere in the volume.

The $I_{\max}$ distribution found for 5,000 such trials is shown in Figure 2. Evidently, in this redshift range, the real Virgo cluster has substantially more power than is found in the synthetic ones and it seems that this excess power can only be ascribed to the periodicity. The hypothesis under test (there is a periodicity in the range 70–75 kms^{-1}) is preferred over the null hypothesis (no such periodicity) at a confidence level of about 50,000 to 1. This must be reduced by a factor $\sim$5 because of several freedoms which the investigators have given themselves (we have excluded dwarf irregulars from the study, the definition of the core of the Virgo cluster is somewhat arbitrary etc.). The result turns out to be very insensitive to the assumed velocity of the sun around the centre of the Galaxy: the differential solar motion correction

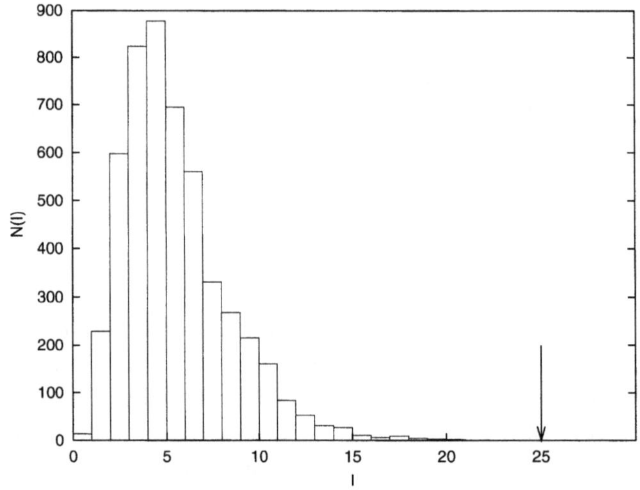

Figure 2. Testing for the significance of the 71 km s^{-1} periodicity. 5000 synthetic Virgo clusters are created as described in the text, and a periodicity in the range under test is sought in a 20 × 20° grid centred on the solar apex. The vertical arrow shows the peak power in the real Virgo dataset.

across the few degrees subtended by the Virgo cluster is small, and uncertainties in this differential correction are second order.

We conclude that, to a high level of confidence, the spiral galaxies of the Virgo cluster show the same periodicity which Tifft (1976) claimed for the Coma cluster.

3. The 36 km s^{-1} Claim

A second major claim, made by Tifft and Cocke (1984), was that there exists a global, galactocentric quantisation of redshifts. For galaxies with narrow HI profiles a periodicity of 24.2 km s^{-1} was claimed, while for broad HI profiles the claimed periodicity was 36.2 km per second. To see these periodicities it was necessary to correct for a solar motion of

$$V_{\odot} = 233.6\text{km s}^{-1}, l_{\odot} = 98.6°, b_{\odot} = 0.2°$$

To test this, Guthrie and Napier (1996) used high-precision catalogued redshift data, simulations having indicated that for the samples sizes employed the effect would only be seen in data of the highest quality. After excluding data employed by Tifft and Cocke (1984), there remained 97 spirals with redshifts measured formally to $\sigma \leq 3$ km s^{-1}. Each redshift had been measured and reduced by at least five groups of observers using five different radio telescopes. These were generally galaxies with broad HI profiles. A remarkably strong periodicity ($I \sim 40$ for power defined so that $\overline{I} = 2$ for white noise) close to that claimed does indeed emerge for vectors in the neighbourhood of the solar motion (Figures 3 and 4). Significance

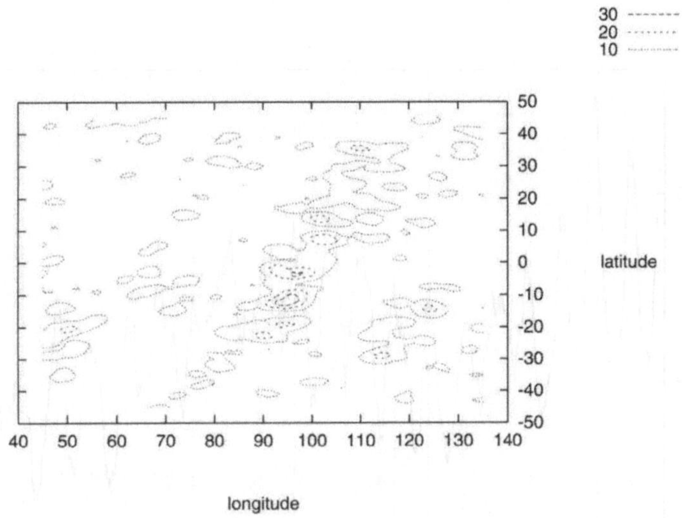

Figure 3. Contour map of power in the sample of 97 high-precision redshifts of field galaxies as the solar apex ranges over the sky. In this example the solar speed $V_\odot = 219$ km s^{-1}.

testing involves the creation of synthetic data sets, as with the Virgo cluster, and exploration of the power they generate by chance. The hypothesis of non-periodicity is thereby rejected at a significance level $\sim 10^{-5}$.

In this case, since the galaxies are scattered over the sky, individual corrections for the solar motion vary considerably and so the peak powers generated vary strongly with the adopted solar apex. A real periodic signal occurring in the frame of reference of a single velocity generates signals at other velocity vectors and identifying the 'real' signal is not a trivial exercise. To find the 'true' solar vector, the sensitivity of the signal to variations in the vector must be decreased. This can be done in a number of ways, but the following approach is particularly instructive. About half the galaxies in the sample belonged to small groups or associations containing two to six companions. By looking at the differential redshifts within these groups, the sensitivity of the signal to $V_\odot$ may be decreased. The resulting dataset is small (50 galaxies) and so was enhanced by adding galaxies obtained from a catalogue by Tifft with measured signal-to-noise ratios greater than 10 and which also belonged to catalogued groups. This enhanced dataset contained 89 galaxies in 28 groups scattered throughout the Local Supercluster. The power distribution in these 28 LSC groups turns out to have a well-defined maximum at $\sim$37.5 km s^{-1} for a solar vector (remarkably!)

$$V_\odot = 220\text{km s}^{-1}, l_\odot = 90.0°, b_\odot = 0.0°$$

Is this periodicity a local phenomenon, peculiar to individual groups? Or is it global, that is, is there phase coherence from one group to another? This question was explored by constructing synthetic local superclusters. The procedure was

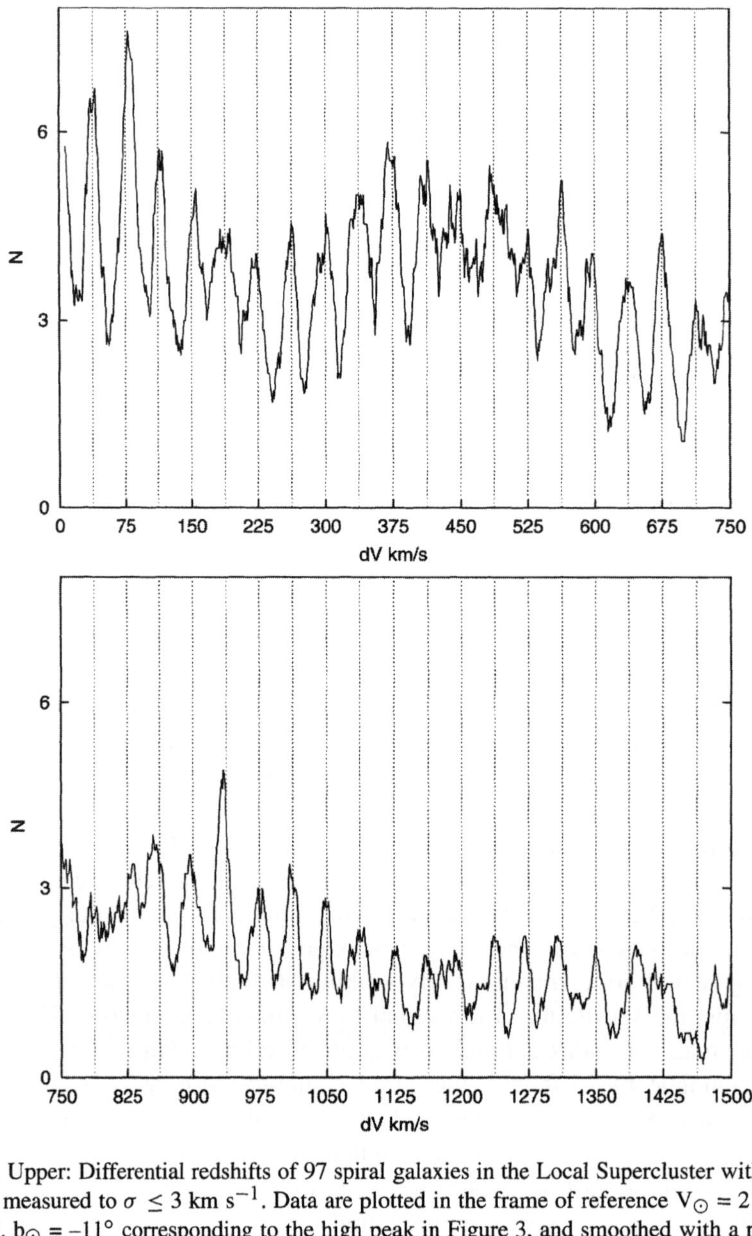

Figure 4. Upper: Differential redshifts of 97 spiral galaxies in the Local Supercluster with systemic redshifts measured to $\sigma \leq 3$ km s^{-1}. Data are plotted in the frame of reference $V_{\odot} = 216$ kms^{-1}, $l_{\odot} = 96°$, $b_{\odot} = -11°$ corresponding to the high peak in Figure 3, and smoothed with a rectangular window of width 13 km s^{-1}. The vertical dotted lines are the best-fit periodicity of 37.5 km s^{-1}. The first 40 cycles are shown, but in fact the periodicity is detectable out to at least 90 cycles within the LSC. Upper: 0–750 km s^{-1}. Lower: 750–1500 km s^{-1}.

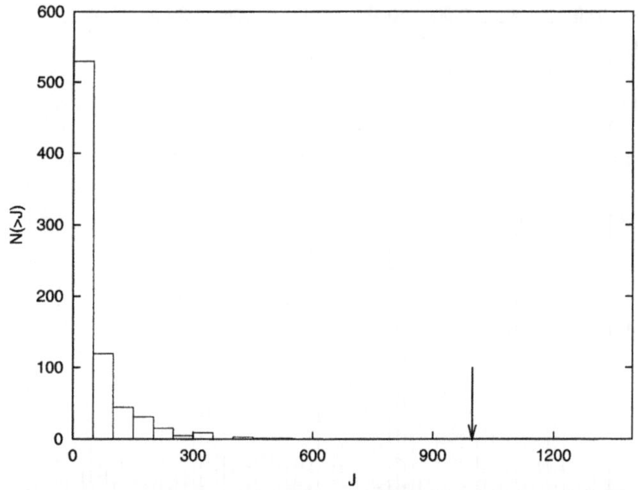

Figure 5. Testing for 'local' vs 'global' periodicity: the power distribution for 1000 synthetic Local Superclusters compared with that for the real LSC (vertical arrow). Each synthetic LSC was generated by shifting the 28 groups randomly while preserving their internal relative velocities, so preserving any internal periodicity but destroying phase coherence between groups. A single global periodicity is preferred at a high confidence level to local, disconnected ones.

to preserve the internal relative redshifts of each group but shift their systemic redshifts bodily by an amount just sufficient to destroy any phase coherence. This is generally equivalent to shifting each cluster radially by one or two diameters and placing its systemic velocity at the position given by the Hubble flow: the overall structure of the LSC is preserved. Thus in this procedure 'new physics' is accepted as part of the null hypothesis, but it is assumed to be a local rather global phenomenon. The statistic J adopted was the overall power measured by the number of peaks exceeding some threshold as the solar vector was varied in the range $V_\odot = 220 \pm 20$ km s^{-1} in speed, $l_\odot = 90 \pm 45°$ in longitude and $b_\odot = 0 \pm 45°$ in galactic latitude, in steps of 5 km s^{-1} and 1°. The outcome of 1000 trials is shown in Figure 5. Clearly, the signal in the real LSC is significantly different from those in the synthetic ones. It follows that the periodicity is a global rather than a local phenomenon, occurring at least throughout the inner regions of the LSC where the clusters are concentrated.

Efforts have been made to determine whether some obscure artefact is responsible for these periodicities, but without success. It turns out that the requirements for an artefact are as follows: (1) the bug must know the sun's galactocentric velocity; (2) it must preserve phase coherence of a falsely generated periodicity over the whole sky; (3) if due to a bug in radio telescopes or their software etc., it must be present independently in all the major radio telescopes; (4) it must operate at photographic and radio wavelengths; and (5) it must behave like a real signal with regard to precision and numbers of data employed and so on.

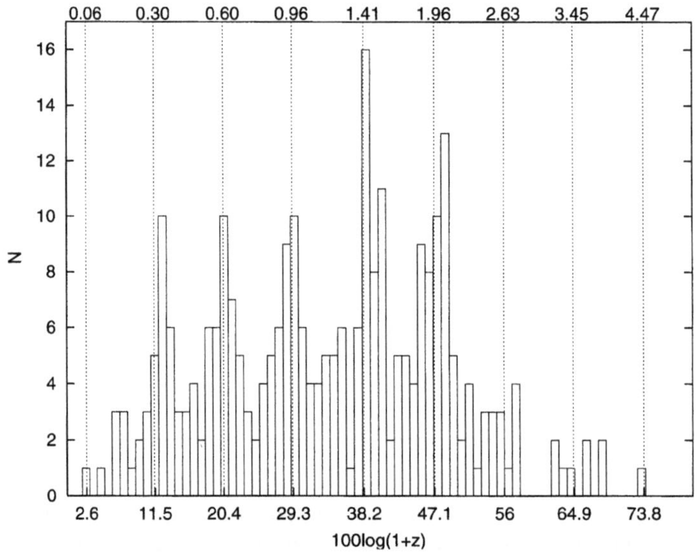

Figure 6. Redshift distribution of 290 QSOs compiled by Karlsson (1990) and Burbidge and Napier (2001) in order to test claims of a periodicity 0.089 in $\log_{10}(1+z)$. Selection procedures etc. are discussed in BN.

4. The QSO Periodicity Claim

The third anomalous redshift claim tested so far is that QSOs in the neighbourhood of bright, nearby, active spirals show a periodicity of 0.089 in $\log_{10}(1+z)$. There is some imprecision in the formulation of this hypothesis (what do we mean by close? By active? What phase should we associate with this periodicity, and with what standard errors?). The hypothesis was tightened up somewhat by bootstrap sampling of 116 QSOs used by Karlsson (1990) to test it, and fresh data were then employed as described in Burbidge and Napier (2001), alias BN: these comprised 57 QSO pairs with separations less than 10 arcseconds, 39 X-ray QSOs near active galaxies (comprising a complete sample) and 78 3C(R) radio QSOs, again comprising a virtually complete sample. Figure 6 is a histogram of the combined Karlsson and BN datasets: the periodicity is clearly present and seems to extend three cycles beyond that originally claimed. Monte Carlo trials yield a formal significance level of a few parts in 100,000, whether the null hypothesis is defined through smoothing of the given data or the z-distribution of QSOs as a whole.

It has often been argued that the QSO periodicity is an artefact of observational selection efects (see the discussion in BN). However the data employed here were selected precisely to avoid such effects. It has also been claimed that the result is a statistical artefact caused by edge effects (Hawkins et al., 2002), but this is demonstrably erroneous: inter alia, edge effects were automatically allowed for in the procedures employed, and the periodicity is easily seen by eye, without any statistical analysis (Figure 6).

5. Summary and Conclusions

We have used new, high precision data and rigorous statistical procedures to test three anomalous redshift claims in an objective fashion, and we confirm all three. Specifically:

1. In the galactocentric frame of reference, the spiral galaxies of the Virgo cluster are distributed with a periodicity of 71 km s^{-1}. This is a straightforward empirical result: the periodicity is obvious to the eye and insensitive to large uncertainties in the solar apex. The finding is akin to that of Tifft (1976) who, using low accuracy photographic data, claimed that the galaxies of the Coma cluster show a redshift periodicity of 72 km s^{-1}.
2. Tifft and Cocke (1984) also claimed a galactocentric periodicity of 36.2 km s^{-1} in wide-profile field galaxies. We found a galactocentric periodicity of 37.5 km/s to be present in such galaxies scattered throughout the LSC. The phenomenon, although global, is found to be stronger in small groups and associations.
3. We have confirmed the claim that QSOs clustered around bright, nearby galaxies show a periodicity 0.089 in $\log_{10}(1 + z)$. The confidence levels associated with each of these findings are high and we can find no statistical or observational artefacts capable of explaining them. While it seems that non-periodicity can be rejected with confidence, it is less clear that periodicity with respect to a fixed galactocentric vector is the only alternative. The solar apex varies with increasing distance and the local periodicities could be due to discrete velocity residuals with respect to it (in the original study of the Virgo Cluster by Guthrie and Napier (1991), a variable solar apex was adopted and a significant periodicity found).

If the periodicities are indeed real, then by Occam's Razor we might expect them to be manifestations of a single, underlying phenomenon, and to be connected with other 'unexpected' features of our neighbourhood such as the remarkable linearity and coldness of the local Hubble flow.

References

Burbidge, G. and Napier, W.M.: 2001, *Astron. J.* **121**, 21.
Guthrie, B.N.G. and Napier, W.M.: 1991, *MNRAS* **243**, 431.
Guthrie, B.N.G. and Napier, W.M.: 1996. *Astron. Astrophys.* **310**, 353.
Hawkins, E., Maddox, S.J. and Merrifield, M.R.: 2002, *MNRAS*, in press.
Karlsson, K.G.: 1990, *Astron. Astrophys.* **239**, 50.
Tifft, W.G.: 1976, *Astrophys. J.* **206**, 308.
Tifft, W.G. and Cocke, W.J.: 1984, *Astrophys. J.* **287**, 492.

REDSHIFT PERIODICITIES, THE GALAXY-QUASAR CONNECTION

W.G. TIFFT
Steward Observatory, University of Arizona, Tucson, Arizona 85721

Abstract. The Lehto-Tifft redshift quantization model is used to predict the redshift distribution for certain classes of quasars, and for galaxies in the neighborhood of $z = 0.5$. In the Lehto-Tifft model the redshift is presumed to arise from time dependent decay from an origin at the Planck scale; the decay process is a form of period doubling. Looking back in time reveals earlier stages of the process where redshifts should correspond to predictable fractions of the speed of light. Quasar redshift peaks are shown to correspond to the earliest simple fractions of c as predicted by the model. The sharp peaks present in deep field galaxy redshifts surveys are then shown to correspond to later stages in such a decay process. Highly discordant redshift associations are expected to occur and shown to be present in the deep field surveys. Peaks in redshift distributions appear to represent the spectrum of possible states at various stage of the decay process rather than physical structures.

1. Introduction

Modern cosmology presumes to understand the cosmic redshift as a simple continuous Doppler-like effect caused by expansion of the Universe. In fact there is considerable evidence indicating that the redshift consists of, or is dominated by, an unexplained effect intrinsic to galaxies and quasars. In this paper we discuss and relate three lines of such evidence including evidence for characteristic peaks in the redshift distribution of quasars, the issue of associations between objects with widely discordant redshifts, and redshift quantization associated with normal galaxies.

Nonuniformities in the quasar redshift distribution, most notably peaks near z of 0.30, 0.60, 0.96, 1.41, 1.96, and 0.061 and its multiples, have been discussed since the 60s (see Burbidge and Napier, 2001). The issue of association between objects with very different redshifts also dates from that period (Burbidge and Sargent, 1970), as does redshift quantization in normal galaxies (Tifft, 1976). Figure 1 illus-

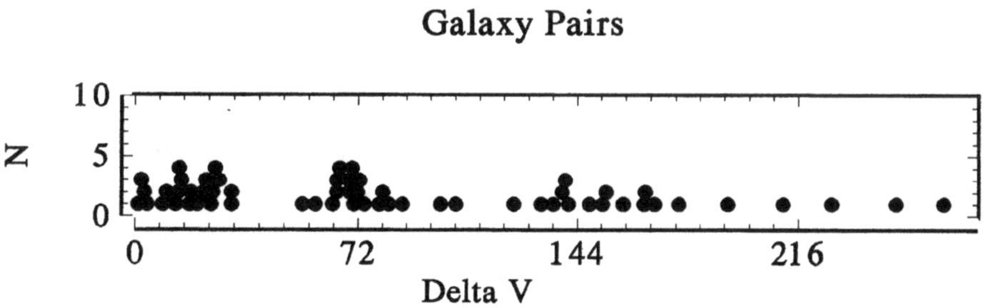

Figure 1. The quantized distribution of differential redshifts for isolated double galaxies.

Astrophysics and Space Science **285:** 429–449, 2003.

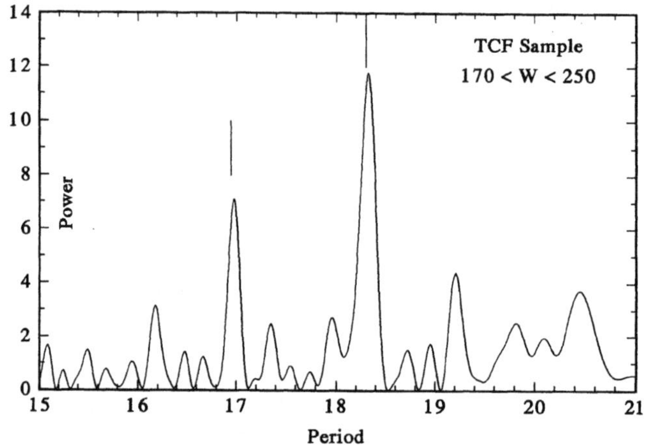

Figure 2. The power spectrum of a global sample of 21 cm galaxy redshifts of a specific type defined by 21 cm profile width W. Specific periods predicted by the Lehto-Tifft model are denoted by vertical lines.

trates the quantized differential redshift distribution found for double galaxies (Tifft and Cocke, 1989); Figure 2 shows characteristic redshift periods found globally (Tifft, 1996) using concepts which predict specific values.

Understanding the relationship between different facets of the intrinsic redshift requires familiarity with the predictive framework of redshift quantization developed using local galaxies. Work prior to 1992 employed empirical periodic intervals studied differentially or globally within a galactic center rest frame. Important independent confirmation work was carried out in the 90s by Guthrie and Napier (1991), and has been continued by Napier (W. Napier, this conference). In 1992 emphasis shifted to the cosmic background rest frame, and in 1993 Ari Lehto, a Finnish physicist, suggested a mechanism for predicting periodicities. Subsequent testing, (Tifft, 1996, 1997), confirmed and extended concepts leading to the current Lehto-Tifft quantization model. The underlying concept, described in the Astrophysical Journal (Tifft, 1996), follows:

Lehto was originally motivated to describe properties of fundamental particles from first principles. Begin with the Planck units, the Planck energy, time, mass and light speed. These units can be related to the Planck time or frequency as defined by the fundamental constants h, c, and G. Observable properties of particles (and redshifts) are then presumed to be generated by a *decay* process beginning at the Planck scale. Lehto proposed that the Planck units decay by a period doubling process, an action doubling process in factors of 2, a process well known in decay of systems toward chaos. If this was the case present day fundamental properties of matter and energy should relate to the Planck units by simple powers of 2. When Lehto explored this idea he found what appeared to be ratios involving cube-root powers of 2. Lehto suggested that doubling could be occurring in a 3-dimensional or 3-parameter space which is perceived as 1-dimensional through the action of a cube-root transformation.

2. The Lehto-Tifft Equations

Lehto's initial form for permitted redshift intervals (periods) follows immediately from the above statement.

$$P = c2^{-N/3}, \tag{1}$$

where N is an integer ≥ 0. A form applicable to particle physics replaces c with the Planck mass or energy. It is relevant here to reflect for a moment on the significance of the equation. The equation is a general form of Kepler's third law where, in ordinary space, a spatial distance relates to the 2/3 power of a time interval. Such a relation is a unique property of 3-d spaces. The existence of such a pattern involving only temporal terms raises the possibility that temporal/frequency/energy space is actually 3-dimensional. If such a 3-d temporal space was flowing relative to 3-d spatial space the flow speed would by definition be the speed of light, a constant. Time would be reduced to 1-d by aberration and generate our perceived 4-space. The model automatically provides constancy of c and preserves special relativity. Temporal space must be quantized since it involves a stepwise decay process from fixed units. Space and flowing time (dynamical 4-spaces) are not so restricted. Quantum and continuous physics can co-exist. We will not pursue such a 3-d temporal cosmology here, but it is useful to introduce the concept. This paper depends on the empirical fact that redshift data are consistent with such a periodicity equation, regardless of its interpretation. In fact it is now possible to explicitly derive the Lehto-Tifft equations. This will be reported in a subsequent paper since it occurred subsequent to the Cardiff conference where this paper was presented.

Equation 1 is rewritten to distinguish cube-root doubling 'families'

$$P = c2^{-N/3} = c2^{-\frac{3D+L}{3}}. \tag{2}$$

D is the number of doublings, and $L = 0$ to 2 specifies which root is involved. Equation 2 accounts for most redshift periods, however, the complete set of observed periods requires a second cube root operation to yield nine ninth-root families distinguished by an index T.

$$P = c2^{-N/9} = c2^{-\frac{9D+T}{9}}. \tag{3}$$

Equation 3 completely and uniquely describes all presently observed redshift periodicities (Tifft, 1996, 1997). T values do not occur randomly. Pure doubling, $T = L = 0$, is dominant, followed by the 'Keplerian' $T = 6$ ($L = 2$) value. Odd T values shifted by 1 from these values ($T = 1, 5, 7$) are present but less common, while the even values ($T = 2, 4, 8$) are rare or absent. Which T family is involved appears to relate to galaxy morphology (Tifft, 1997). Equation 3 appears to be necessary to describe structure found in the redshift distribution of local galaxies. Deeper in space, at high redshift, the $T = 0$ family becomes increasingly dominant. As noted in the previous paragraph, the origin of the Lehto-Tifft equations is now believed to be known, including why particular T values are preferred. This will be reported in a subsequent paper.

A second basic relationship is required to assess redshift quantization. Redshift intervals dilate with distance due to relativistic and geometric effects. Such effects must be removed to evaluate underlying quantization. In classical cosmology geometry, or 'curvature', is described by a 'deceleration' parameter q_o. The Hubble 'constant' is a function of time, $H = H(t) = f(z, q_o)$ where $q_o = 1/2$ corresponds to Euclidian geometry. Such a relationship must exist in any cosmology. q_o need not be uniquely associated with gravitation but $H(t)$ will be described much the same way. The removal of z-dependent distortion in the redshift is referred to as the 'cosmological' correction.

The cosmological correction was first investigated by Tifft and Cocke (1984) for application in global quantization studies. A correction assuming that redshift intervals dilate in proportion to $\sqrt{H(t)}$ was shown (Tifft, 1996) to linearize quantization in galaxy redshifts to well beyond 10,000 km/s if q_o is set exactly equal to 1/2. In this paper no significant deviation in the form of the correction is seen out to $z = 1$ or 2. The relationship is derived using the classical $H(t)$ formulation to find an expression for redshift intervals as a function of z and q_o which is then integrated in a Taylor expansion about $q_o = 1/2$. For q_o exactly equal to 1/2 the integration gives a closed relationship between z(observed) and z(Lehto-Tifft).

$$z_{obs} = \left[\frac{z(LT)}{4} + 1\right]^4 - 1 \qquad z(LT) = 4[(1 + z_{obs})^{1/4} - 1]. \tag{4}$$

This form empirically fits all data examined (Tifft, 1996, 1997). Current tests therefore use Equation 4 to convert observed redshifts to $z(LT)$ for quantization evaluation.

Equation 4 is consistent with the 3-d temporal concept previously mentioned. In such a model energy is associated with temporal volumes which vary as t^3. If photon redshifts relate to energy density, which varies as volumes evolve, the rate of change will be perceived as H. Volume will evolve as t^2 so H should depend on t^2 and redshift intervals vary as $\sqrt{H(t)}$ as observed.

3. Looking Back In Time

In the Lehto-Tifft model redshifts evolve through a systematic doubling decay process, now quite advanced, to yield the small redshift intervals observed locally. Looking back in time should reveal earlier decay stages. High redshift data should

TABLE I
Quasar redshift peaks

z(LT)	D=2	z(obs)	z(peak)
c/16		0.064	0.061
c/8		0.131	
c/4	1/4 c	0.274	0.30
c/2	2/4 c	0.601	0.60
	3/4 c	0.989	0.96
c	4/4 c	1.441	1.41
	5/4 c	1.968	1.95

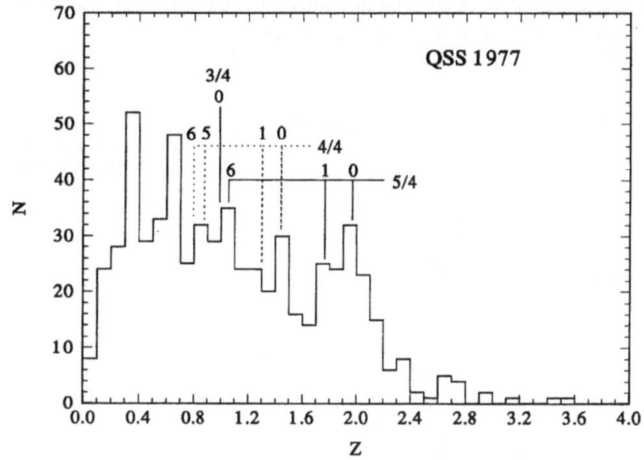

Figure 3. The redshift distribution for quasars, mostly radio sources, known by the mid 1970s. The distribution shows peaks ascribed to an intrinsic component. Redshifts predicted by the Lehto-Tifft model are denoted by lines and identified with specific doubling stages and T values.

be increasingly structured and consistent with predictable early doublings and mixing. Are redshifts associated with early levels related to the peaks in the quasar redshift distribution? Table I shows that there is a very close agreement. The table is constructed by converting the earliest doublings and mixings, basic fractions of c, to observed values using Equation 4, to compare with the quasar peaks. The quasar redshift distribution is indeed consistent with predicted early redshifts generated in a universe evolving in a doubling decay process. Peak locations also match an empirical logarithmic series described by Karlsson (1977).

The issue of specific distinct peaks in the quasar redshift distribution is in dispute, however it is unlikely that the match in Table I would occur if the peaks were

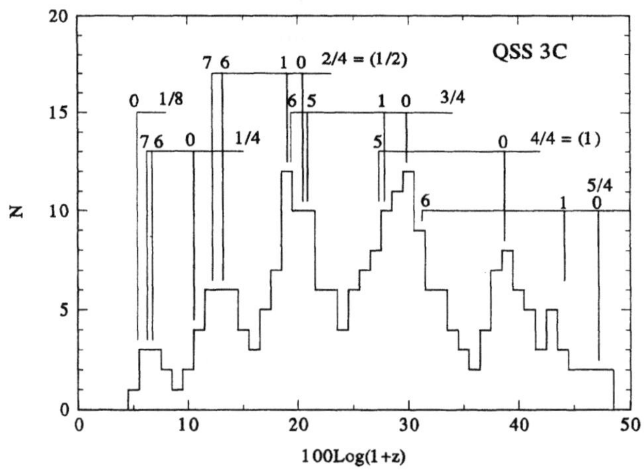

Figure 4. The redshift distribution for objects in the 3C radio survey. The distribution shows peaks ascribed to an intrinsic component. Redshifts predicted by the Lehto-Tifft model are denoted by lines and identified with specific doubling stages and *T* values.

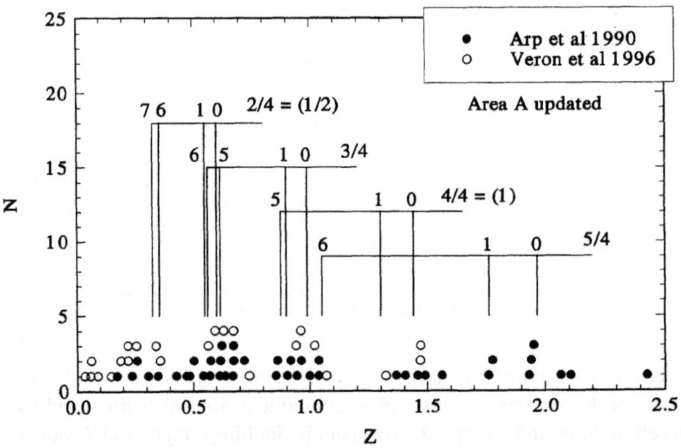

Figure 5. The redshift distribution of quasars and active galaxies in a south galactic cap field. The study was augmented using additional data to compare with redshift peaks predicted by the Lehto-Tifft model.

simply random fluctuations. It seems quite likely that the distribution depends upon the type of quasar involved. Burbidge (1980) and Burbidge and Napier in Hoyle, Burbidge and Narlikar (2000) indicate that peaks are clear when quasars which are strong radio sources are involved. This is illustrated in Figure 3, where quasars known in 1977, virtually all radio sources, are shown and in Figure 4 which shows the complete set of 3c quasar radio sources. The figures also indicate other most likely *T* states permitted in the Lehto-Tifft model. The gaps associate quite clearly

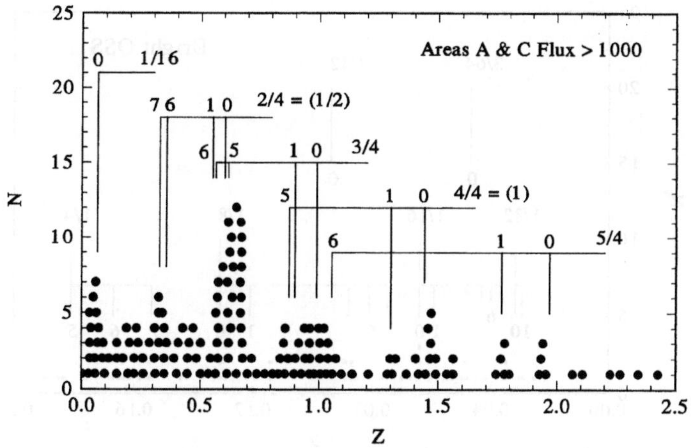

Figure 6. The redshift distribution, shown in Figure 5, further augmented with objects in a large portion of the south galactic cap to further confirm predicted redshift peaks.

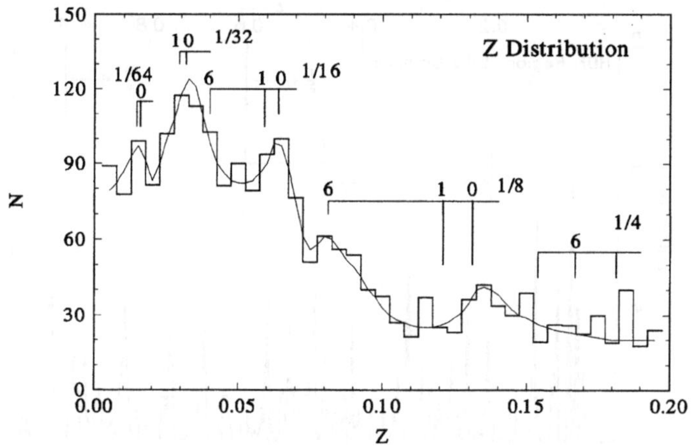

Figure 7. The general redshift distribution for quasars and active galaxies, as known in 1996, to illustrate the relationship between the observed distribution and redshift peaks predicted by the Lehto-Tifft model.

with the absent or least likely *T* states. Burbidge and Napier (2001) also associate the peaks with quasars close to low redshift galaxies, however other recent studies based upon large optically selected samples do not confirm the effect (Hawkins et al., 2002). This may not be surprising since the early samples contained more radio selected objects. In the Lehto-Tifft model the peak redshifts represent unique potentially active stages. Subsequent smaller scale decay toward lower redshifts may move objects out of the active stage and smear out the distribution. We will specifically illustrate this idea using high redshift galaxies, presumed to be evolutionary

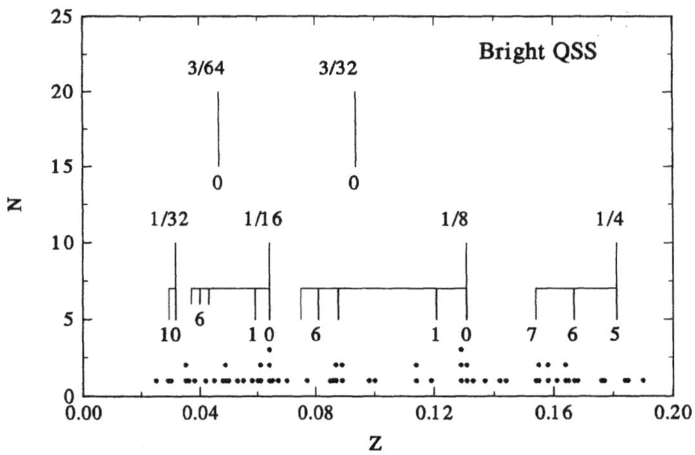

Figure 8. The redshift distribution for bright quasars, in the Schmidt-Green survey, to illustrate the relationship between the observed distribution and redshifts predicted by the Lehto-Tifft model.

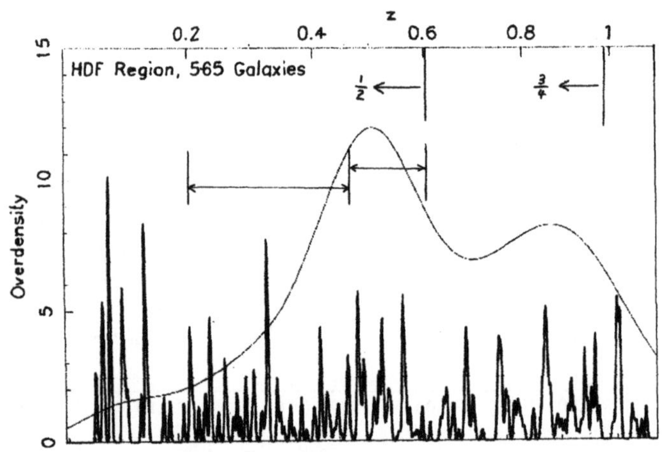

Figure 9. The smoothed galaxy density (smooth line) and local overdensity fluctuations associated with redshift clumping in the Hubble Deep Field region according to Cohen et al. (2000). The association of broad density peaks with the $c/2$ and $3c/4$ quasar peaks is shown as are intervals used for detailed comparisons with predicted redshifts.

products of quasars, later in this paper. Whether or not an excess of quasars near galaxies indicates physical association remains an open question, but association and periodicity need not be directly related.

Burbidge and Napier (2001) also invoke a new category of quasars, close associations. They claim that such a sample of redshifts, which includes a significant number with $z > 2.5$, extends the Karlsson (1977) $\log(1 + z)$ periodicity to higher redshifts. In fact the number of higher redshifts is quite small and the distribution rather noisy. The redshifts are also well described in the Lehto-Tifft model. Above $z(LT) = 1$ the Lehto-Tifft model proceeds by frequency mixing which should yield the well known sharp drop in redshift frequency at high z. Specific z values are predictable and appear to match the associated quasar redshifts very well. Close associations in quasars seem to be analogous to close binary galaxy systems where the quantized pattern shown in Figure 1 was found. Further details discussing matches to high redshift values will be presented in a subsequent paper.

To further investigate the association of radio emission with the quasar peaks we extended a study by Arp et al. (1990) based upon quasars selected by radio flux. Their study found the Karlsson peaks in a study of several specific fields. Figure 5 shows, with filled symbols, the distribution found in a south galactic hemisphere field. With the assistance of N. Jobson and L. Lippiello the sample was augmented (open symbols) using the Veron-Cetty and Veron (1996) quasar catalog. The sample was then extended to the entire south hemisphere south of the initial field to yield Figure 6. A peaked distribution related to the Karlsson and specific Lehto-Tifft predictions persists. We note the gaps where rare or absent T values fall, and a possible association with $T = 1$ and perhaps $T = 6$ values not predicted in any other model.

Finally, to extend studies to possible lower redshift peaks, especially the $c/8$ peak predicted to fall at $z(obs) = 0.131$, N. Jobson extended the quasar and active galaxy redshift distribution to low redshifts using the Veron-Cetty and Veron (1996) catalog. The sample was limited to objects with three significant figures in the redshift value to eliminate many rough or preliminary values listed. Figure 7 shows that the $c/8$ peak is apparently present. This peak is the x2 multiple of the $c/16$ peak at $z = 0.06$ consistent with both the Karlsson and Lehto-Tifft model. The $c/8$ peak is not directly predicted by Karlsson, however, Burbidge has noted several multiples of the 0.06 period, compatible with the Lehto-Tifft model. Figure 8 shows the Schmidt and Green (1983) bright quasar sample which also shows small concentrations at the $c/8$ and $c/16$ locations. These results at low redshift are intended only to show broad consistency. The overall pattern becomes very complex at high doubling numbers (Tifft 1996, 1997). We will therefore now turn to much simpler early doubling patterns predicted to be present *for galaxies* at high redshift

4. The Galaxy-Quasar Connection

To preserve sharp quantization of redshift and the Hubble law, as observed, objects must evolve by cascading down through a series of specific redshift states. We have suggested that the principal peaks in the quasar redshift distribution represent an active stage relating to the earliest doublings and mixings associated with decay from the Planck unit c. In such a model subsequent decay can be expected to produce less active quasars and ordinary galaxies distributed in a redshift 'spectrum'

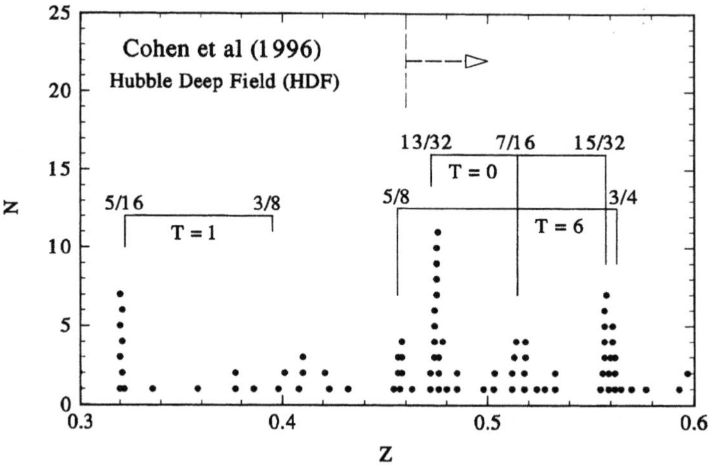

Figure 10. The redshift distribution in the Hubble Deep Field region, below the $z = 0.6$ quasar peak, used to illustrate association of redshift peaks with specific values predicted by the Lehto-Tifft model. The analysis is divided into two parts, above and below $z = 0.46$.

between the primary peaks. Since the spectrum of states is far advanced locally the obvious place to look for such a process is at high redshift. Decay from the first doubling, $c/2$ associated with the $z = 0.6$ quasar peak, may be expected to be especially clear. We would expect the redshift spectrum to show a strong $T = 0$ dominance, multi-periodic 'phased' decay products, and discordant redshift associations where physically related objects have decayed into different but related states. Local decay has proceeded into the $D = 12$ to 16 range, but at $z = .5$ we would expect to see periods in the $D = 4$ to 9 range corresponding to $c/16$ - $c/512$ (20,000+ to 500 km/s).

To test this hypothesis we turn to the excellent accurate redshift surveys recently done in and around the Hubble deep field (HDF) and a southern deep field (SDF) by Cohen et al. (1996, 1999, 2000). Figure 9, adopted from Cohen et al. (2000), shows the overdensity redshift spectrum for the HDF region. The sharp spikes are classically interpreted as remarkably low velocity dispersion clouds or 'walls' in a developing 'cell' structure. The smooth broad curve shows the heavily smoothed overall galaxy density distribution. The general distribution shows peaks displaced below $z = .6$ and $z = .96$ consistent with being evolved remnants of early $c/2$ and $3c/4$ doublings. To investigate this we examined the redshift pattern in intervals immediately below $c/2$ and around $c/4$.

The first study, done in 1997, utilized the original HDF region study by Cohen et al. (1996). A redshift frequency diagram is shown in Figure 10. Predicted locations are shown for the simplest fractions of c expected to fall in the interval, including $T = 1$ and 6 values. Peaks are present very close to predicted locations, especially three strong peaks directly below $c/2$ associated with $c/16$ and $c/32$ intervals. Figure 11 shows how precise the fit to these three peaks is. Redshift 'phase' at the

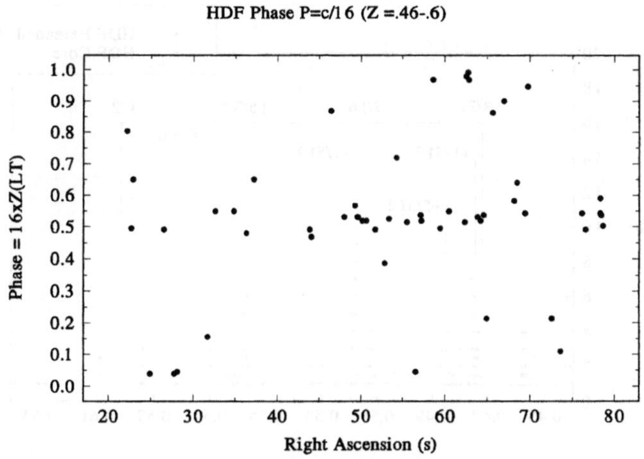

Figure 11. The redshift phase distribution (at the $c/16$ period) for the 1996 Hubble Deep Field sample with $0.46 < z < 0.6$ shown in Figure 10.

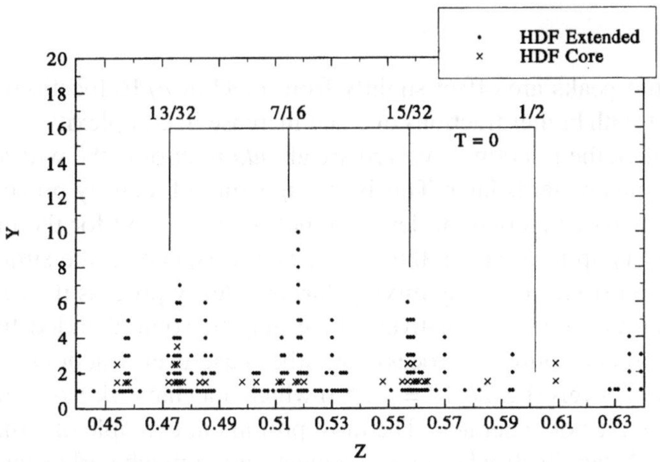

Figure 12. The redshift distribution in the Hubble Deep Field region below the $z = 0.6$ quasar peak using the complete catalog (Cohen et al., 2000). The three major $c/32$ periodic peaks immediately below the $z = 0.6$ quasar peak are indicated.

$c/16$ period (the decimal part of $cz(LT)$ divided by $c/16$) concentrates strongly at 0.5 (odd $c/32$ values) and less strongly at 0.0 ($c/16$ values). Redshift values appeared to be periodic and phased as predicted. Further analysis requires a larger sample, which became available with the publication of a general catalog of HDF data by Cohen et al. (2000).

Figure 12 shows the redshift distribution for the complete HDF catalog in the interval immediately below the $c/2$ location. The three primary concentrations and finer structures are apparent. Figure 13 shows that the lower two main peaks and

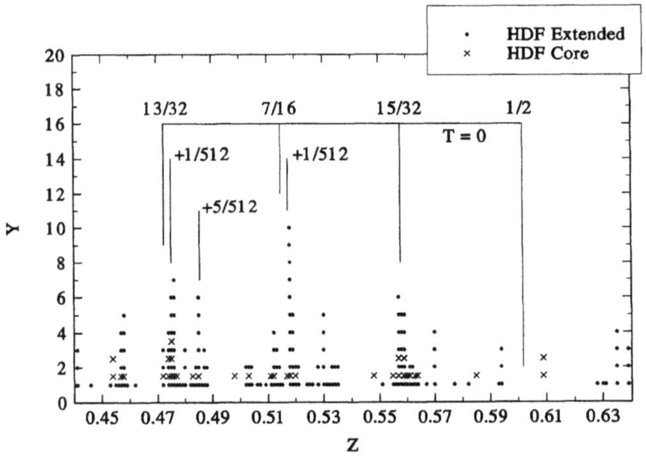

Figure 13. The redshift distribution in the Hubble Deep Field region below the $z = 0.6$ quasar peak using the complete catalog (Cohen et al., 2000). Fine structure around the three major $c/32$ periodic peaks is indicated.

certain satellite peaks are offset slightly from $c/32$ or $c/16$ locations. The offsets are consistent with higher fractions of c and increase in complexity with decreasing redshift. Further, the fractions involved are all *odd* fractions. We will see additional examples of such patterns later. This is the signature of a decay process which has proceeded to a specific doubling level but not beyond, $c/32$ for the main distribution with offsets up to $c/512$ in this case. Small offsets near the simpler fractions can be expected from frequency mixing during a decay process if a range of doublings exists at any time. Alternatively there may be some detailed finer structure within the primary doubling process. In either case even fractions correspond to earlier doubling levels (i.e. 2/512 = 1/256) which may have already decayed (1/256 → 1/512) or were not generated. The most prominent example of a missing higher state is the $c/2$ state itself which we associate with a much earlier quasar era. The distance to any particular aggregate of galaxies seems to take us back to a time period when decay had proceeded only so far, to the $D = 5$ or $c/32$ level for the main peaks in this HDF example. This is a normal lookback expectation, however D not redshift measures lookback time; redshift may distribute over a range of *specific* levels at a given distance.

Just as we see that the small offsets above the $c/32$ peaks correspond to the first few odd offsets of higher doublings, so too the offsets of the triple pattern below the $c/2$ location correspond to offsets of 1/32. 1/16, and 3/32 of c. The smallest offsets (1/32 and 1/512) are fit very accurately in both cases. We will see other examples of this. The states closest to the simple fractions appear to be the most persistent, while the simple fractions themselves are usually absent or infrequent above the highest doubling present.

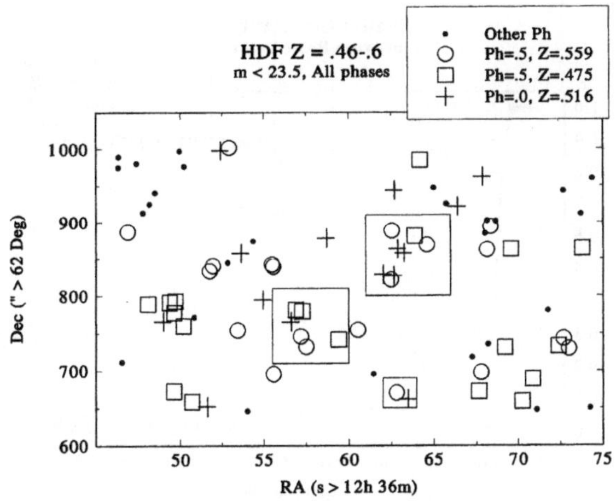

Figure 14. Positions for Hubble Deep Field objects, with $0.46 < z < 0.6$ and $m < 23.5$, distinguished by redshift peak membership. Large boxes indicate concentrations comparable in scale to local galaxy groups. The small box indicates a discordant redshift pair.

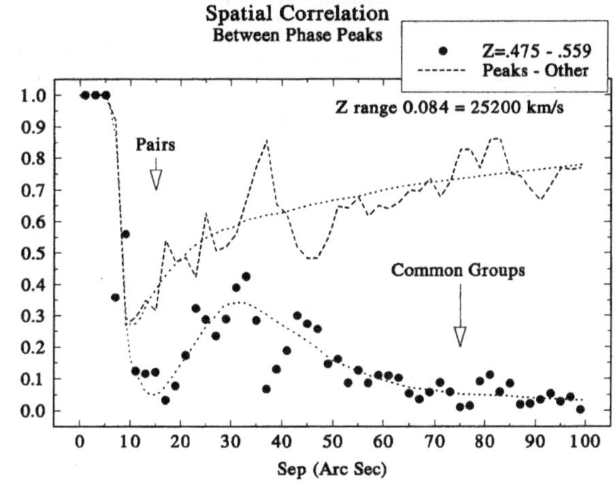

Figure 15. A Monte Carlo evaluation of the significance of the discordant redshift associations between the $z = 0.475$ and $z = 0.559$ redshift peaks in the Hubble Deep Field, and between these peaks and objects with redshifts in between the three major peaks. Physical association is indicated for both discordant pairs and groups.

As noted above, a common decay level of galaxies for the three $c/32$ peaks suggests a common spatial (distance) association despite the wide redshift spread. To test this we looked at the correlation between galaxy position and redshift peak membership. Figure 14 is a plot of galaxy positions with $0.46 < z < 0.6$ for a rectangular region in the HDF extended study. Symbols distinguish the three dominant redshift peaks and objects that are in between. The sample is cut at a

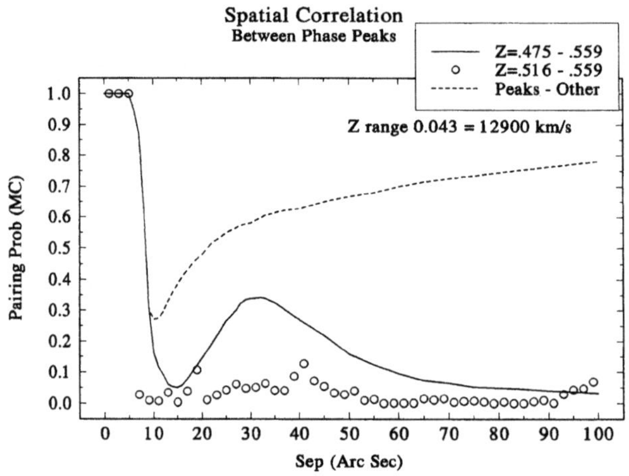

Figure 16. A Monte Carlo evaluation, similar to Figure 15, of discordant physical associations between the $z = 0.516$ and $z = 0.559$ peaks. Results from Figure 15 are shown for comparison.

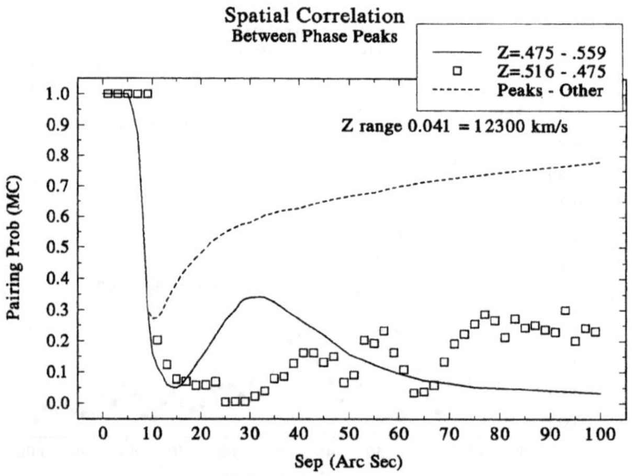

Figure 17. A Monte Carlo evaluation, similar to Figure 15, of discordant physical associations between the $z = 0.516$ and $z = 0.475$ peaks. Results from Figure 15 are shown for comparison.

uniform magnitude to account for deeper sampling in the central HDF although this affects few galaxies in the range of z involved. Galaxies appear to be clumped. At the distance associated with $c/2$ 1 Mpc has an angular scale of about 100 arc seconds. The larger boxes in Figure 14 therefore correspond with typical galaxy group sizes locally (although the two scales in the figure are different). We see that what looks like spatial clumps are composites of the three redshift peaks. We further see numerous pairings of galaxies from different peaks. One is shown with a small box in Figure 14.

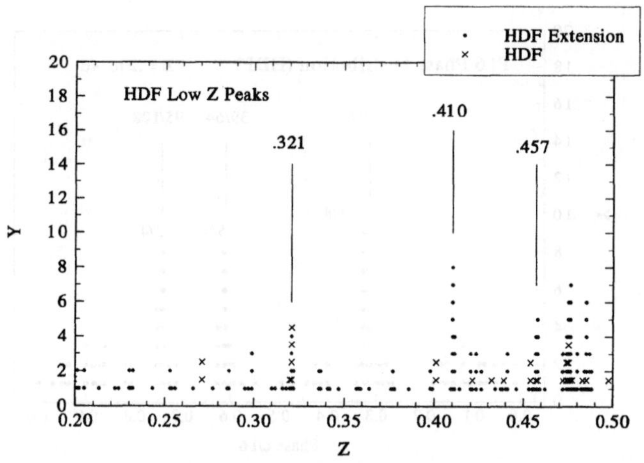

Figure 18. The redshift distribution in the Hubble Deep Field region below $z = 0.5$ using the complete catalog (Cohen et al., 2000). Three peaks are present below $z = 0.46$.

To assess the significance of the discordant associations D. Christlein carried out Monte Carlo evaluations of the angular separation distribution between pairs which combine different redshift peaks. The number of pairs between specific peaks, in ranges of separation, were counted for the observed distribution and compared with 1000 samples generated by random wrapped displacements in RA and Dec of the objects in one peak relative to the other. Block displacements were used to preserve any real clumpy structure. Figure 15 shows the result for associations between the two extreme peaks separated by 25000 km/s in redshift, and for associations with objects which are not in the peaks. The peaks which phase together show a clear excess of associations for both close pairs and groupings. Objects which do not phase together do not show significant associations. Figures 16 and 17 show the result of comparisons between adjacent peaks which have redshift differences in the 12000–13000 km/s range. Again we find clear evidence for physical association of galaxies widely discordant in redshift which phase together in periods corresponding to simple fractions of the speed of light.

5. Extensions To Lower Redshift

To further investigate the deep field work we examined extensions to lower redshift. Figure 18 shows three prominent narrow redshift peaks in the $0.2 < z < 0.46$ interval below the region previously discussed. Many other redshifts distribute throughout the interval but the pattern is not random. Figure 19 examines *phase* relationships using the $c/16$ period. The three prominent peaks are well isolated but appreciably enhanced. The peaks have been significantly reinforced by addition of points at different redshifts which phase together as seen previously in the region

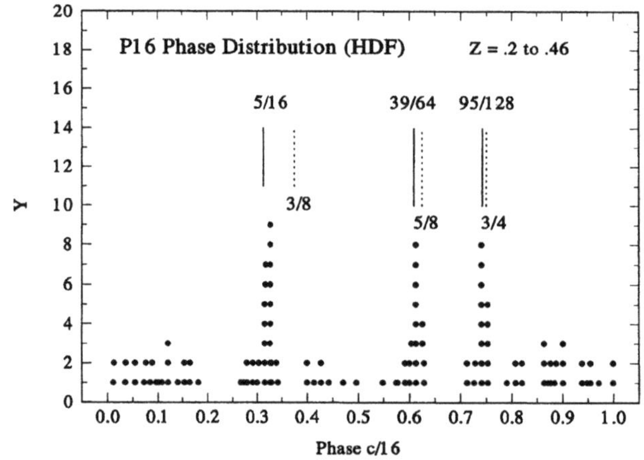

Figure 19. Redshift phase at the $c/16$ period for the Hubble Deep Field objects in the $0.2 < z < 0.46$ interval. The redshift peaks in Figure 18 fall at 1/8 or 1/16 fractions of the $c/16$ interval and are reinforced by objects at other redshifts in phase with them.

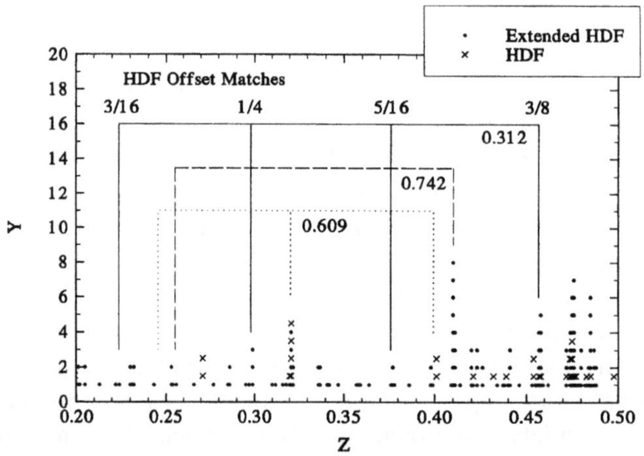

Figure 20. The redshift distribution in the Hubble Deep Field region below $z = 0.5$, as shown in Figure 18, with $c/16$ phase linkages marked for the three reinforced phase peaks in Figure 19.

below $z = .6$. Figure 20 shows the $c/16$ couplings for the three peaks. The 17 points in the $z = .457$ peak, which falls at phase 0.312, are enhanced by 11 points from $z = 0.376$, 0.298, and 0.224 to generate a sharp clear peak. The other peaks are also enhanced. The $c/16$ period pervades the region. Except for the linearity correction from Equation 4, relatively small at low z, this period is essentially a phase shifted version of the $z = 0.06$ period noted previously by Burbidge (1968).

What is new is the appearance of several phase shifted versions within which one particular cycle is especially enhanced. The phase diagram in Figure 19 shows that the peaks associate with specific eights, 5/8 and 3/4, or sixteenths, 5/16, of the $c/16$ cycle. There may be small offsets below the simple fractions but more precise z values will be required to examine any detailed substructure. An overall periodicity of 8×16 ($c/128$) will superimpose the two upper phase peaks and weaker peaks at phase 7/8 and 1/8 which associate with $c/16$ periods matching redshifts at $z = 0.421$ and 0.441. The phase 0.312 family of redshifts is one doubling higher in the $c/256$ series. Formulae describing $z(LT)$ for the three phase peaks are $(16n+5)/256$, $n = 3-6$, and $(8n+5)/128$ or $(4n+3)/64$ for $n = 3-5$. These are consistent with forms expected if doubling decay occurs for values offset from simpler primary fractions by mixing.

As noted in the region just below $c/2$, only odd fractions of intervals are present. The 2/8, 4/8 and 8/8 fractions are missing. As noted above this is consistent with a decay sequence displayed in distance. The galaxies in Figure 19 are at doubling level $D = 8$ ($8 \times 16 = 128 = 2^8$). Even fractions reduce to smaller D values which may have existed earlier and have subsequently decayed if they formed at all. They should not be found in a cloud associated with a distance corresponding to $D = 8$. The odd-integer characteristic of the patterns is a strong confirmation that a time dependent doubling decay model has produced the structure.

The fact that only odd multiples of a given D level are to be expected within a given cloud of objects may be related to the absence of even T values in the Lehto-Tifft doubling equation for redshift families, or more precisely the absence of T values which are direct powers of two, 2, 4, and 8. Six divides only once to the odd value of 3. It is 1, 5, 7, with 6, and sometimes 3 T values that are seen in addition to the pure $T = 0$ doubling sequence. (As noted earlier, the origin of the Lehto-Tifft equations and the generation of particular T values within a doubling process now appears to be understood and will be published separately.)

The final step in the examination of the lower redshift sample involves looking at spatial associations. Figure 21 shows the spatial location of galaxies in the $0.2 < z < 0.46$ range using superimposed symbols to identify objects connected by the $c/16$ period incorporating the $z = 0.457$ peak. Clumping appears to be present, on a larger angular scale as expected from the lower redshifts involved. The concentration in the northeast part of the figure and the general contrast between the east and west halves also suggest physical associations of discordant redshift values; a much larger sample will be required to properly test this possibility in the lower redshift range.

6. The Southern Deep Field

To test the findings associated with the HDF investigations we looked at an independent SDF study by Cohen et al. (1999). Figure 22 shows the redshift distribution in the $0.3 < z < 0.6$ range. There is a striking periodicity at $c/64$ and $c/32$ aligned with predicted $T = 0$ redshift states as indicated. A phase diagram in Figure 23,

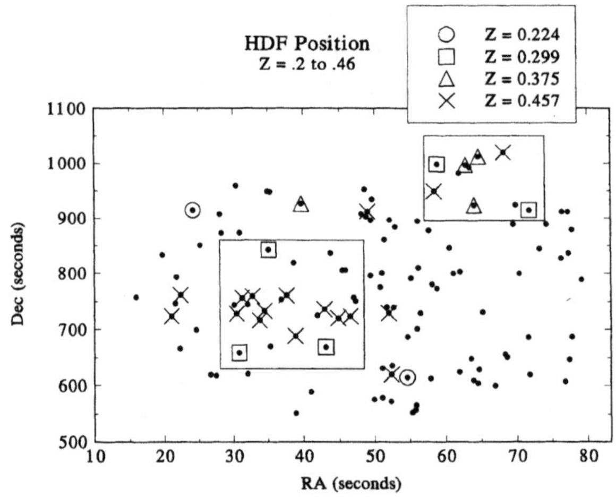

Figure 21. Positions for Hubble Deep Field objects with $0.2 < z < 0.46$, distinguished by redshift, for objects in the 0.3125 phase peak of Figure 19. Boxes indicate possible physical concentrations including possible discordant redshift associations.

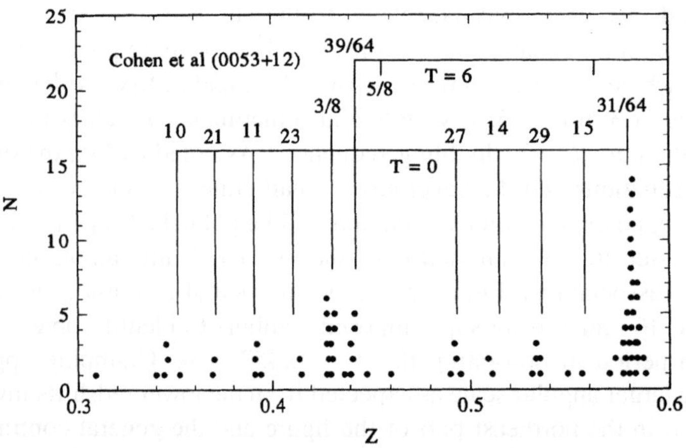

Figure 22. The redshift distribution in a southern deep field region (Cohen et al., 1999) for $0.3 < z < 0.6$ to illustrate the association of redshift peaks with specific values predicted by the Lehto-Tifft model.

constructed for the $c/32$ period, shows a strong concentration around 0.5 (odd $c/64$ states) and a weaker concentration around 0.0 ($c/32$ states). This pattern is expected for objects at or approaching the $D = 6$ ($c/64$) doubling level. The pattern is remarkably consistent with expectations. The phase distribution around 0.5 may contain a spatially dependent stepped pattern at still higher fractional intervals, as seen in the HDF data, but more accurate data are needed.

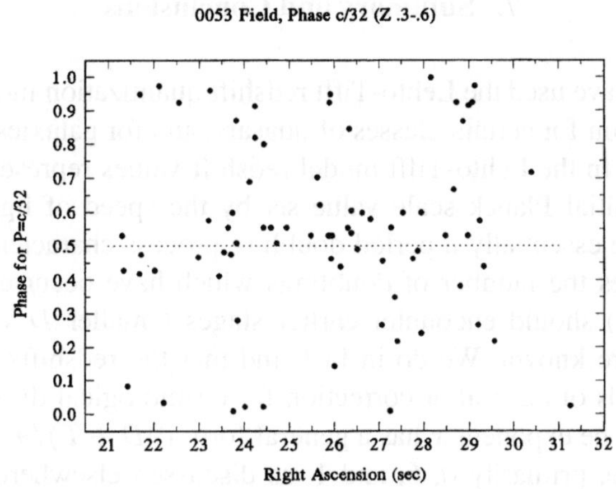

Figure 23. The redshift phase distribution (at the $c/32$ period) for the 1999 southern deep field sample shown in Figure 22.

A detailed examination suggests that the z range above and below $z = 0.46$ should be separated as was done for the HDF. Above this level the $c/64$ period is very strong with a trace of $c/32$ and perhaps the 7/16 state. The smaller fractions show the largest offsets. As noted in the HDF the state adjacent to the empty $c/2$ state is strong and accurately matched. Our model suggests this is a single aggregate at a distance corresponding to $D = 6$ decay, closer than the similar HDF association in the $D = 4$ to 5 range.

The lower redshift region may represent more than one aggregate. The $c/64$ states are barely represented while the $c/32$ states are stronger. The 5/16 level is present and a leading 3/8 state is strong. The 3/8 state is the first eighth state below $c/2$, consistent with the smallest offset being strong and well fit. This may associate it with a higher doubling pattern. Figure 22 shows a possible $T = 6$ state also in the $c/8$ pattern at 5/8. The 5/8 state is empty but the state 1/64 below it is populated. The range of D involved suggests that the pattern could represent more than one structure, much as spectral lines from different multiplets overlap. The objects below the 3/8 state associate primarily with the $D = 5$ level since few occupy the $D = 6$ level. If D is associated directly with distance these objects could be more distant than the higher redshift group. Individual redshifts may have little to do with actual distance. Much more study will be required to examine such a picture, map possible structures and understand if and how D levels relate to actual distance. The SDF appears to be consistent with the HDF, showing redshift patterns consistent with large associations at various stages of doubling decay.

7. Summary and Conclusions

In this paper we have used the Lehto-Tifft redshift quantization model to predict the redshift distribution for certain classes of quasars, and for galaxies in the neighborhood of $z = 0.5$. In the Lehto-Tifft model redshift values represent specific decay states from an initial Planck scale value set by the speed of light, c. The decay is presumed to be essentially a period doubling process characterized by an index D which indicates the number of doublings which have occurred. Looking back in time (distance) should encounter earlier stages (smaller D values) for which redshift values are known. We do in fact find that the redshifts are periodic and in phase at periods of $c2^{-x}$ after correction for cosmological distortion associated with the model. The exponent x has a general form $(9D + T)/9$, with only certain T values possible, primarily 0, 6, and 1, as discussed elsewhere. Pure doubling, $T = 0$, is normally dominant and at high redshift little else was expected or found.

We find that peaks in the quasar distribution at $z = 0.061$, 0.3, 0.6, 0.96, 1.41, and 1.95 correspond very closely to Lehto-Tifft predictions corresponding to 1/16, 1/4, 1/2, 3/4, 1.0, and 5/4 of c (0.064, 0.274, 0.601, 0.989, 1.441, and 1.968). These peaks appear to associate with an active stage where radio emission or other indications of activity are apparent. The doubling level associated with this stage of quasar activity is $D = 0$ to 4, with $T = 0$ dominant and perhaps some traces of $T = 1$ and 6 which are believed to be the next most likely values.

We find that galaxy redshifts in the Hubble Deep Field, and a similar field in the southern galactic hemisphere, concentrate sharply at higher fractional values of c displaced below the primary quasar peak at $z = 0.6$. Close to the $z = 0.6$ peak doublings range around $D = 4$ to 8 and increase further at lower z as expected. A fine structure at still higher fractions of c is present. The patterns of D present are consistent with doubling evolution from decreasing D values as a function of increasing lookback. A series of phase shifted $c/16$ periodicities is especially prominent around z of 0.4.

Galaxies which are periodically phased, but in different sharp redshift peaks, appear to be strongly spatially associated as both pairs and small groups. The objects in specific associations have similar D values but may distribute over a wide redshift range corresponding to periodically spaced redshifts at a given D level. The redshift pattern appears to be a 'spectrum' of related redshifts associated with individual large galaxy clouds which have evolved to different D levels. Monte Carlo evaluations of the discordant redshift patterns around $z = 0.5$ in the HDF show strong physical associations between redshift peaks which differ by 12,000 and 25,000 km/s ($c/32$ and $c/16$).

We conclude that more extended studies using precise redshifts, especially near $z = 0.5$ below the $z = 0.6$ active quasar peak, can provide critical, perhaps definitive, information relating to the nature of the redshift. This window corresponds to the region directly below the first doubling ($c/2$) where strong $T = 0$ states at low doubling levels are predicted to be prominent. Photometric redshifts are not accurate enough for such a test. Equivalent local studies require much higher resolution since doubling appears to have proceeded to $D = 16$ or beyond. Precise 21-cm redshifts or equivalent are required for most local work on quantization.

This conference is being held in honor of Fred Hoyle, a man who unhesitatingly pursued new and exciting ideas, generally well ahead of many of the scientific views of his time. I am pleased to have known such a man, and have attempted to follow his path by using this opportunity to present some new ideas here that just might help us better understand the cosmos that Fred so loved. Preprints of this paper, related future publications, and links to most previous publications by the author can be accessed on the internet at http://SASTPC.org

References

Arp, H., Bi, H.G., Chu, Y. and Zhu, X.: 1990, *A&A* **239**, 33.
Burbidge, E.M. and Sargent, W.: 1970, in: D. O'Connel (ed.), *Nuclei of Galaxies*, North Holland, p. 351.
Burbidge, G.: 1968, *ApJ* **154**, L41.
Burbidge, G.: 1980, *Texas Symp. Proc. Ann. New York Acad. of Sci.*
Burbidge, G. and Napier, W.M.: 2001, *AJ* **121**, 21.
Cohen, J.G. et al.: 1996, *ApJ* **471**, L5.
Cohen, J.G. et al.: 1999, *ApJS* **120**, 171.
Cohen, J.G. et al.: 2000, *ApJ* **538**, 29.
Guthrie, B.N.G. and Napier, W.M.: 1991, *MNRAS* **253**, 533.
Hawkins, E., Maddox, S.J. and Merrifield, M.R.: 2002, *MNRAS*, in press.
Hoyle, F., Burbidge, G. and Narlikar, J.V.: in: *A Different Approach to Cosmology*, Cambridge, p. 332.
Karlsson, K.G.: 1977, *A&A* **58**, 237.
Napier, W.M.: in this conference.
Schmidt, M. and Green, R.F.: 1983, *ApJ* **269**, 352.
Tifft, W.G.: 1976, *ApJ* **206**, 18.
Tifft, W.G.: 1996, *ApJ* **468**, 491.
Tifft, W.G.: 1997, *ApJ* **485**, 465.
Tifft, W.G. and Cocke, W.J.: 1984, *ApJ* **287**, 492.
Tifft, W.G. and Cocke, W.J.: 1989, *ApJ* **336**, 128.
Veron-Cetty, M.P. and Veron, P.: 1996, *A Catalog of Quasars and Active Galaxies*, ESO Scientific Report – electronic form.

RESEARCH WITH FRED

HALTON ARP
Max-Planck-Institut für Astrophysik, 85741 Garching, Germany

A small memoir about his response to a discovery that has turned out to have important relevance to cosmology and cosmogony.

Pasadena 1971

The phone rang in my office at Santa Barbara St.

'This is Fred Hoyle calling from the Cal Tech campus. Can I come up and see some of your pictures of connected objects of different redshifts?'

I was thrilled that a person of such eminence would come to see my observations. As he sat across the desk from me he pushed his glasses up on his forehead and brought the glass plate almost touching his face, I started to explain the features of the image but he immediately said:

'No, don't tell me anything I just want to look.'

I sat for a long time in silence while he looked at a smaller galaxy attached to one of much lower redshift. Finally he gave me back the plate without saying anything, thanked me, got up and left.

Seattle 1972

The American Astronomical Society had scheduled Fred Hoyle to give its prestigious Russell Lecture in the spring of 1972. I did not know what it would contain but I was looking forward to hearing his assessment of the state of astronomical theory. I myself gave a short observational paper in the general session, the last paragraph of which summarized the empirical conclusions and is reproduced below.

I present the paper heading and the last paragraph as it appeared in the Society (BAAS) publication. I see that I had somehow gotten the idea from Fred that particle masses growing with time could explain my evidence that the intrinsic redshift of young objects was high and then diminished as they aged. I do not remember discussing this with him but then I would not even remember the events of that meeting if it were not for an incident which happened after his talk.

Astrophysics and Space Science **285:** 451–457, 2003.

From the Bulletin of the American Astronomical Society, Spring 1972:

05.01.10 Morphology and Redshifts of Galaxies. HALTON ARP, Hale Observatories, Carnegie Institution of Washington, California Institute of Technology.

————-three paragraphs omitted————

"The present observations are used inductively to conclude that the compact objects originate in the nuclei of large galaxies where the physical conditions approach singular values and that their excess redshifts are related to their young age as measured from this event. In my opinion, of the kind of explanations that the current observations require, one of the simplest is one along the lines of Hoyle's suggestion that electrons and other atomic constituents can be created with initially smaller mass. Then smaller $h\nu$ emissions result from a given atomic transition, and radiation from all objects in the new galaxy is shifted to the red. As the galaxy ages, its atomic parameters asymptotically approach that of older matter."

On the opposite page I reproduce the title page of Fred's Russell Lecture. I think the juxtaposition on facing pages of the observational conclusions compared to the theoretical analysis *given at the same AAS meeting* emphasizes the critical nature of the moment. It seems even more critical now, looking back on it and considering the importance accorded the Russell lecture. I had no idea what the Lecture would contain but clearly I sensed it would be of importance in attempting to get the observations of discordant redshifts seriously considered by a larger number of astronomers. There is even the amusing irony that the 'Developing Crisis in Astronomy' was 'Supported in part by the National Science Foundation'.

The Lecture

I was enthralled at Fred's presentation. He opened by saying,

'It is sometimes said that nothing is known from astronomy which goes outside the range of currently known physics ... if one accepts ... the origin of the universe at some moment of time ... then it is this phenomenon of "origin" which lies outside currently-known physics.'

Following this shot across the bows he went on to say '... t = 0 represents a discontinuity not an origin ... The emergence of a discontinuity suggests a quantum transition ... Physical systems cease to be classical when the action becomes small, as it certainly does near t = 0 because all masses are zero at t = 0.' There was the key I had been waiting for. And in the mathematical equations he presented he derived a Hubble law solely from a non-velocity, particle mass growth as a function of t^2.

He continued, 'Under conditions of weak local gravitational fields particle masses are dominated by ... distant interactions.' Ah, Machian not Einsteinian physics. On the other hand Hoyle pointed out '... particle masses within [dense] aggregates

THE DEVELOPING CRISIS IN ASTRONOMY*

FRED HOYLE

Institute of Theoretical Astronomy, Cambridge, England

and

California Institute of Technology, Pasadena, California

Henry Norris Russell Lecture

Presented at

The Seattle Meeting of the American Astronomical Society

April 8-12, 1972

*Supported in part by the National Science Foundation [GP-27304, GP-28027].

arise only only from internal interactions' which led him to remark, '... I have little faith in the usual treatment of the 'black hole' problem.

Then came the part that electrified me – after showing how the particle masses as a function of cosmic time could be related by strict mathematical transformations to the usual cosmological models – he then said, *'This concept appears necessary if we are to understand the result reported by Arp for the galaxy NGC 7603 and its appendage.'* So I finally knew, *that* is what he had concluded when he silently returned the plate of this object to my desk!

After his talk we were standing outside the lecture room discussing the relation of the observations to what he had said when we were approached by one of the leading astronomers of the day, Martin Schwarzschild. Martin stood there looking up at us and, in his inimitable way, slowly contorting his expression in an effort to say something of evident importance. Finally he blurted out, 'You are both crazy'.

I was startled, but flattered to be linked together with Fred. Fred just looked blank. It is interesting for me to reflect that this is the incident that remained foremost in my memory. It was only after recalling it, that I then made an effort to remember, and, rereading material of that time, I was able to recall all the further details that I am now reporting.

Publication Problems with the Russell Lecture

The tradition had always been for the Astrophysical Journal to routinely publish the Russell Prize Lecture. So Fred sent his manuscript in shortly after the meeting. To everyone's utter amazement, sometime later he received a referee's report from the editor! Fred was angry and simply never replied to the editor. When I asked where the paper would appear he merely indicated that he was interested in other matters.

About that time a debate on the reality of discordant redshifts had been organized in Washington D.C. by the American Association for the Advancement of Science. I was asked to represent the affirmative side and a leading conventional advocate was sought unsuccessfully among many candidates for the opposing side. Finally John Bahcall seized the opportunity to defend the status quo. The debate was published in a book titled 'The Redshift Controversy' (Frontiers in Physics, W.A. Benjamin Inc. 1973, ed. George Field). Both sides of the debate were invited to republish papers that supported their side of the argument. I suddenly realized that here was an opportunity to get Fred's Russell lecture published. He consented. The title page of his lecture as published in the book is reproduced here. The Lecture is reproduced in the book just as Fred had submitted it to the ApJ but which was never published there. I would urge all people interested in the history of this subject to read the paper and realize at the same time that this seminal paper would never have been available in published form if it had not been for the above, fortunate happenstance.

Subsequent Developments

In his Russell Lecture Fred already mentioned how the association of quasars with low redshift, active galaxies had made it improbable at the much less than 10^{-5} level that their apparent physical association was accidental. Jayant Narlikar, a former Hoyle student, made an elegant solution of the field equations with particle masses variable with time which showed how the discordant redshifts could be physically understood (Narlikar, 1977) – completely in the spirit of Hoyle's Russell Lecture. In the following years detailed evidence has built up demonstrating how quasars are ejected from active galaxies and how their redshifts evolve from high to low with cosmic time. (see for example Narlikar and Arp, 1993 and Arp, 1998.) (Recent reviews of new evidence may be referenced also in Hoyle, Burbidge and Narlikar, 2000; Arp, Burbidge, E., Chu et al., 2002.)

But the climaxing observation appeared in year 2002. How exceedingly ironic that 30 years after Fred Hoyle pointed to NGC 7603 as a crucial system which must force our acceptance of the existence of discordant redshifts – after a generation has passed – the luminous link between this active Seyfert and its appendage is observed to have two high redshift, quasar-like objects in it.

Two young spanish astronomers with the modest aperture Nordic Optical Telescope on La Palma took spectra of the two knots imbedded in the optical filament which joins the cz = 8,000 km/sec Seyfert with the 16,000 km/sec companion galaxy. They found compact, emission line objects with redshifts of z = .243 and z = .391. (López-Corredoira and Gutiérrez, 2002). Their results are summarized here in Figure 1.

Unfortunately it will come as no surprise to researchers in this field to learn that despite a probability of less than $8x10^{-10}$ of being accidental this decisive observation with fundamental consequences was rejected by Nature and by the Astrophysical Journal and has only now appeared in Astronomy and Astrophysics.

An even more recent discovery supports this kind of ejection origin for quasars. The large, X-ray and hydrogen ejecting galaxy NGC 3628, has been shown to contain *two quasars of z = .995 and z = 2.15 in an X-ray filament which emerges directly from the nucleus of the galaxy.* (See Arp, Burbidge, E.M. Chu et al., 2002.)

What Have We Done to Science?

During the years I visited with Fred from time to time to show him the newest observational results which were struggling to get published. He would instantly size up the results and say something like, ‘ Well chip, they will certainly have to admit now that their assumptions are wrong.’ After a while we both knew that it would not be accepted in the foreseeable future. He never dwelt on the lost effort, money or the dismal state of the science. He was always trying to think ahead to the next insight, the next synthesis of physics. It will always be a pleasure and

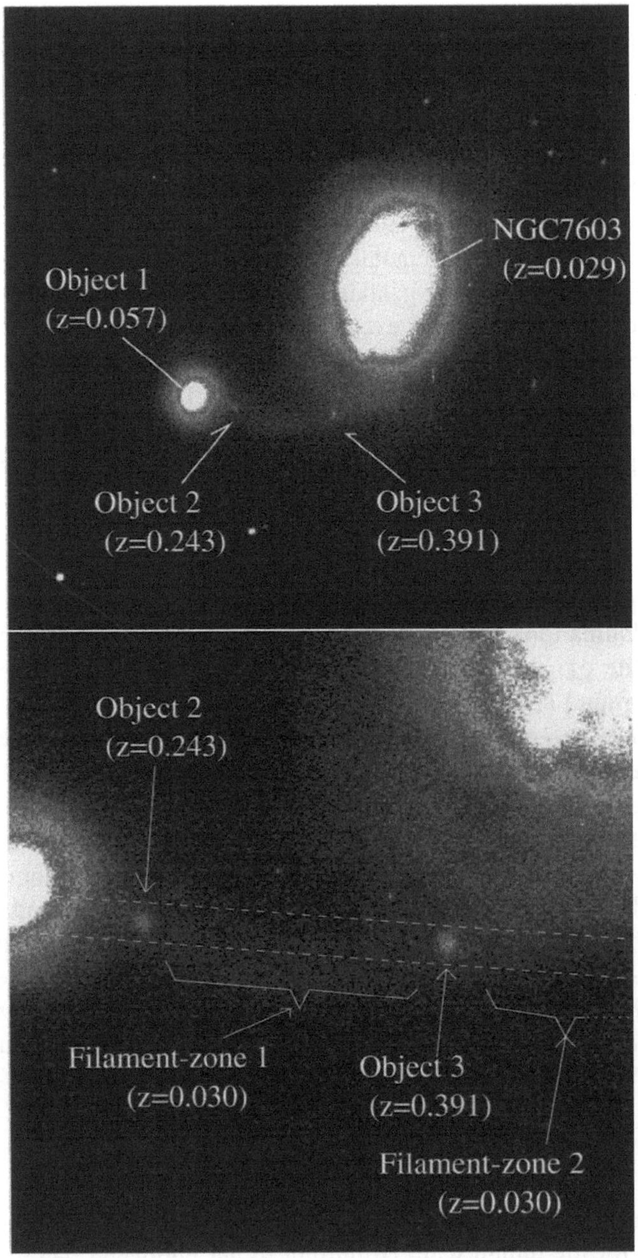

Figure 1. The main galaxy, NGC 7603 is an active, X-ray bright Seyfert with a redshift of 8,000 km/sec. The companion is smaller with a redshift of 16,000 km/sec and a bright rim where the filament from the Seyfert enters it. The recent measures indicate the filament is drawn out of the low redshift parent and contains the two emission line, high redshift, quasar like objects. From López-Corredoira and Gutiérrez (2002).

inspiration however to look back and read his clear, courageous logic and also sad to think how far ahead we might be now if more people had joined in the discovery of new understandings instead of insisting on complexifying and patching up their commitment to old dogma. I can still hear him saying, 'They defend the old theories by complicating things to the point of incomprehensibility.'

We should have crossed over that bridge to a more correct physics that Fred pointed to so clearly more than three decades ago.

References

Arp, H.: 1998, *Seeing Red: Redshifts, Cosmology and Academic Science*, Apeiron, Montreal.

Arp, H., Burbidge, E.M., Chu, Y., Flesch, E., Patat, F. and Rupprecht, G.: 2002, *A&A* **391**, 833.

Hoyle, F.: 1972, Russell Lecture in 'The Redshift Controversy', Frontiers in Physics, W.A. Benjamin Inc., Reading, Mass., 1973, pp. 294–314.

Hoyle, F., Burbidge, G. and Narlikar, J.: 2000, *A Different Approach to Cosmology*, Cambridge Univ. Press.

López-Corredoira, M. and Gutiérrez, C.: 2002, *A&A* **390**, L15.

Narlikar, J.: 1977, *Ann. Phys.* **107**, 325.

Narlikar, J. and Arp. H.: 1993, *ApJ* **405**, 51.

THE DISCOVERY OF MAJOR NEW PHENOMENOLOGY IN SPIRAL DISCS AND ITS THEORETICAL MEANING

D.F. ROSCOE
University of Sheffield, School of Mathematics and Statistics, Sheffield, UK

Abstract. The work described in this paper has its origins in a speculation on the realization of Mach's Principle – specifically, in the formulation of two questions: *(a) Can a globally inertial space & time be associated with a non-trivial global matter distribution? (b) If so, what are the general properties of such a global distribution?* This analysis (Roscoe, [2002a] (astro-ph/0107397)), led us to conclude that a globally inertial space & time can be associated with a non-trivial global matter distribution, and that this distribution is necessarily fractal with $D = 2$; that is mass $\sim R^2$. Gravitational processes are then to be understood in terms of perturbations of this equilibrium space-matter-time structure. We give a very brief overview of these gravitational processes, specifically applied to spiral galaxies. These considerations led directly to the discovery of a completely new phenomenology in spiral discs reported in detail in Roscoe [2002b] (astro-ph/0107300). We review this phenomenology, and give a brief account of its theoretical explanation.

1. Introduction

Fred Hoyle believed that, whenever major problems arose in our understanding of any physical processes (be it astrophysical, biological or whatever) then one should not be afraid to engage in radical ways of thinking – so long as one remained generally faithful to the scientific method that has served so well for the past three centuries. For example, it was adherence to this philosophy, and a belief in the anthropic principle (he existed), which led him to predict the existence of an excited state at an energy of 7.6 million electron volts in the nucleus of carbon-12 – in this way the long-standing problem of the creation of the heavy elements was effectively solved. The same philosophy informed his approach to cosmology. However, many of the cosmological problems which Fred Hoyle sought to solve remain unsolved. For example, one could cite questions concerning the nature of quasar redshifts (possibly stimulated by the ambiguous nature of the Hubble Diagram for quasars) – are they really cosmological, or what? The motivation for the work of the paper presented here can be understood from this perspective – that radical thought must, sooner or later, be deployed if our understanding of much astrophysical phenomenology is to advance.

For no other reason than that many of these problems have proven to be surprisingly long-standing, this author was led, finally, to question the way in which Mach's Principle is (or is supposed to be) incorporated into GR (General Relativity). GR allows an internally consistent discussion of an *empty inertial spacetime* – that is, the empty spacetime of special relativity is a particular solution of the field equations. But, if one takes the view that some strong form of Mach's Principle is

Astrophysics and Space Science **285:** 459–470, 2003.

fundamental to cosmology – so that for any truly fundamental theory the idea of the *inertial frame* is irreducibly connected to the *existence of matter* – then this latter circumstance appears to suggest that, perhaps, General Relativity is insufficiently primitive to allow a fundamental understanding of large-scale astrophysical processes in general. This was the starting point for the work which is briefly described in this paper, written to celebrate *Fred Hoyle's Universe*.

2. The Starting Point

The ideas underlying what is now known as 'Mach's Principle' can be traced to Berkeley (1710; 1990) for which a good contemporary discussion can be found in Popper (1953). Berkeley's essential insight, formulated as a rejection of Newton's ideas of absolute space, was that the motion of any object had no meaning except insofar as that motion was referred to some other object, or set of objects. Mach ((1960), reprint of 1883 German edition) went much further than Berkeley when he said *I have remained to the present day the only one who insists upon referring the law of inertia to the earth and, in the case of motions of great spatial and temporal extent, to the fixed stars*. In this way, Mach formulated the idea that, ultimately, inertial frames should be defined with respect to the average rest frame of the visible universe.

It is a matter of history that Einstein was greatly influenced by Mach's ideas as expressed in the latter's *The Science of Mechanics* ... (see for example Pais (1982)) and believed that they were incorporated in his field equations so long as space was closed (Einstein (1950)). The modern general relativistic analysis gives detailed quantitative support to this latter view, showing how Mach's Principle can be considered to arise as a consequence of the field equations when appropriate conditions are specified on an initial hypersurface in a closed evolving universe. In fact, in answer to Mach's question asking what would happen to inertia if mass was progressively removed from the universe, Lynden-Bell, Katz and Bicak (1995) point out that, in a *closed* Friedmann universe the maximum radius of this closed universe and the duration of its existence both shrink to zero as mass is progressively removed. Thus, it is a matter of record that a satisfactory incorporation of Mach's Principle within general relativity can be attained when the constraint of closure is imposed.

However, there is a hardline point of view: in practice, when we talk of physical space (and the space composed of the set of all inertial frames in particular), we mean a space in which *distances* and *displacements* can be determined – but these concepts only have any meaning insofar as they refer to relationships within material systems. Likewise, when we refer to elapsed physical time, we mean a measurable degree of ordered change (process) occurring within a given physical system. Thus, all our concepts of measurable 'space & time' are irreducibly connected to the existence of material systems and to process within such systems –

which is why the closed Friedmann solutions are so attractive. However, from this, we can also choose to conclude that any theory (for example, general relativity notwithstanding its closed Friedmann solutions) that allows an internally consistent discussion of an empty inertial spacetime must be non-fundamental at even the classical level.

To progress, we take the point of view that, since all our concepts of measurable 'space & time' are irreducibly connected to the existence of material systems and to process within such systems, then these concepts are, in essence, *metaphors* for the relationships that exist between the individual particles (whatever these might be) within these material systems. Since the most simple conception of physical space & time is that provided by inertial space & time, we are then led to two simple questions:

Is it possible to associate a globally inertial space & time with a non-trivial global matter distribution and, if it is, what are the fundamental properties of this distribution?

In the context of the simple model analysed, definitive answers to these questions are provided in Roscoe (2002a), and these answers can be summarized as:

- *A globally inertial space & time can be associated with a non-trivial global distribution of matter;*
- *This global distribution is necessarily fractal with $D = 2$ or, in other words mass $\sim R^2$*

3. A Quasi-Fractal Mass Distribution Law, $M \approx r^2$: The Evidence

A basic assumption of the *Standard Model* of modern cosmology is that, on some scale, the universe is homogeneous; however, in early responses to suspicions that the accruing data was more consistent with Charlier's conceptions of an hierarchical universe (Charlier, (1908; 1922; 1924)) than with the requirements of the *Standard Model*, de Vaucouleurs (1970) showed that, within wide limits, the available data satisfied a mass distribution law $M \approx r^{1.3}$, whilst Peebles (1980) found $M \approx r^{1.23}$. The situation, from the point of view of the *Standard Model*, has continued to deteriorate with the growth of the data-base to the point that, (Baryshev et al. (1995))

> *...the scale of the largest inhomogeneities (discovered to date) is comparable with the extent of the surveys, so that the largest known structures are limited by the boundaries of the survey in which they are detected.*

For example, several recent redshift surveys, such as those performed by Huchra et al. (1983), Giovanelli and Haynes (1986), De Lapparent et al. (1988), Broadhurst et al. (1990), Da Costa et al. (1994) and Vettolani et al. (1994) etc have discovered massive structures such as sheets, filaments, superclusters and voids, and show that large structures are common features of the observable universe; the most significant conclusion to be drawn from all of these surveys is that the scale of the largest inhomogeneities observed is comparable with the spatial extent of the surveys themselves.

In recent years, several quantitative analyses of both pencil-beam and wide-angle surveys of galaxy distributions have been performed: three recent examples are give by Joyce, Montuori and Labini (1999a) who analysed the CfA2-South catalogue to find fractal behaviour with $D = 1.9 \pm 0.1$; Labini and Montuori (1998a) analysed the APM-Stromlo survey to find fractal behaviour with $D = 2.1 \pm 0.1$, whilst Labini, Montuori and Pietronero (1998b) analysed the Perseus-Pisces survey to find fractal behaviour with $D = 2.0 \pm 0.1$. There are many other papers of this nature in the literature all supporting the view that, out to medium depth at least, galaxy distributions appear to be fractal with $D \approx 2$ so that mass $\approx R^2$.

This latter view is now widely accepted (for example, see Wu, Lahav and Rees (1999)), and the open question has become whether or not there is a transition to homogeneity on some sufficiently large scale. For example, Scaramella et al. (1998) analyse the ESO Slice Project redshift survey, whilst Martinez et al. (1998) analyse the Perseus-Pisces, the APM-Stromlo and the 1.2-Jy IRAS redshift surveys, with both groups finding evidence for a cross-over to homogeneity at large scales. In response, the Scaramella et al. analysis has been criticized on various grounds by Joyce et al. (1999b).

So, to date, evidence that galaxy distributions are fractal with $D \approx 2$ on small to medium scales is widely accepted, but there is a lively open debate over the existence, or otherwise, of a cross-over to homogeneity on large scales.

To summarize, there is considerable debate centered around the question of whether or not the material in the universe is distributed fractally or not, with supporters of the big-bang picture arguing that, basically, it is not, whilst the supporters of the fractal picture argue that it is with the weight of evidence supporting $D \approx 2$. This latter position corresponds exactly with the picture predicted by the present approach.

4. Gravitating Systems

Theories of gravitation are derived by assuming some conception of an inertial background (for example, Newton's absolute space, or the flat spacetime of Einstein's special relativity) and perturbing it in someway to generate the required theory. Clearly, the assumed nature of the inertial background has a fundamental influence on the structure of the final theory. So, we can expect that perturbations of an inertial background which is irreducibly associated with a fractal $D = 2$ matter distribution will produce a gravitation theory that is quite distinct from existing theories.

We proceeded as follows:

- Perform a point-source perturbation of the inertial background to obtain a point-source theory, and confirm that it reproduces Newtonian gravitation in the point-source case;
- Extend to a two point-source perturbation, and confirm that the Newtonian two-body theory is reproduced;
- Use the insights gained to write down the N-body theory – which is generally intractable;
- Note that in conditions of very high symmetry, and in the limit from an N-body theory to a continuum theory, there exists the possibility of a tractable form emerging;
- As an example, apply conditions of cylindrical symmetry and use the resulting equations as a model of an idealized spiral galaxy (ie, a perfect disc with no lumps and bumps.)

5. Spiral Galaxies: An Overview of the Theory

When applied to model an *idealized disc*- that is, a mass distribution with perfect cylindrical symmetry with no bulges or irregularities of any kind – the theory says that rotational velocity must satisfy the following differential equation:

$$\frac{1}{V_{rot}}\frac{dV_{rot}}{dR} = \left[\frac{(\lambda \pm p)^2 - q\,(\lambda^3 - \lambda - q)}{(\lambda \pm p)^2 + q^2}\right]\frac{1}{R} \equiv \frac{\alpha}{R} \tag{1}$$

where (p, q, λ) are dimensionless constants satisfying the quartic (actually a pair of quadratics) condition

$$p\,\lambda^2 \pm (1 + p^2 + q^2)\,\lambda + p = 0. \tag{2}$$

Thus, the rotational velocity is found to be given by

$$V_{rot} = A\,R^{\alpha} \quad \text{where} \quad \alpha \equiv \left[\frac{(\lambda \pm p)^2 - q\,(\lambda^3 - \lambda - q)}{(\lambda \pm p)^2 + q^2}\right] \tag{3}$$

where, by virtue of the quartic condition (2), the exponent α must satisfy one of the conditions:

$$\alpha = F_i(p, q)\,, \quad i = 1, 2, 3, 4 \tag{4}$$

In this latter expression, p and q are uninterpreted dimensionless dynamical parameters, and the F_i functions arise when the four solutions of (2) for λ in terms of p and q are substituted into (3). That is, according to the theory, disc dynamics for idealized spirals are confined to one of four distinct surfaces in the (α, p, q) parameter space.

In the following, we firstly show that the power-law model (without reference to the quartic conditions (2)) gives a good statistical resolution of bulk optical

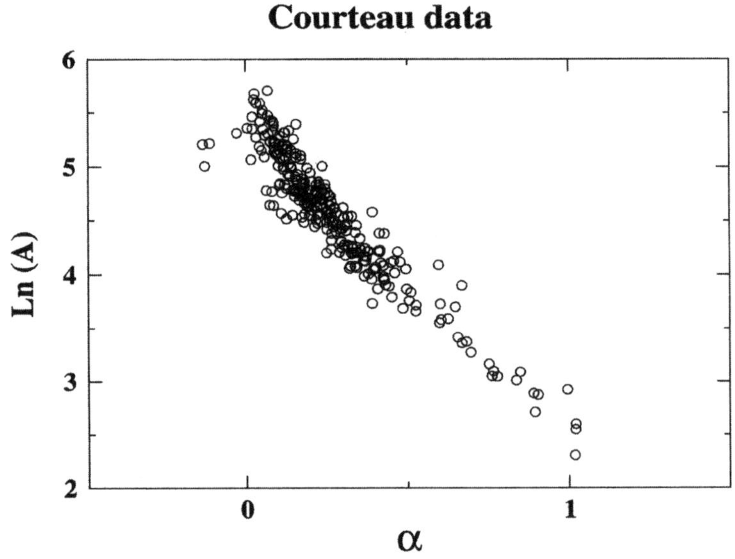

Figure 1. Plot of ln A against α for Courteau sample.

rotation curve (ORC) data. Secondly, we consider the evidence supporting the idea that disc dynamics *are* confined to four distinct surfaces in the parameter space of spiral as predicted above.

6. Spiral Galaxies: The Basic Power-Law Model

The power-law model given at (3) was derived from a theory of the *idealized disc* and, consequently, can only be considered to have a *statistical* validity in those parts of optical discs that are exterior to the central bulgy regions of spirals. The following overview is based on the analyses of Roscoe (1999a; 2002b).

We give a quick demonstration of the effectiveness of the power-law model in resolving the dynamics in bulk samples by reference to the data from a sample of 305 spiral galaxies published by Courteau (1997). Figure 1 gives the basic scatter plot of the model parameters ($\ln A$, α) computed (by regression techniques) for all the galaxies in the sample. A detailed investigation shows that the variation is this plot can be virtually all accounted for by variations in the luminosity properties of the galaxies used in the sample. In particular, we find that the model

$$\ln A \approx -2.8701 - 0.3796\, M + \alpha\, (7.6475 + 0.5027\, M + 0.0118\, S)$$

accounts for about 98% of the variation in Figure 1. Here M represents the *absolute magnitude* and S represents the *surface brightness* of each object in the sample. Figure 2 plots $\ln A$ (as calculated on each ORC by regression on the dynamical

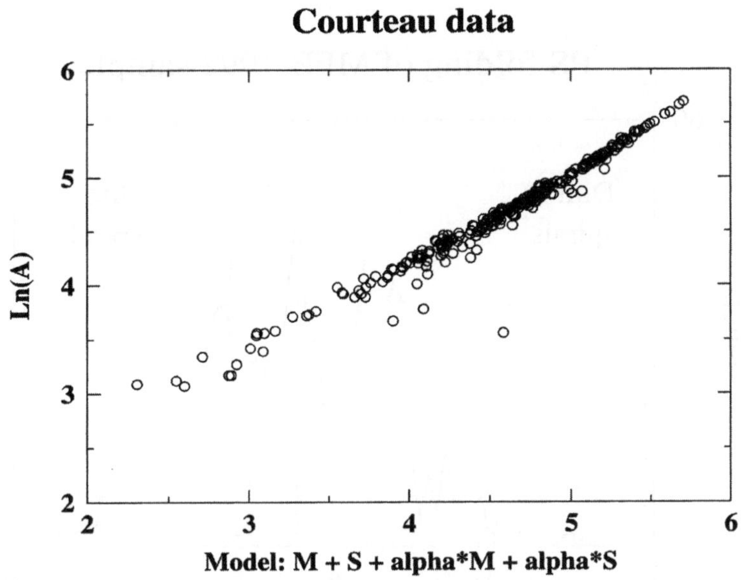

Figure 2. Plot of ln A against model for Courteau sample.

data) against its value as predicted by the model given above for ln A in terms of luminosity properties, and we see an almost perfect fit.

Similar results can be obtained from any of the four samples that we have analysed. Thus, we can consider the power-law description of circular velocities in idealized spiral discs to be virtually perfect when applied to large ensembles.

7. Discrete Dynamical Classes: The Observations

Of considerably more interest, the theory also makes an extremely strong prediction about the nature of the exponent, α, of the power-law model at equation (4). Specifically, it predicts that α should be constrained to occupy one of four distinct surfaces in (α, p, q) space, where (p, q) are dynamical parameters in the model.

This prediction has interest on many levels – but specifically because nothing like it features in any other extant theory of disc dynamics. The effect was first noticed – tentatively – during a pilot study of a small sample of Rubin et al. (1980) ORCs, and this initial identification was used to define a hypothesis which was subsequently tested on the Mathewson et al. (1992) sample and reported in Roscoe (1999a). The effect was subsequently confirmed in Roscoe (2002b) on three further large samples (Dale et al (1997) et seq, Mathewson and Ford (1996) and Courteau (1997)). For completeness, we give a brief review of this evidence here.

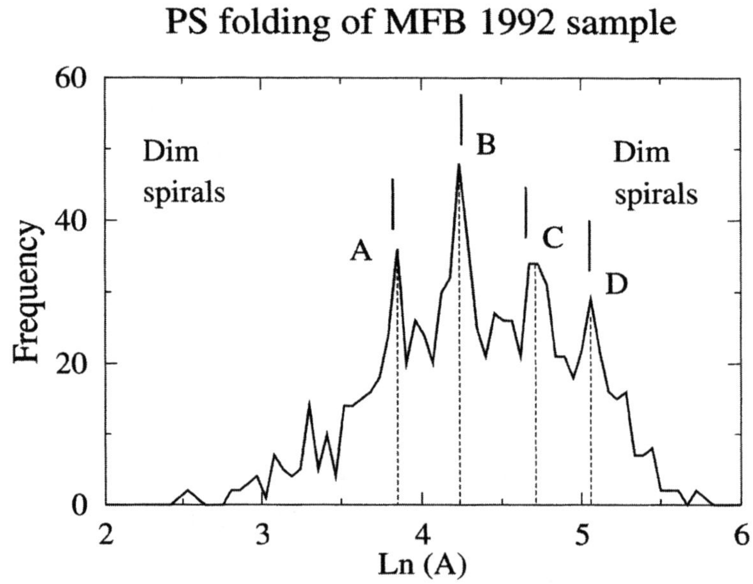

Figure 3. Frequency plot of ln A for MFB sample with Persic and Salucci folding solution.

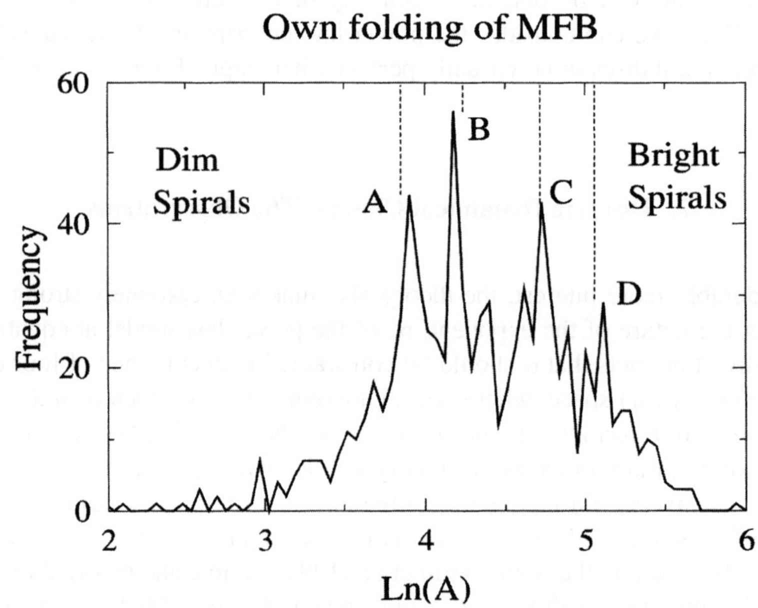

Figure 4. Frequency plot of ln A for MFB sample with own folding solution.

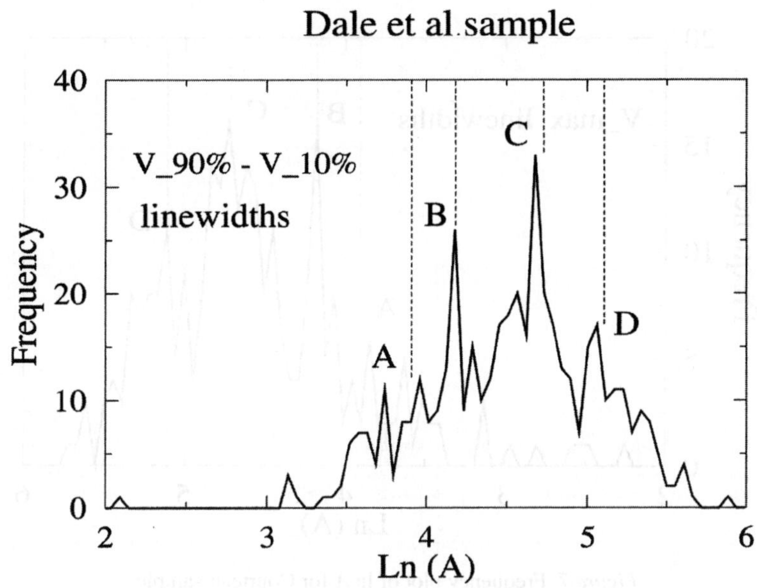

Figure 5. Frequency plot of ln A for Dale et al. sample.

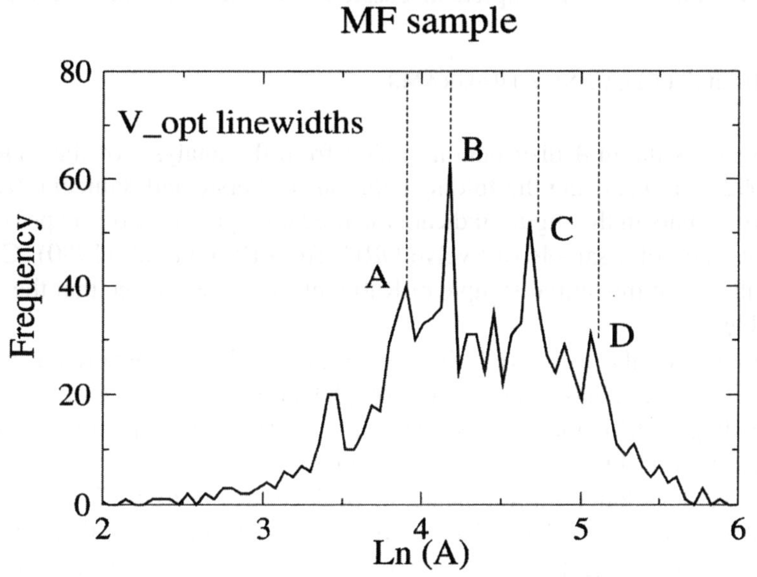

Figure 6. Frequency plot of ln A for MF sample.

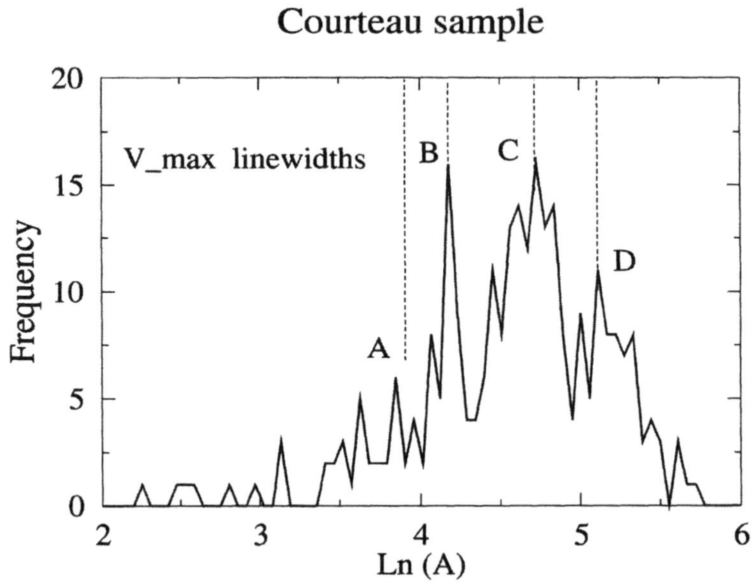

Figure 7. Frequency plot of ln A for Courteau sample.

In practice, the evidence takes the form of the ln A frequency diagrams for each of the four samples, and then interpreting the meaning of these diagrams. See Roscoe (2002b) for a complete discussion of the ln A parameter computation.

7.1. THE ln A FREQUENCY DIAGRAMS

Figure 3 shows the ln A distribution arising from the analysis of the Mathewson et al. (1992) sample, using the folding solutions of Persic and Salucci (1995). The short vertical bars in that figure indicate the predicted positions of the peaks, based on a pilot study of a sample of twelve ORCs from Rubin et al. (1980). Given the very small size of this initial sample of Rubin objects, we can see that the match is remarkable.

Figure 4 is similar to figure 3, but uses folding solutions generated by the author's own software, and is shown for completeness. Figures 5, 6 and 7 show the corresponding distributions for the Dale et al. ((1997) et seq) sample, the Mathewson and Ford (1996) sample and the Courteau (1997) sample. In each of these cases, the vertical dotted lines indicate the peak centres of figure 3.

The A peak in figures 5 and 7 are more-or-less absent because this peak corresponds to very dim objects, and these are very much under-represented in the two samples concerned.

The joint probability of the observed peaks in the four samples arising by chance alone, given the original hypothesis raised on the small Rubin et al. sample, has been computed in Roscoe (2002b), using extensive Monte-Carlo simulations, to be vanishingly small at $< 10^{-20}$.

7.2. Relationship of the ln A Frequency Diagrams to the Theory

It is clear from the four diagrams that $\ln A$ has a marked preference for one of four distinct values, say $\ln A = k_1, k_2, k_3, k_4$. However, we also know, from §6, that $\ln A$ is a strongly defined function of the galaxy parameters (α, M, S), so that $\ln A = F(\alpha, M, S)$. Putting these two results together gives

$$F(\alpha, M, S) = k_i\,, \quad i = 1, 2, 3, 4\,. \tag{5}$$

Consequently, the $\ln A$ frequency diagrams imply that spiral discs are confined to one of four distinct surfaces in (α, M, S) space.

But M is a measure of absolute galaxy luminosity, and therefore a measure of the corresponding total galaxy mass, m say (assuming no dark matter). Similarly, the surface brightness parameter, S, which is a measure of the density of absolute galaxy luminosity, can be considered as a measure of mass density, ρ, in a galaxy. That is, $(M, S) \approx (m, \rho)$. If the dynamical parameters (p, q) in (4) are now identified as dimensionless proxies for (m, ρ), then we see that (5) is essentially identical to (4).

We can reasonably state, therefore, that theory and observation are in complete accordance in respect of the phenomenon of discrete dynamical classes for disc galaxies.

8. Conclusions

By taking a particular strong form of Mach's Principle seriously, we have been led to a theory which, in its turn, has led to the discovery of a major new phenomenology in spiral discs. This phenomenology says that, in effect, the dynamics of spiral galaxies are constrained to occupy one of four distinct surfaces in a particular parameter space. We then indicated that, from the point of view of the theory which led to the discovery in the first place, these constraints can be understood as the physical manifestation of an algebraic quartic condition which must be satisfied if the equation of motion for rotational velocities is to be consistently solved.

In a wider context, the new phenomenology provides strong circumstantial evidence that the assumed form of Mach's Principle contains within it at least a seeds of reality, and gives further justification to Fred Hoyle's philosophy that, from time to time, science needs radical thinking.

References

Baryshev, Yu, V., Sylos Labini, F., Montuori, M. and Pietronero, L.: 1995, *Vistas in Astronomy* **38**, 419.

Berkeley, G.: 1710 *Principles of Human Knowledge*.

Berkeley, G.: 1721, *De Motu*.

Broadhurst, T.J., Ellis, R.S., Koo, D.C. and Szalay, A.S.: 1990, *Nature* **343**, 726.

Charlier, C.V.L.: 1908, *Astronomi och Fysik* **4**, 1 Stockholm.

Charlier, C.V.L.: 1922, *Ark. Mat. Astron. Physik* **16**, 1 Stockholm.

Charlier, C.V.L.: 1924, *PASP* **37**, 177,

Courteau, S.: 1997, *AJ* **114**, 6, 2402–2427.

Da Costa, L.N., Geller, M.J., Pellegrini, P.S., Latham, D.W., Fairall, A.P., Marzke, R.O., Willmer, C.N.A., Huchra, J.P., Calderon, J.H., Ramella, M. and Kurtz, M.J.: 1994, *ApJ* **424**, L1.

Dale, D.A., Giovanelli, R. and Haynes, M.: 1997, *AJ* **114**(2), 455–473.

Dale, D.A., Giovanelli, R., Haynes, M.P., Scodeggio, M., Hardy, E. and Campusano, L.E.: 1998, *AJ* **115**(2), 418–435.

Dale, D.A., Giovanelli, R. and Haynes, M.P.: 1999, *AJ* **118**(4), 1468–1488.

Dale, D.A. and Uson, J.M.: 2000, *AJ* **120**(2), 552–561.

Dale, D.A., Giovanelli, R., Haynes, M.P., Hardy, E. and Campusano, L.E.: 2001, *AJ* **121**, 1886–1892.

De Lapparent, V., Geller, M.J. and Huchra, J.P.: 1988, *ApJ* **332**, 44.

De Vaucouleurs, G.: 1970, *Science* **167**, 1203.

Einstein, A.: 1950, *The Meaning of Relativity*, 3rd edn, Princeton Univ. Press.

Giovanelli, R., Haynes, M.P. and Chincarini, G.L.: 1986, *ApJ* **300**, 77.

Huchra, J., Davis, M., Latham, D. and Tonry, J.: 1983, *ApJS* **52**, 89.

Joyce, M., Montuori, M. and Labini, F.S.: 1999, *ApJ* **514**, L5.

Joyce, M., Montuori, M., Labini, F.S. and Pietronero, L.: 1999, *A&A* **344**, 387.

Labini, F.S. and Montuori, M.: 1998, *A&A* **331**, 809.

Labini, F.S., Montuori, M. and Pietronero, L.: 1998, *Phys. Lett.* **293**, 62.

Lynden-Bell, D., Katz, J. and Bicak, J.: 1995, *MNRAS* **272**, 150.

Mach, E.: 1960, *The Science of Mechanics – A Critical and Historical Account of its Development* Open Court, La Salle.

Mathewson, D.S., Ford, V.L. and Buchhorn, M.: 1992, *ApJS* **81**, 413.

Mathewson, D.S. and Ford, V.L.: 1996, *ApJS* **107**, 97.

Martinez, V.J., PonsBorderia, M.J., Moyeed, R.A. and Graham, M.J.: 1998, *MNRAS* **298**, 1212.

Pais, A.: 1982, *Subtle is the Lord* – The Science and Life of Albert Einstein, OUP.

Peebles, P.J.E.: 1980, *The Large Scale Structure of the Universe*, Princeton University Press, Princeton, NJ.

Persic, M. and Salucci, P.: 1995, *ApJS* **99**, 501.

Popper, K.R.: 1953, *A Note on Berkeley as precursor of Mach, J. Phil. Sci.* **4**, 26.

Roscoe, D.F.: 1999, *A&A* **343**, 788–800.

Roscoe, D.F.: 1999, *A&A* **343**, 697–704.

Roscoe, D.F.: 2002, *GRG* **34**, 5, 577–602.

Roscoe, D.F., 2002, *A&A* **385**, 431–453.

Rubin, V.C., Ford, W.K. and Thonnard, N.: 1980, *ApJ* **238**, 471.

Scaramella, R., Guzzo, L., Zamorani, G., Zucca, E., Balkowski, C., Blanchard, A., Cappi, A., Cayatte, V., Chincarini, G., Collins, C., Fiorani, A., Maccagni, D., MacGillivray, H., Maurogordato, S., Merighi, R., Mignoli, M., Proust, D., Ramella, M., Stirpe, G.M. and Vettolani, G.: 1998, *A&A* **334**, 404.

Vettolani, G. et al.: 1994, in: *Proc. of Schloss Rindberg Workshop: Studying the Universe With Clusters of Galaxies*.

Wu, K.K.S., Lahav, O. and Rees, M.J.: 1999, *Nature* **397**, 225.

THE PREHISTORY OF THE STEADY STATE THEORY

ROBERT TEMPLE

Astronomers and cosmologists may well assume that the controversy between adherents of the steady state and the big bang theories of the Universe is a modern one. However, it has a prehistory. Of course, this prehistory is not formulated in the same terms. It goes back to ancient times when little was understood about the true nature of the stars or of the cosmos. But the underlying question is the same, which comes down to this: Did the Universe come into existence, either by spontaneous generation or creation, or has it always existed? And of course, the question that goes with this is: Will the Universe ever cease to exist, or will it go on existing forever?

Who was it who first suggested that the Universe might always have existed, and indeed might always go on existing without end? It was the philosopher Aristotle, who died in 322 BC at the age of 63. We all know that Aristotle was a remarkable man. He founded the sciences of modern logic, zoology, anatomy, political science, and psychology, while he co-founded with his colleague Theophrastus the science of botany. He also discovered the Eustachian Tubes of the human ear 1900 years before Bartolommeo Eustacchi, known as Eustachius (1520–1574), in the sixteenth century. He scientifically dissected more than 300 species of creatures including humans, he collected 158 constitutions of states as part of his project for comparative political science studies (and he even wrote a new constitution for his home town), he attempted to reduce all of the data of the world of sense into comprehensible and ordered form and to formulate the principles of both logic and thought in order to deal with it coherently. He carried out scientific experiments of an early sort, however much they may have fallen short of our standards of today. He accepted nothing on trust and challenged all fixed ideas. In 1970 an article in Scientific American by Imre Toth revealed that Aristotle was a forerunner of non-Euclidian geometry because he had overtly challenged the parallel postulate in geometry. The prehistory of Riemann, Lobachevski and the others, thus lies with Aristotle, though it was not carried further from the fourth century BC until the nineteenth century AD.

But for us here today, the most remarkable thing is that Aristotle was the first person to challenge the theory that the Universe had come into being at all. He makes this absolutely clear in the First Book of his work *De Caelo*, 'On the Heavens', when he says: 'That the world was generated all are agreed' (Aristotle, 1970, 279b13) By 'all', he means all who preceded or were contemporary with him. He then proceeds to distinguish between two schools of thought: 'having been generated, however, some say it is eternal, others say that it is destructible like any other natural formation'. He opens his lengthy discussion about the eternity of the

Astrophysics and Space Science **285:** 471–477, 2003.

Universe by saying, in his typical way, 'Let us start with a review of the theories of other thinkers ...' (Aristotle, 1970, 279b6). Aristotle always did this, including in all of his fundamental arguments a historical survey of variant opinions. In this, too, he behaved as a true scientist.

There is no need for us to survey all of the views summarized by Aristotle. He makes mincemeat of them. He says, for instance, that 'to assert that the Universe was generated and yet is eternal is to assert the impossible; for we cannot reasonably attribute to anything any characteristics but those which observation detects in many or all instances. But in this case the facts point the other way: generated things are seen always to be destroyed.' (Aristotle, 1970, 279b18–21). And so on. He proceeds to demolish the views of his teacher Plato, who held this position.

Aristotle's own views are strikingly and vividly maintained: 'the word "duration" possessed a divine significance for the ancients, for the fulfilment which includes the period of life of any creature, outside of which no natural development can fall, has been called its duration. On the same principle the fulfilment of the whole heaven, the fulfilment which includes all time and infinity, is "duration" – a name based upon the fact that it *is always* – duration immortal and divine. From it derive the being and life which other things, some more or less articulately but others feebly, enjoy.'

In lengthy and painstaking arguments, Aristotle demolishes the arguments that the Universe could ever have come into being at all, or ever can cease to exist. In his famous and widely published work 'On Philosophy', which was the best known of his writings for the public during his lifetime, and which has been largely pieced together from extensive surviving fragments, Aristotle strongly advocated his theory of what is generally translated as 'the Eternity of the World', but which was really the Eternity of the Universe. His views thus became widely known throughout the Greek world of his day (Aristotle, 1952).

The later writer Philo says in his own treatise on the Eternity of the World: 'Aristotle was surely speaking piously and devoutly when he insisted that the world is ungenerated and imperishable, and convicted of grave ungodliness those who maintained the opposite ... he used to say in mockery (we are told) that in the past he had feared lest his house be destroyed by violent winds or storms beyond the ordinary, or by time or by lack of proper maintenance, but that now a greater danger hung over him, from those who by argument destroyed the whole world'(Aristotle, 1952, Fragment 18 of *On Philosophy*, pp. 88–9).

The colleague and immediate successor of Aristotle as head of his school at Athens was Theophrastus. He also wrote a treatise on the Eternity of the Universe, which contained painstaking analyses and dismissals of the contrary arguments, which he attributes to unnamed people whom he calls 'the sophists'. He was very worried about the reading public being persuaded of a created Universe, and said: 'It is necessary to counter so much special pleading, in case anyone who lacks experience should submit to its authority', and he calls the arguments of these 'sophists' an actual 'deception' (Theophrastus, 1992, pp. 342–57). So as we can see, the debate was already pretty hot within the generation immediately after Aristotle, at the end of the fourth century BC.

The debate continued to rage throughout the following centuries. The Stoic School of philosophy was pretty passionately in favour of a strange theory that the Universe was periodically destroyed and recreated, which is not unlike one of our modern theories today. Opponents argued that this was logically indefensible, because the fire into which the Universe was said to be dissolved continued to exist, so that the Universe didn't really end after all. And then how could it be recreated only from fire? It is therefore not surprising that two prominent Stoic philosophers dissented from their School's doctrine, and supported the theory of un uncreated and unending Universe. The earliest of these was Boethus of Sidon, in the third century BC, who was a translator of Aristotle despite being a Stoic, and later the much better known Stoic thinker Panaetius of Rhodes, who lived c. 185–c. 109 BC. They said that if the Universe were created and destructible, we would then have something created out of the non-existent, and that that was preposterous. They also said that 'there is nothing outside the Universe except possibly a void,' and no cause existed either within or without the Universe to destroy it. 'And if it were to be destroyed without a cause, clearly the origin of the destruction will arise from what does not exist,' and this any reasonable person must reject as absurd (Philo, 2001, 15–16, pp. 239–241).

Another important treatise in the prehistory of the Steady State Theory is called 'On the Eternity of the Universe', and it was attributed to an early Pythagorean philosopher of the 5th century BC called Ocellus Lucanus. However, it was really written by a later Pythagorean or Aristotelian in the second or first century BC. This treatise advocates an eternal Universe in very strong terms. The anonymous author states: 'It appears to me that the Universe is indestructible and unbegotten, since it always was, and always will be ... there is not anything external to the Universe, since all other things are comprehended in the Universe, and it is *the whole* and *the all*. ... the Universe is without a beginning, and without an end ...' (Ocellus Lucanus, 1831, pp. 1–2).

There is a close affinity between this famous treatise and one written by Critolaus of Phaselis, who was Head of the Aristotelian School at Athens in the second century BC., and it may be that he was the true author of the treatise attributed to Ocellus. Like his predecessor Theophrastus, Critolaus was worried about the public being misled, and he boasts of having refuted 'those who strengthen falsehood against truth'. He insists that 'the Universe must surely be uncreated and therefore is indestructible'. He also echoes the founder of his school, Aristotle, who had criticized 'those who would destroy the world' by expressing the same sentiments exactly in attacking 'those who propound the destruction of the world' (Philo, 2001, 14–15, pp. 233–8). We thus see that, for centuries, the Aristotelian School argued with great passion about this subject, just as their founder had done.

A lengthy treatise 'On the Eternity of the Universe' was written by the author Philo the Jew of Alexandria, who was born in 20 BC. He actually wrote two separate treatises on this subject, and the only translation of both of them appeared in 1855 and has never been reprinted (Philo, 1855A and 1855B), but the main one is readily available in translation today (Philo, 2001). He surveys the entire debate from Aristotle's day to his own time. Philo, as a pious Jew, strongly maintained that although the Universe had been created by God, it would never end. Fred Hoyle was aware of the general trend of this historical debate, because he specifically stated in 1965 in his book *Galaxies, Nuclei, and Quasars*: '... the concept of a world with beginning but without end is a part of Judaism and of the derivative Christian culture.' (Hoyle, 1966, p. 88) Fred was not a historian of science, as he was too busy with other things, and so I do not find any mention anywhere in his books of any of these ancient authors specifically. But it is important for us to realize that Fred was always aware of the wider philosophical context of his work.

The Roman author Cicero also entered into the debate, and wrote (in *Lucullus*), in commenting on the Stoic views that the Universe had been created: 'When your wise Stoic has said all these things to you syllable by syllable, Aristotle will come with the golden flow of his speech, to say that the Stoic is talking nonsense; he will say that the world never came into being, because there never was a new design from which so noble a work could have taken its beginning, and that it is so well designed in every part that no force can effect such great movements and so great a change, no old age can come upon the world by lapse of time, so that this beauteous world should ever fall to pieces and perish' (Aristotle, 1952, Fragment 20 of *On Philosophy*, pp. 92–3).

Christians did not at all like an eternal Universe, as it was against their religion. The Christian bishop Lactantius (in his *Institutes*, Book II) wrote in the third century AD: 'If the world can perish as a whole because it perishes in parts, it clearly has at some time come into being; and as fragility proclaims a beginning, so it proclaims an end. If that is true, Aristotle could not save the world itself from having a beginning' (Aristotle, 1952, Fragment 20 of *On Philosophy*, p. 93).

A very lengthy account of Aristotle's theories of the eternity of the Universe were preserved by the late Greek philosopher Proclus in the fifth century AD (Proclus, 1820, Vol. I, pp. 246–9), but these texts have been left out of the standard collections of Aristotelian fragments, and no attempt appears ever to have been made to integrate them into a reconstruction of the lost work 'On Philosophy'. Here some work may be left to do by classicists. One key observation made by Proclus about Aristotle's theory is the statement that according to him 'eternity is stable infinite power'. (*Ibid.*) Although this falls far short of any notion of continuous creation, it is clear that the reason why no continuous creation was ever postulated by Aristotle was his total ignorance of the fact that the Universe is expanding. If he had known of the expanding Universe, I have no doubt that he would have postulated continuous creation to go on filling it up forever as it expanded forever. All of his ideas point to this. Because the concept of expansion was lacking, stability was thought to prevail. If Aristotle were alive today, he would realize that stability can be achieved in an expanding Universe through continuous creation. Indeed, it is the only way to have stability in the long term. He would have jumped at the idea.

In 'On the Heavens', Aristotle states: 'Anything which always exists is absolutely imperishable. It is also ungenerated, since if it was generated it will have the power for some time of not being' (Aristotle, 1970, 281b26–27). But Aristotle was thinking of the whole. In an expanding Universe, Aristotle would have wished to preserve the whole as having been ungenerated and eternal, and would have come to the concept that the parts must be generated and perishable, in the same way that individual beings like humans and animals are. As a modern cosmologist, Aristotle would have grasped very quickly the notion that a star and a man are not that different except in scale, as both are transients in an eternal Universe.

To conclude, then, Aristotle's views on the eternity of the Universe are expressed very clearly in the first sentence of Book Two of 'On the Heavens': '... the heaven as a whole neither came into being nor admits of destruction, as some assert, but is one and eternal, with no end or beginning of its total duration, containing and embracing in itself the infinity of time ...' (Aristotle, 1970, 283b26–30).

More than eight hundred years later these views were to be rebutted with ferocious intensity by the Christian philosopher Johannes Philoponus, in the sixth century AD. He wrote a treatise, substantial fragments of which survive, entitled 'Against Aristotle on the Eternity of the World' (Philoponus, 1987). Philoponus was absolutely determined to have a Creator God, and he wrote his gigantic work (which his opponents claim defeated all attempts to read it because of its great length, its verbosity and its mock-profundity) in order to defend his God against the outrageous suggestion by Aristotle that there might never have been a Creation of the Universe. The bulk of the comments preserved, as well as many of the arguments used by Aristotle himself, are extremely tedious and of no real interest to us today, dealing as they do with the nature of the supposed perfect circular movements of stars and so forth. There is no need for us to survey such details in our brief survey.

A very strenuous rebuttal of Philoponus's attack on Aristotle was made by another philosopher of the sixth century AD, Simplicius. He was an Aristotelian, and his work was called 'Against Philoponus on the Eternity of the World'. He ridiculed Philoponus for being long-winded and having feeble arguments. This work is lost but also survives in fragments. Simplicius criticised Philoponus by saying that he 'regarded it as a matter of great importance if he could entice large numbers of laymen to disparage the heavens and the whole world as things that are just as perishable as themselves. ... this gentleman not only dared to write against Aristotle's arguments in the first book of "On the Heavens" concerning the eternity of the heavens and the world – without understanding what the text says, as I have attempted to demonstrate ... He also opposed the arguments at the beginning of Aristotle's "Physics", Book 8, which show that motion and time are eternal; his objections are beside the point, as one can clearly grasp from my replies to him' (Simplicius, 1991, pp. 107–8).

The real reasons for the flaring up of this debate in the early centuries of Christianity, and for Philoponus's attack on Aristotle, were probably to do with religious dogma interposing itself. Philoponus was a Christian. Christianity has no room for an eternal Universe, as it dictates as part of its dogma that there must be a Creator, and hence a Creation. For Christianity there is and can only be a Big Bang, the biggest possible bang, in fact, where God does his stuff in true Hollywood style and gives us the greatest production of all time, the Universe as we know it, to which trees, animals and man are added in swift array, so that what wasn't there a moment ago is suddenly there, and history can commence without an awkward development period.

We thus see that the debate over the eternity of the Universe commenced not in the 1940s, but more than 2300 years ago. This is the prehistory of the dispute. And I believe it helps to put the whole matter in some perspective. It also strengthens the suspicion which some have had that psychological attitudes are factors in the debate. It may be that there are some people who 'want' the Universe to end, who fear endlessness. On the other hand, people of an open disposition may hate the very idea of a Universe which is not open like themselves, which might end, or which might once not have existed. Whereas some people fear eternity, there are doubtless others who crave it. Do the former support the big bang and the latter support the steady state? This is an interesting field for study. I don't believe any psychological profiles have ever been prepared of the psychological types who take prominent positions on either side of the debate. Having a close acquaintance with two of the three founders of the Steady State Theory, I can call attention to the obvious by pointing out that both Fred Hoyle and Tommy Gold have long been renowned for their open, frank, and candid personalities. Is this a psychological profile of someone who wants an open-ended Universe? Some of the early supporters of the big bang theory at the time of the hot debates in earlier decades were noted for being what I can tactfully call 'less open' personalities. Does a cautious, inward-looking person tend to favour a Universe which will draw to an end? Does a retentive person want a retentive Universe? Can he not tolerate openness in the cosmos any more than he likes it in human beings?

Aristotle, who started all of this debate, was an open personality. This is clear in countless ways, and from what we know of his biography. I believe he would have had a lot in common with the man we have come here to celebrate, Fred Hoyle. Fred often spoke of the conflict between open and closed people, or what he called Type A and Type B personalities, so he was well aware of this very issue, and of how personality traits could dominate attitudes towards cosmology. They seem to have done so for nearly two and a half millennia, and they will doubtless go on doing so for as long as we perishable commodities known as human beings are allowed to go on existing in this Universe which may or may not end, depending on who or what is right, Aristotle or Philoponus, Steady State or Big Bang.

References

Aristotle: 1952, *The Works of Aristotle* translated under the editorship of Sir David Ross, Vol. XII, *Select Fragments*, Oxford, 1952. The surviving fragments of *On Philosophy* are found on pp. 78–99.

Aristotle: 1970, *The Works of Aristotle* translated under the editorship of Sir David Ross, Vol. II, Oxford, 1970. *De Caelo* (*On the Heavens*) is contained within this volume, which also includes *The Physics* and *On Generation and Corruption.*

Hoyle: 1966, Fred Hoyle, *Galaxies, Nuclei and Quasars*, Heinemann, London, 1966.

Ocellus Lucanus: 1831, *Ocellus Lucanus On the Nature of the Universe, Taurus . . . On the Eternity of the World, Julius Firmicus Maternus on the Thema Mundi . . . Select Theorems on the Perpetuity of Time by Proclus*, translated by Thomas Taylor, London, 1831.

Philo: 1855A, 'A Treatise on the Incorruptibility of the World', in *The Works of Philo Judaeus*, translated by C.D. Yonge, Henry G. Bohn, London, 1855, Vol. IV, pp. 21–61.

Philo: 1855B, 'A Treatise concerning the World', in *The Works of Philo Judaeus*, translated by C.D. Yonge, Henry G. Bohn, London, 1855, Vol. IV, pp. 180–210.

Philo: 2001, 'On the Eternity of the World', in *Philo*, translated by F. H. Colson, Harvard University Press, USA, Vol. IX, 2001, pp. 172–291.

Philoponus: 1987, Johannes Philoponus, *Against Aristotle on the Eternity of the World*, translated by Christian Wildberg, Duckworth, London, 1987.

Proclus: 1820, Proclus, *Commentaries of Proclus on the Timaeus of Plato*, translated by Thomas Taylor, 2 vols., 1820.

Simplicius: 1991, *Place, Void, and Eternity: Philoponus: Corollaries on Place and Void* translated by David Furley, with *Simplicius: Against Philoponus on the Eternity of the World*, translated by Christian Wildberg, Cornell University Press, Ithaca, New York, USA, 1991.

Theophrastus: 1992, *Theophrastus of Eresus: Sources for His Life, Writings, Thought & Influence*, ed. By Wiliam W. Fortenbaugh, Part One, E.J. Brill, Leiden, Netherlands, 1991.

PART IV: INTERSTELLAR MATTER

INTERSTELLAR MATTER AND STAR FORMATION

P.M. SOLOMON
Department of Physics and Astronomy, State University of New York,
Stony Brook, NY 11794, USA

1. Introduction

I first met Fred Hoyle in the summer of 1967 as a visitor to the brand new Institute of Theoretical Astronomy IOTA, the precursor to the Institute of Astronomy. Fred had invited me to spend the summer in Cambridge at the suggestion of Peter Strittmatter. Although I was not working on either nuclear astrophysics or cosmology, his major research topics at the time, Fred took an interest in my work on interstellar molecules and we developed a friendship. His interests were very broad and he asked me to organize a meeting at the institute in 1969 on the then new topic of infrared astronomy.

Over the years we discussed a huge variety of topics ranging from infrared astronomy, the microwave background, cosmology, stellar mass loss, interstellar matter, comets, Stonehenge, the navigation system of birds, cricket and politics, particularly Watergate. When he came for a visit you never knew what the topic would be; Fred was not only stimulating and brilliant he was also fun to be with.

Fred Hoyle's work on interstellar matter and star formation can be divided into four periods spanning the years from 1940 until the beginning of the twenty first century. The second period includes his famous work on star formation (and galaxy formation) through fragmentation of gas clouds in the early 1950's. The third period beginning ten years later addresses the origin and composition of interstellar grains in the 1960's and 70's. The early work on grains (almost all with N.C. Wickramasinghe) included the first suggestion that grains formed in stellar atmospheres and were composed primarily of graphite and organic compounds including aromatic hydrocarbons rather than ice. Beginning about 1978 in a long series of papers and books (also with N.C.W.) he worked on the identification of interstellar grains with biological material and the influence of interstellar grains on the origin of life on earth (panspermia).

Even though I knew Fred Hoyle pretty well, there were still a few surprises in the literature which I found in the process of preparing for this meeting. Long before his 1953 paper on the collapse and fragmentation of interstellar clouds, Fred Hoyle wrote a stunning article on the physics of interstellar gas in 1940 emphasizing the importance of molecular hydrogen. This was one of his first few astronomical publications and his first on the physics of interstellar matter.

Astrophysics and Space Science **285:** 481–495, 2003.

In addition to Fred Hoyle's contributions to the literature on interstellar matter and star formation, he was also actively interested in the early 1970's in the new field of millimeter astronomy and the study of molecular clouds. In fact in 1972 while still at the Institute, in a little known effort he almost succeeded in getting funding for a UK millimeter wave telescope to be located in Australia for the study of the galactic center region and the southern Milky Way. Much later after substantial modification I believe this led to the funding of the JCMT submillimeter telescope in Hawaii.

Here I will discuss highlights from his published work on interstellar matter from 1940 until about 1977 (elsewhere in this volume, N. Wickramasinghe gives a more complete summary of the work on interstellar grains including work during the last 20 years).

2. The 1940 Paper: On the Physical Aspects of Accretion by Stars

This paper (Hoyle and Lyttleton, 1940) contains the most complete treatment of the heating and cooling of interstellar gas published at that time. The context is the physical state of a pure hydrogen interstellar cloud accreting onto a massive star. In part this is was a reply to a criticism of their first (1939) accretion paper by (Atkinson (1940)). For a pure H cloud the temperature had been previously estimated to be 10,000 K, too high for substantial accretion. The problem is posed in terms of cooling mechanisms and the solution, cooling by molecular hydrogen H_2, is almost 30 years ahead of its time. The paper considers cooling by H_2, the photoionization and dissociation of H_2 and the formation of H_2. It is fascinating to see how Fred Hoyle formulated the problem more than 60 years ago even in the absence of any laboratory (or good theoretical) cross section or radiative lifetimes.

2.1. Cooling by H_2

> The estimate of 10,000°K depends on a calculation by Eddington based on the assumption that, apart from the radiation emitted in the capture of electrons by positive ions, the interstellar material has effectively no opportunity for further emission. It is therefore necessary to investigate whether the cosmical cloud is capable of emitting radiation by any process other than the inverse photoelectric effect. It will be shown that there is another process by which it can get rid of practically all the internal energy supplied by selective absorption...
>
> ... it can be seen that the main radiative processes are as follows:
>
> (1) Free-free transitions or Bremsstrahlung, in electron–proton collisions.
>
> (2) Infra-red emission by excited hydrogen molecules.
>
> A consideration of the cross section for Bremsstrahlung shows that the process (1) is quite ineffective in reducing the temperature of the cloud. ... On the other hand, it can readily be shown that any appreciable percentage

> of hydrogen molecules must be highly effective in radiating the energy supplied by selective absorption. This can be seen by considering an assembly of hydrogen (protons and $H_2{}^+$) illuminated by a source of temperature T.

The discussion goes on to calculate the net heating per cc from photoionization followed by recombination as the recombination rate times kT with a recombination cross section $\sigma = 4 \times 10^{-17}/T'$, where T' is the gas temperature. They then consider cooling by H_2 rotational and vibrational transitions.

> Consider now the excitation of a hydrogen molecule by collisions with electrons. Interchange of energy between the translatory energy of the electron and the rotational states of the hydrogen molecule will take place freely. But the interchange with the vibrational states is somewhat more intricate and the cross section for complete equipartition between the electron and the molecule may be roughly estimated as 10^{-2} times the kinetic cross section or... $\sigma' = 2 \times 10^{-18}$ cm^2.

The energy 'supplied by the electrons to the gas', or the energy loss per excitation is taken as equal to kT' times the collisional excitation rate. After equating the net heating and cooling rates they show that the gas temperature T' must be small compared with T unless the ratio [of electrons to hydrogen molecules] is very large.

In the next paragraph we see recognition of the importance of collisional de-excitation to cooling or rather the absence of collisional de-excitation even for highly forbidden transitions.

> The foregoing calculation assumes that between successive excitations of a particular molecule, the molecule has sufficient time to radiate away the energy acquired in the first excitation. In connexion with this point it may be noticed that the excitation involved here concerns only the lowest electronic state of the molecule. The corresponding transitions are forbidden ones, since the electric dipole moment of the hydrogen molecule in its lowest electronic state is zero. Accordingly, the required transitions are due to a quadrupole moment, and since the emission of the radiation takes place in the infra-red the probability of these transitions may be expected, by analogy with the atomic case, to be smaller than the probability of an allowed transition by a factor of about 10^{-8}. If, therefore, we take 10^{-6} sec. as the lifetime for an allowed transition we may write 10^{-2} sec. as the approximate lifetime for the forbidden transitions. This lifetime is very short compared with the average time between successive excitations of the molecule. This may easily be seen; for if we suppose that the number of electrons per cm^3. is about 10^3 and if we take their velocity as 5×10^7 cm. per sec., the probability per molecule of exciting collisions per sec. is
>
> $$2 \times 10^{-18} \times 10^3 \times 5 \times 10^7 = 10^{-7}$$

> (the factor 2×10^{-18} being the cross-section for this process), and the time between successive excitations of any molecule is accordingly of the order of 10^7 sec. Thus in spite of the fact that in the above calculation all the estimates made involve a considerable margin of safety, the lifetime is nevertheless only 10^{-5} of the interval between successive excitations.

This is the first realization in the literature of the importance of cooling by molecular hydrogen.

2.2. The Lifetime of Molecules of Interstellar Hydrogen

Hoyle and Lyttleton formulated the problem of the photoionization and photodissociation of H_2 in terms of gas flowing through the interstellar medium at about 5 km/s and encountering the ultraviolet radiation from an O or B star.

> ... for a hydrogen molecule to undergo ionization and dissociation which requires radiation of energy greater than 15.6 e.V., the molecule must approach the critical sphere surrounding some star. This sphere will in general be largest for O stars, but the density of O stars in space is so much smaller than for B stars that the B stars may altogether have more effect.

They then go on tho show that lifetime of molecular hydrogen in interstellar space under these conditions would be extremely long and that any molecules which formed would remain until they actually entered the HII region. Even in that case they would have time to do some cooling.

2.3. Existence of Interstellar Molecules

Near the end of this paper is a section addressing the question of the formation of interstellar molecules. This section was written and submitted for publication shortly before the first observations of interstellar CH lines in stellar spectra, a point that the authors comment on at the end of this section. Again they are considering a pure hydrogen gas. They clearly recognize that normal three body reactions are not possible in the interstellar medium and that some special two body formation process is required. The mechanism they suggest, two body radiative association of hydrogen atoms in a relatively dense cloud is not the correct mechanism even in the absence of dust. They argue that given enough time H_2 will form. What is most interesting and valid is the argument that in a gas with no heavy elements, some hydrogen molecules must form to enable cooling before star formation can take place.

> It is now appropriate to investigate shortly the question of how interstellar molecules may come to be formed. In this connexion it may be noticed that in order to calculate the time required for a cloud of known density and temperature to form a comparable density of hydrogen molecules, some estimate has at the present time to be made for the cross-section for the formation of H_2 by two body collisions... If the stars are regarded as forming before

> the molecules the cosmical cloud may be taken to have a temperature of some 10,000°K in accordance with Eddington's estimate. On the other hand it seems probable that it is by accretion that the stars form in the first place, and in this case it would appear that the molecules must form before the condensation of the stars can begin.

The cooling problem addressed in 1940 by Hoyle and Lyttleton in the context of accretion is analogous to the cooling problem in the early universe where the gas contains no heavy elements and even a small amount of molecular hydrogen is critical for cooling and formation of the first generation of stars. Their suggestion that some H_2 will form by pure gas phase processes is correct although the specific mechanism 2 body association is wrong.

Another aspect of this work which is not correct in detail is the assumption that molecular hydrogen can be photodissociated only by being exposed to very hard ultraviolet photons with energy greater than 15.6 eV inside an HII region. However H_2 can be photodissociated by absorption at energies greater than only 11.1 eV in the Lyman Bands followed by emission into the continuum in a process I described in 1966 (see Field, 1966). Thus H_2 formed at a low rate would be subject to photodissociation even outside an HII region. However H_2 formed on grains (in a metal rich environment) at a high rate will be self shielded by line opacity in all clouds with a hydrogen density greater than about 100 cm^{-3} resulting in clouds that are almost completely molecular hydrogen (Solomon and Wickramasinghe, 1969; Solomon and Wickramasinghe, 1969).

3. The Fragmentation of Gas Clouds into Galaxies and Stars

This paper (Holye, 1953) presents a theory for the formation of galaxies and stars in a grand scheme of isothermal gravitational contraction followed by fragmentation then further contraction and fragmentation into smaller units in a repeating series which stops when the cooling becomes inefficient due to the opacity of the fragments. At this stage the process ends since the cooling time becomes much greater than the collapse time. Among the questions addressed are:

> Why are the masses of galaxies mainly confined in the range 3×10^9 to $3\times10^{11}\odot$?
>
> Why is the typical mass of a type II star of order $\odot$?
>
> Why do type II stars apparently form almost simultaneously with the origin of a galaxy?

Hoyle also speculates on the formation of type I stars near the end of the paper where there is a prescient discussion of processes which may inhibit star formation and enable a spiral galaxy disk to maintain a large gas content.

Here I concentrate on the star formation aspects of this paper although the general process is applied to galaxies and stars. The model of 'hierarchical fragmentation' is based on thermal stability of the gas at a temperature of 10,000 K and the gravitational instability of a fragment with more than the Jeans mass. Using the notation of the time the collapse criterion was given as

$$\frac{GM}{V^{1/3}} > 5\Re T \tag{1}$$

where V is the volume of the cloud. The Jeans mass expressed in current notation is approximately

$$M \approx (\pi k/Gm_p\mu)^{3/2}\, T^{3/2}\rho^{-1/2}$$

The key to the mechanism is the decrease in collapse time for each successive generation of fragmentation which leads to a runaway process as the collapse time t $\approx (G\rho)^{-1/2}$ decreases with decreasing Jeans mass resulting from isothermal contraction.

The first part of the paper is a detailed discussion of the temperature of a pure hydrogen gas heated by turbulence with collisional ionization. The turbulence is assumed to be at velocities > 10 km/s appropriate to the formation of galaxies and type (population) II stars. After considering the cooling curve and ionization equilibrium he concludes that there will be 2 stable temperature regimes '*either the temperature lies in the range 10,000°–25,000° or the temperature must be high, of the order of* $3\times10^{5\circ}$ *or more.*' Fragmentation is then considered in detail for the 10,000 K regime leading to the formation of ever smaller subunits until stellar masses are reached. The process is described beginning with a galactic mass.

MODEL FOR THE HIERARCHY STRUCTURE

Step 1.—A spherical galactic condensation of mass $3.6\times10^9\odot = M_0$ say, temperature $10^{4\circ}$ K, initial density $\rho_0 = 10^{-27}$ gm/cm^3, initial radius $R_0 = 1.2\times10^{23}$ cm, compacts by a factor $k^{2/3}$ and then divides into k equal masses, each of radius R_0/k, density $\rho_0 k^2$, and temperature $10^{4\circ}$ K.

It will be noticed that the obvious requirements are satisfied by these assumptions. Thus each of the subunits satisfies the contraction condition. Moreover, with their assumed radii the subunits can be fitted into the contracted volume of the galaxy itself. Remembering that in the first shrinkage the galaxy decreases in dimensions by a factor of about 3, it follows that we must put $k^{2/3}\sim 3$, which gives $k\sim 5$.

Step 2.—The subunits of step 1 themselves contract by a factor of $k^{2/3}$ and then divide into k equal fragments, each of radius R_0/k^2, density $\rho_0 k^4$, and temperature $10,000^\circ$K. Once again these smaller fragments satisfy the contraction condition.

Step 3.—The subunits of step 1 themselves contract by a factor of $k^{2/3}$ and then divide into k equal still smaller fragments, each of radius R_0/k^3, density $\rho_0 k^6$, and temperature $10,000^\circ$K. Once again these still smaller fragments satisfy the contraction condition.

And so on for further steps.—Now, since the time scale for condensation at each step may be expected to follow the inverse square root of the density,

> it follows that the ratio of the time required for an infinity of such steps to the time required for just the first step is given by
>
> $$1 + \frac{1}{k} + \frac{1}{k^2} + \frac{1}{k^3} + \ldots = \frac{k}{k-1}, \tag{2}$$
>
> which for $k = 5$ gives a ratio close to unity. Thus we see that the main expenditure of time lies in the first step and that further steps require comparatively little time. It is indeed this rapid convergence of the steps that provides the strongest reason for the opinion that dissipation arises from the development of the sort of hierarchy structure postulated above.

The importance of the isothermal nature of the process is stressed and an explanation is given for the inability of the pure hydrogen gas to cool below 10,000 K.

> Now since the time scale for each successive step decreases as $1/k$, it follows that the rate at which thermal energy is made available per unit mass increases by a factor of k at each step. On the other hand, the rate of radiation per unit mass increases by a factor k at each step. Accordingly, radiation is favored by the changes that take place as fragmentation proceeds. There is, therefore, a tendency for the temperature to fall, *but the temperature cannot fall much below* $10,000°$ *K, since otherwise the hydrogen would become insufficiently ionized for effective radiation to occur.* This conclusion is subject to the implicit assumption that the fragments remain optically thin to radiation. The entire nature of the condensation process is altered if the fragments become optically thick to the whole of the radiation spectrum of the hydrogen. A discussion of this issue forms the main topic of the following section.

The paper goes on to consider the conditions which might terminate the process and result in the fragmentation of a galaxy into type II stars.

> For a transition to the adiabatic condition, we require L to become less than the rate at which gravitational energy is being released in our problem. The gravitational energy released per fragment in the nth step of the hierarchy process is of the order of
>
> $$\frac{GM_0^2}{R_0 k^n}(k^{2/3} - 1), \tag{3}$$
>
> the release occupying a time of order $k^{-n}(G\rho_0)^{-1/2}$. In this time the energy radiated is $Lk^{-n}(G\rho_0)^{-1/2}$. Hence a transition from the isothermal to the adiabatic condition begins to occur at the stage where
>
> $$L = \frac{G^{3/2} M_0^2 \rho_0^{1/2}}{R_0}. \tag{4}$$

The opacity of the gas is the limiting factor which determines the luminosity and the mass of the final fragments. The remainder of the treatment of the formation of type II stars consists of a discussion of the opacity and the strong effect of a small admixture of metals on the opacity coefficient. In particular for temperatures of about 3,000–5,000 K the electrons supplied by metals form H^- which is the dominant opacity source. At lower temperatures heavy molecules may raise the opacity further. Using the negative hydrogen ion as the main source of opacity with a metal abundance of about 5×10^{-6} in the predominantly neutral gas, Hoyle arrives at a limiting mass for the fragments of between 0.3 and 1.5 $M_\odot$.

In the 1953 paper there is no mention of the importance of cooling by H_2 in a pure hydrogen gas, in contrast to the 1940 paper. While H_2 cooling was the basis for a paper 13 years earlier, this paper considers only atomic hydrogen modified by the addition of some metals. I can only guess at the reason for this. In between these two papers the 21 cm line of HI was discovered and the galaxy was found to be full of HI. This may have led Fred, along with most other astronomers, to forget about hydrogen molecules. I do know that when I discussed my own work involving the abundance and self-shielding of H_2 with him in 1968-69 and evidence for huge quantities of H_2 in dark clouds from CO observations and particularly our observations of the Milky Way molecular ring in 1974, he was very interested. He remarked that the 21 cm astronomers including Oort had been very confident that they had found almost all of the hydrogen.

There were many papers in the 1960's and 70's that improved the Hoyle fragmentation picture by including the influence of other cooling agents, e.g. H_2, a more accurate treatment of heavy elements, molecules and dust in order to arrive at a more accurate picture. However, (Rees (1976)) showed that the final mass of the fragmentation process is insensitive to the final temperature or detailed opacity and can be expressed as a fraction of the Chandrasekhar mass which explains why fragmentation results primarily in objects of stellar mass and not galactic or planetary mass.

The last section of the fragmentation paper is in some ways the most modern. Hoyle speculates on the formation of Type I stars, that is stars forming from a gas with high metallicity and substantial dust in a spiral galaxy disk. This cloud is denser and much cooler than the huge low density cloud which starts the fragmentation in a collapsing galaxy. As a result the timescale for fragmentation is much smaller which presents a serious problem. Hoyle takes as a starting point for the fragmentation leading to the formation of disk stars an interstellar cloud of mass 10^4 $M_\odot$, T = 50 K and density $\rho_0 = 10^{-23}$ gm/cm^3

> ... comparable with the masses of the larger interstellar clouds in the galaxy. The fragmentation time for such a cloud would be less than 4×10^7 years,... On this basis we should at first sight expect the residuum of gas to become entirely condensed into stars in only 4×10^7 years, reckoned from the time that its density rises high enough for tidal shearing to be unimportant. But such an exhaustion of the remaining gas certainly has not occurred in the galaxy or in other spirals. The question is: Why?

There is a clear realization, that in the Milky Way disk and other spirals there must be a process at work which limits the efficiency or regulates star formation. Unimpeded star formation in the disk will lead to what would now be called a starburst. Although it is fashionable to discuss star formation induced star formation or supernova induced star formation, the more important process is the inhibition of star formation. In normal spiral galaxies, star formation on the scale of the whole disk is a slow and inefficient process although it may be occasionally efficient on very small scales. Here are Fred Hoyle's thoughts on this problem, which still sound current 50 years later although the problem is still far from a quantitative solution. He makes 3 'tentative' suggestions; the last 2 are the most interesting.

> b) ... If very luminous stars of high temperature are formed – and there is, of course, observational evidence that this is so – a considerable heating must occur in the gas. (at this point there is a footnote to the forthcoming paper by Oort and Spitzer and an acknowledgement to Lyman Spitzer for a private communication on the subject of heating) It is readily possible for a few highly luminous O stars to heat a gas cloud of Mass $10^4\odot$ to such a degree that the cloud is forced to expand under the action of gas-pressure forces – not of radiation pressure. Thus the following situation may well arise: When there are no luminous high-temperature stars, condensation occurs, and O stars are formed. But as soon as O stars are formed, the clouds are expanded and condensation ceases. After the bright stars become exhausted, the clouds then re-form and more stars condense. On this picture a sort of cyclic process is set up – clouds condense; stars form, including O stars; clouds blow up, and star formation ceases; O stars die; clouds re-form; more stars condense, including O stars; etc. The interstellar clouds of small mass would represent bits of clouds left over from the blowing-up process.
>
> c) It has in recent years been thought that the galaxy may possess a substantial magnetic field. It has been suggested that the magnetic field owes its origin to the general motions of ionized interstellar clouds. But this suggestion has the disadvantage, from the point of view of explaining the polarization properties of the interstellar medium, that the magnetic vector cannot apparently maintain its direction over sufficiently great distances. This difficulty is much alleviated if it be supposed that the magnetic field is built up, not in the motions of the interstellar clouds, but by the motions of the original gas cloud out of which the galaxy originally formed. In this case the magnetic vector would also alter in direction from point to point, but alignment might be expected over far greater distances than on the earlier view. Moreover, the concentration of the residuum of gas to a disklike shape would tend to align such a magnetic field parallel to the galactic plane, in accordance with the requirement of Davis and Greenstein in their theory of the polarization properties of the medium.
>
> Now it may occur, after a sufficient fraction of the gas in the disk has become condensed into stars, that the self-gravitation of the remaining gas is insufficient to overcome the magnetic pressure, in which case there is no tendency for large-scale condensation into stars. In exceptional regions, self-gravitation may indeed overcome the magnetic pressure – for example, in the

Orion nebula – and extensive condensation may occur there. In this way, it might be possible to explain why a gaseous disk has in the main managed to survive in the galaxy and in other spirals for several billion years.

4. The Composition of Interstellar Grains

Fred Hoyle produced a large body of work in conjunction with Chandra Wickramasinghe addressing the question of the composition of interstellar grains. Their first paper in 1962 challenged the prevailing view that grains were composed primarily of ice and instead were composed of graphite formed in carbon stars and ejected into the interstellar medium by radiation pressure. In 1970 the idea of a stellar origin of grains was extended to include supernova (with D. Clayton). It is now generally accepted that interstellar grains originate in stars. In the mid 1970's in a series of papers Hoyle and Wickramasinghe suggested that complex organic compounds were a major component of grains. This included the then radical idea that heterocyclic hydrocarbons could explain features in the interstellar dust spectrum. During the 1980's they suggested that interstellar grains were not only organic but biological material, including bacteria. In all of this work they used the spectrum of interstellar dust (absorption and emission) from the ultraviolet to the mid infrared in order to identify the constituents of grains. I will briefly discuss some of the highlights from their early work (up until about 1977) below. A much more complete review including the later work is presented by Chandra Wickramasinghe elsewhere in these proceedings.

4.1. Graphite particles as interstellar grains

In this important paper Hoyle and Wickramasinghe examine both the observational evidence for graphite and develop a theoretical model for the formation of graphite in cool carbon rich stellar atmospheres and the consequent expulsion into the ISM. Graphite came to be regarded as an important constituent of the ISM for the next decade. They changed their view on graphite in favor of complex hydrocarbons in 1977–78 based on the new ultraviolet observations of the extinction curve near 2200Å.

Their convincing arguments advanced in favor of graphite were developed in detail; they can be briefly summarized as follows:

1. The high average extinction requires that grains be composed primarily of the abundant elements carbon and oxygen. Water ice is possible but a carbon compound such as graphite should be considered.
2. The theoretical extinction curve for graphite fits the observed extinction (see Figures 1 & 2).
3. Graphite has a sufficiently high albedo to explain reflection nebula.

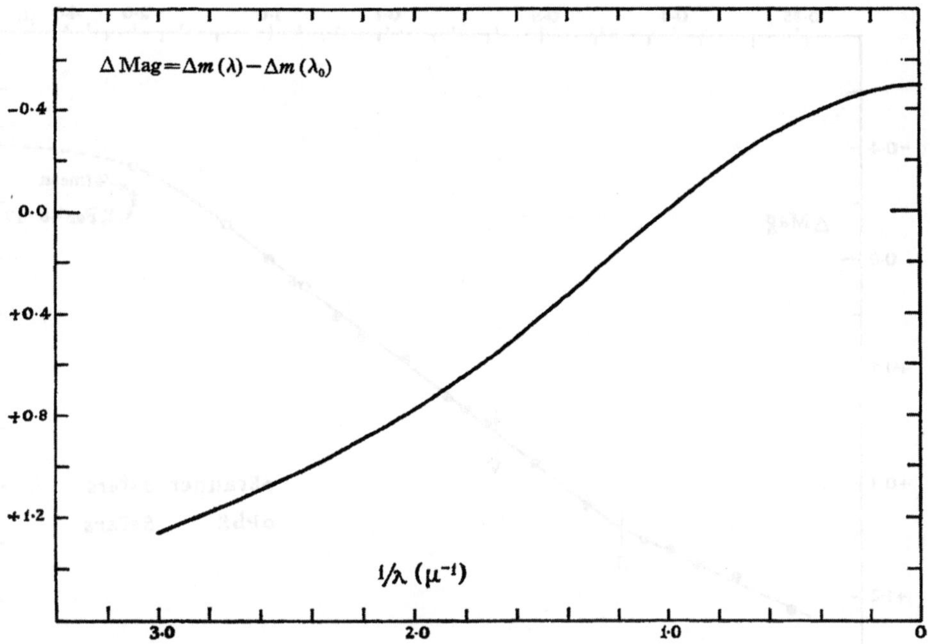

Figure 1. Theoretical curve of interstellar reddening (for Graphite). [Originally Figure 1 in Hoyle and Wickramasinghe (1962).

4. Graphite cannot form in the ISM due to the very low growth rate at low density. Stars forming graphite must have an excess concentration of C over O since CO will take up almost all of the less abundant constituent. Graphite can form in pulsating late type carbon stars during the coolest part of the pulsation cycle when the partial pressure of gaseous carbon will exceed the vapor pressure of bulk graphite by a small margin.
5. Radiation pressure in the outermost layers can eject the grains into the ISM.

The summary of the paper in the abstract shows the full range of factors considered in favor of graphite.

> The interstellar reddening curve predicted theoretically for small graphite flakes is in remarkable agreement with the observed reddening law, suggesting that the interstellar grains may be graphite and not ice. This possibility is not in contradiction with the high albedos of reflection nebulae at photographic wave-lengths, provided the particles have sizes of order 10^{-5} cm.
>
> The origin of graphite flakes at the surfaces of cool carbon stars is considered, about 10^4 N stars in the galaxy being sufficient to produce the required density of interstellar grains in a time of 3×10^9 years.
>
> Grains tend to be formed in the pulsation cycle of an N star at temperatures $< 2700°$ K. The grains have an important effect on the photospheric opacity, causing the photospheric density to decrease very markedly as the temperature

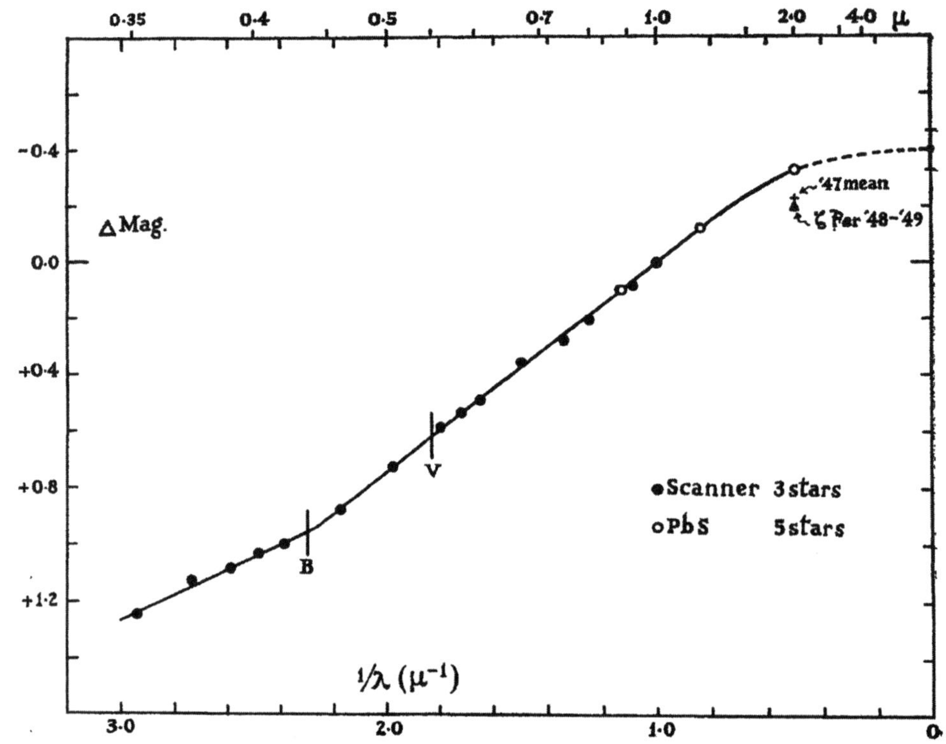

Figure 2. Normalized interstellar reddening curve derived from photoelectric scanner observations, and from infra-red filter observations with a lead sulphide photo-conductive cell. [Originally Figure 2 in Hoyle and Wickramasinghe (1962).

falls towards 2000° K. It is this fall of density that allows the grains to be repelled outwards by radiation pressure and to leave the star altogether in spite of the frictional resistance of the photospheric gases. The grains do not evaporate as they leave the atmosphere of the star. . . .

. . . Graphite is highly refractory and would not evaporate in HII regions. Graphite chemisorbs hydrogen and is therefore an effective catalyst in the production of interstellar H_2. Indeed graphite grains would be highly efficient in the production of interstellar molecules in general. . . .

4.2. Organic compounds as interstellar grains

In the 1970's new spectroscopic data from space based ultraviolet and ground based infrared observations led to a large increase in the data available for the identification of interstellar grains. In a series of papers in 1976–77 Hoyle and Wickramasinghe shifted their attention to organic compounds and polymers as the source of ultraviolet and infrared features. For example they compared the infrared emission expected from polysaccharide grains with the observed spectra from a

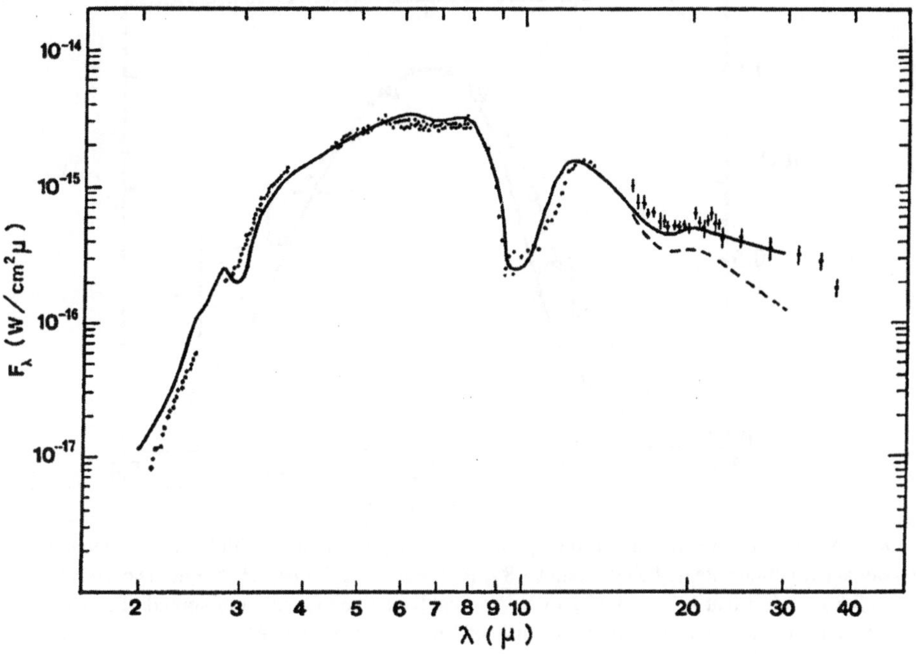

Figure 3. Points represent flux data in the 2–40 μm region for OH 26.5 + 0.6 (Forrest et al., 1978). Solid curve is the best-fitting-normalized flux from a polysaccharide grain model with $T = 430K, \alpha = 2.6$. (Dashed curve omits contribution to longwave flux from cooler cloud – see text.) [Originally Figure 2 in Hoyle and Wichramasinghe (1977a).

wide variety of sources (Hoyle and Wickramasinghe, 1977b). For some sources such as OH 26.5+06 they showed a comparison between the observed spectrum from 2 μ to 30 μ and the expected emission from an optically thin source of polysaccharides at 400K (Hoyle and Wickramasinghe, 1977a. See Figure 3.

The goodness of fit over a wide spectral range is taken as evidence of a correct identification, although they leave open the possibility that other organic compounds are important.

> While it is true that the absorption peaks at approximately 3, 3.4 and 10μm are common to many organic compounds we could find no other condensed molecular solid with a spectrum to match a wide range of astronomical data.

At first they suggested that the polysaccharides might be produced from gaseous formaldehyde in interstellar space but a few months later they suggested (Hoyle and Wickramasinghe, 1977d) that polysaccharides could form in an outflow from a very young O star with a huge mass loss rate and an optically thick cooler envelope, based on an outflow model for strong infrared sources without HII regions. (Hoyle et al., 1973).

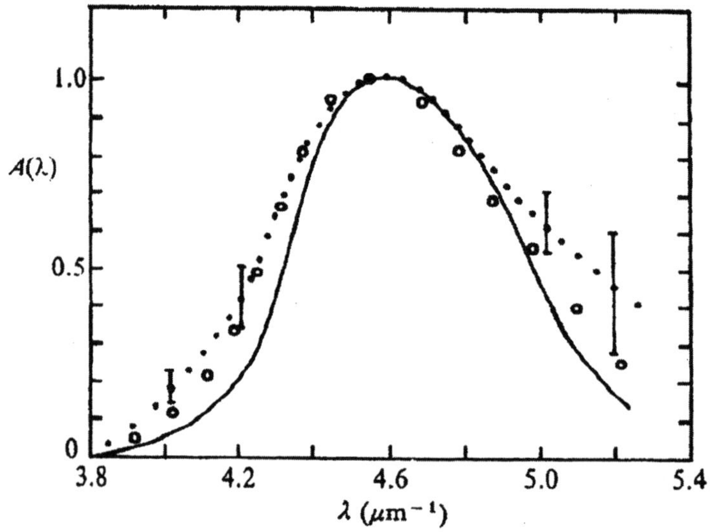

Figure 4. Normalised average molar absorptivity for $C_8H_6N_2$ isomers (solid curve) compared with interstellar extinction data in the waveband 3.8 $\mu m^{-1} < \lambda^{-1} < 5.4\ \mu m^{-1}$. Normalisation is to 0.0 at $\lambda^{-1} = 3.8\ \mu m^{-1}$, 1.0 at $\lambda^{-1} = 4.55\ \mu m^{-1}$. Vertical bars give indication of spread of astronomical data. Dotted curve is the mean extinction curve of Bless and Savage (1972). Open circles give mean extinction $\langle E(\lambda - V)/E(B - V)\rangle$ relative to extinction data for θ-Orionis, and normalised as above. [Originally Figure 1 in Hoyle and Wickramasinghe (1977c).]

At about the same time they realized that graphite by itself could not produce the shape of the important ultraviolet feature in interstellar extinction at 2200Å. Rather than pure graphite as the source of the UV feature they suggested hydrocarbons in the form of nitrogenated **heterocyclic hydrocarbons** (Hoyle and Wickramasinghe, 1977c). They compared the mean observed extinction near 2200Åwith measured absorptivity for the hydrocarbons showing that the peak wavelengths agreed very well and the shape was close although not perfect (see Figure 4). Hoyle and Wickramasinghe also comment that the observed strength of this feature would require only about 10% of interstellar C and N in the form of the hydrocarbons.

Although these specific hydrocarbons may not be the dominant form in the interstellar medium it is now generally accepted that some type of hydrocarbon possibly polycyclic aromatic hydrocarbons or PAH's, are an important component of interstellar matter.

As is well known Hoyle and Wickramasinghe went on to suggest that even more complex compounds are present in interstellar matter and in comets including life itself in the form of bacteria. Fred Hoyle firmly believed that life could not have originated on earth. He felt that the probability of life forming from non-living material was simply too small to happen on earth. Although it is generally thought that he adopted his views on the effect of interstellar matter on the earth late in his career, at least outside his science fiction, this is not the case.

In 1939 when Fred was 24 years old he wrote a paper with R. Lyttleton titled 'The Effect of Interstellar Matter on Climatic Variation' (Hoyle and Lyttleton, 1939). This was an application of their accretion mechanism which suggested that when the sun passed through a dense interstellar cloud the accretion rate would be sufficient to temporarily increase the solar luminosity and warm the earth. Ice ages would correspond to periods of no interaction with interstellar clouds. Accretion is not that effective and most interstellar clouds are not that dense but this paper shows the breadth of his thinking throughout his career. Although not correct on the origin of the earth's climate variation he was right a decade later in developing a theory for the origin of the elements on the earth, one of the great scientific accomplishments of his time.

References

Atkinson, R. d'E.: 1940, *Proc. Cambridge Phil. Soc.* **36**, 314.

Bless, R.C. and Savage, B.D.: 1972, Ultraviolet photometry from the Orbiting Astronomical Observatory. II. Interstellar extinction, *Ap. J.* **171**, 293–308.

Field, G.B.: 1966, Hydrogen molecules in astronomy, *Ann. Rev. Astro. and Astrophys.* **4**, 207–244.

Hoyle, F.: 1953, On the fragmentation of gas clouds into galaxies and stars, *Ap. J.* **118**, 513–528.

Hoyle, F. and Lyttleton, R.A.: 1939, The effect of interstellar matter on climatic variation, *Proc. Cambridge Phil. Soc.* **35**, 405–415.

Hoyle, F. and Lyttleton, R.A.: 1940, On the physical aspects of accretion by stars, *Proc. Cambridge Phil. Soc.* **36**, 424.

Hoyle, F., Solomon. P.M. and Woolf, N.: 1973, Massive stars and infrared sources, *Ap. J. (Letters)* **185**, 89–93.

Hoyle, F. and Wickramasinghe, N.C.: 1962, On graphite particles as interstellar grains, *MNRAS* **124**, 417–433.

Hoyle, F. and Wickramasinghe, N.C.: 1977a, Polysaccharides and the infrared spectrum of OH 26.5 + 0.6., *MNRAS* **181**, 51p–55p.

Hoyle, F. and Wickramasinghe, N.C.: 1977b, Polysaccharides and infrared spectra of galactic sources, *Nature* **268**, 610–612.

Hoyle, F. and Wickramasinghe, N.C.: 1977c, Identification of the λ2,220Å interstellar absorption feature, *Nature* **270**, 323–324.

Hoyle, F. and Wickramasinghe, N.C.: 1977d, Origin and nature of carbonaceous material in the galaxy, *Nature* **270**, 701–703.

Rees, M.J.: 1976, Opacity-limited hierarchical fragmentation and the masses of protostars, *MNRAS* **176**, 483–486.

Solomon, P.M. and Wickramasinghe, N.C.: 1969, Molecular hydrogen in dense interstellar clouds, *Ap. J.* **158**, 449–460.

ELEVATING THE STATUS OF DUST

CHANDRA WICKRAMASINGHE
Cardiff Centre for Astrobiology, Cardiff University, 2 North Road, Cardiff CF10 3DY, UK
E-mail: Wickramasinghe@cf.ac.uk

Abstract. The subject of interstellar grains has grown enormously in stature over the past 40 years, the duration of my collaboration with Fred Hoyle. These developments are in large measure due to Fred's prescient inputs into this subject and his encouragement.

If you can look into the seeds of time,
And say which grain will grow and which will not,
Speak then to me, who neither beg nor fear
Your favours nor your hate.

— William Shakespeare, (Macbeth)

1. Introduction

Throughout his long career Fred Hoyle had the rare gift of spotting important problems in science. He was able to make connections between vastly different scientific disciplines, often resulting in major breakthroughs of knowledge. An idea rarely sat in isolation in head: he would unceasingly seek wide-ranging connections. This is the manner in which I perceived Fred's encouragement for me to enter the field of interstellar grains. In 1960 there was little happening in this field, nor did it appear likely that any major developments were about to transpire.

I recall vividly the first moments of my introduction to this subject, walking with Fred in the hills of Cumbria in the late summer of 1961. It was an exceptionally fine morning when we set out on our trek but by noon, as we sat down with our lunch packs, cumulous clouds floating overhead half threatened to rain. Fred looked up to the heavens and remarked prophetically: 'Those clouds could be fully saturated with water vapor, but it will not come down as rain unless the clouds are seeded in some way...' There was more to be said, but the connection he sought then was between a well-recognised meteorological fact and what he considered to be an unsolved problem in astronomy. The problem was the condensation of ice grains in interstellar clouds. If the nucleation of ice particles ice or rain drops was so difficult in terrestrial clouds, how could ice grains nucleate under the exceedingly tenuous conditions of interstellar space where the maximum density is no more than a few million atoms per cubic centimeter.

Astrophysics and Space Science **285:** 497–509, 2003.

2. From Dirty Ice Grains to Graphite

At the time that I started researching this subject, the firmly held view of astronomers was that the interstellar grains were comprised of dirty ice (a mixture of volatile ices dominated by H_2O with traces of metals) that condensed in the tenuous clouds of interstellar space (van de Hulst, 1946; Oort and van de Hulst, 1946). Astronomers were content to accept that this model without question, and only worry about grains when it came to correcting the colours, temperatures and distances of stars. Fred's conjecture that nucleation of ice grains was difficult in space was easy enough to verify by applying standard results of nucleation theory. Within two months of my fateful walk with Fred I found myself challenging the ice-grain paradigm and exploring alternative astronomical environments for the condensation of grains.

We found that the ice grain theory presented other difficulties when confronted with observational data. The accurate interstellar reddening curve obtained by Nandy (1964) revealed a slight but nevertheless significant discrepancy with respect to the ice model in the near ultraviolet region of the spectrum. And searches for a diagnostic spectroscopic signature of water ice at 3.1μm in the spectra of highly reddened stars led consistently to negative results (Danielson et al., 1965). We argued therefore that water-ice could not be the major component of interstellar dust.

The first alternative class of model we discussed was the possibility of grains condensing in the photospheres of cool giant stars. Carbon stars, particularly the variable N-stars, appeared the strongest candidate at the outset, leading to the formation of refractory graphite grains (Hoyle and Wickramasinghe, 1962). These giant stars have a photospheric C/O ratio exceeding unity and are variable over a timescale of $\sim$100 days. We showed that during the minimum phase when the temperature fell below 2000K carbon grains will nucleate and grow to radii of a few hundredths of a micrometer and be expelled from the star due to the action of radiation pressure. Our paper on graphite particles as interstellar grains led to several years of controversy. The matter seemed to be settled once and for all by ruling out ice as a dominant grain component when the first ultraviolet spectra of stars became available (Stecher, 1965). The extinction curve of starlight in the mid-ultraviolet showed a conspicuous feature centred at about 2175A, which was impossible to explain with ice grains (Wickramasinghe and Guillaume, 1965). This difficulty is seen clearly in the curves plotted by Stecher and Donn (1965) and reproduced in Figure 1.

3. Refractory Grain Mixtures

A decisive shift away from volatile icy grains to grains comprised of a refractory material seemed well established by 1970. In addition to our original proposal of

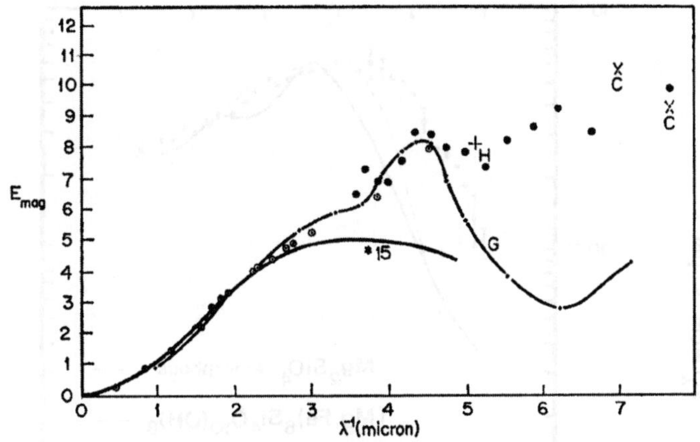

Figure 1. The theoretical extinction curve calculated by Stecher and Donn (1965) for a size distribution of graphite grains (G) compared with Van de Hulst's model #15 for ice grains, and the first set of extinction data points obtained by Stecher (1965). (Reproduced from 1965, *Astrophys. J.* **142**, 1681)

graphite grains from carbon stars we also considered protoplanetary discs, oxygen rich Mira variables, supernova ejecta and nova envelopes as prospective locations for grain formations (Hoyle and Wickramasinghe, 1968, 1969, 1970).

The observed interstellar extinction at 5500A amounts to about 1.8 mag per kpc near the galactic plane in the solar neighbourhood. For any type of submicron sized grain this implies a requirement of an average mass density of grains in the range 1 to 3×10^{-26} g cm^{-3}, about one percent of the interstellar hydrogen density. Since the average composition of interstellar matter must reflect roughly the solar relative abundances, this means that the grain material must be dominated by the elements C, N, O. For each of the mechanisms we proposed for grain formation we were able to present arguments for a reasonable fraction of the total grain mass that is demanded by the extinction data.

With the discovery of a broad emission feature over the waveband 8-12 μm in many astronomical sources a case was made out for a significant contribution to dust from silicate material (Woolf and Ney, 1969; Knacke et al., 1969). On the basis of detailed thermochemical calculations Fred Hoyle and I had earlier found SiO_2, MgO and Fe to be the favoured primary condensates in mass flows from oxygen rich Mira variable stars, so under realistic conditions minerals like $MgSiO_3$ were not unlikely to form (Hoyle and Wickramasinghe, 1968). However the contribution of silicates to the overall infrared emission of dust in extended nebulae like the Trapezium began to look rather dubious in view of a mismatch that showed up between data for laboratory silicates and astronomical spectra obtained by Forrest et al. (1975, 1976). The mismatch appeared to be endemic to all forms of silicates, as determined from laboratory studies of samples under a wide range of ambient conditions. Figure 2 illustrates the extent of the discrepancy that began to worry us in the early 1970's.

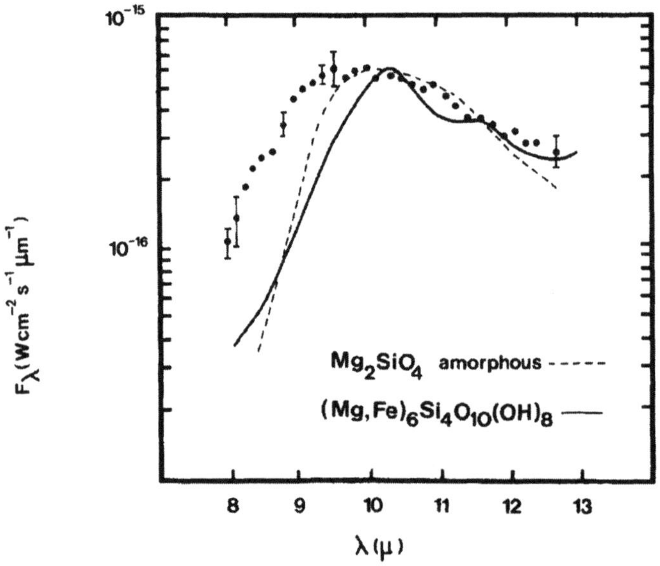

Figure 2. Data for emission by dust in the Trapezium nebula (Forrest et al., 1975, 1976) that was discordant with both hydrated and amorphous silicates.

In these circumstance the astronomical community pursued a somewhat unusual route. The observational data points were used to infer the optical properties of a fictional 'astronomical silicate', and thereafter this hypothetical 'silicate' was adopted as a standard for astronomical comparison, and identifications of 'silicates'. Fred Hoyle and I protested strongly that this procedure represented an unacceptable inversion of logic (Hoyle and Wickramasinghe, 1984). It was our view that the lack of correspondence between data and models in Figure 2 was telling us something important about grains and that was a message that had to be read.

4. The Shift to Organic Grain Models

The first explicit proposal for an organic composition of both interstellar and cometary dust was made in 1974/75 (Wickramasinghe, 1974a,b; Vanysek and Wickramasinghe, 1975). The motivation was mainly the lack of detailed agreement between the astronomical data for heated dust in the Trapezium nebula and the calculated behaviour of real silicates over the infrared waveband 8–13 micrometres as shown in Figure 2. The question that we set out to answer in the summer of 1974 was whether the lack of adequate absorptivity/emissivity over the 8-9.5 μm waveband, that seemed to be an endemic problem for silicates, could be explained with contributions from organic grains. A perusal of spectral atlases soon revealed that polymers involving C-O, C-N, C=C and C-O-C linkages offer the prospect of contributing over precisely the required infrared waveband A formaldehyde-based

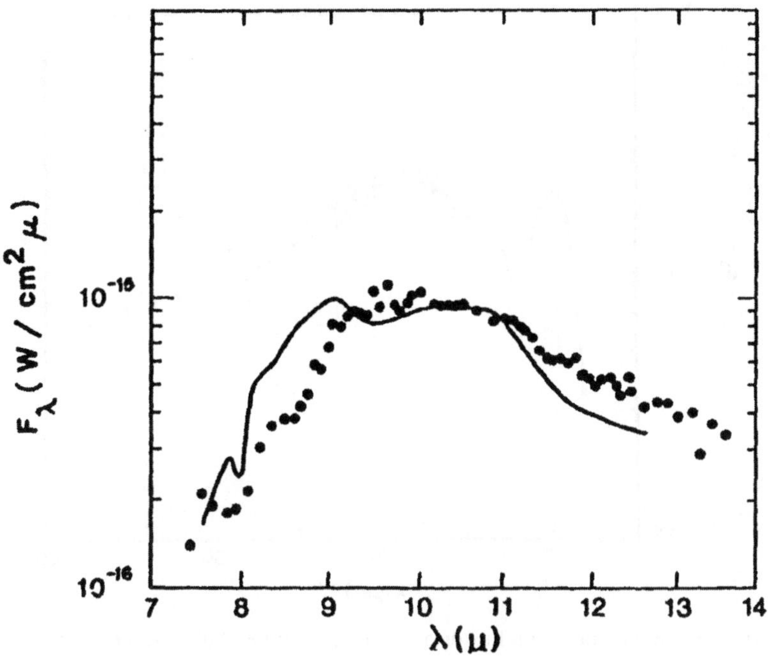

Figure 3. Comparison of emission from polyoxymethylene (POM) grains heated to 190K with the Trapezium data (Wickramasinghe, 1974a,b).

polymer first suggested itself in view of the ubiquitious occurrence of the gas phase molecule H_2CO in interstellar space (Wickramasinghe, 1974a,b). The first comparison of an organic polymer (POM or polyoxymethylene) with the Trapezium data (points) is shown in Figure 3, where it is clear that the 8–9.5 micrometre waveband is particularly well served with an enhanced source of opacity. This comparison represented the first suggestion in the astronomical literature of organic polymeric grains in space.

Encouraged by the initial success of this new class of model as evidenced in Figure 2 we proceeded to consider other organic polymers. We moved progressively through various co-polymers of formaldehyde (Cooke and Wickramasinghe, 1977), and thence to sporopollenin (Wickramasinghe et al., 1977) and other biochemicals such as polysaccharides (Hoyle and Wickramasinghe, 1977). The fits to astronomical data improved systematically along this progression, but the best fits came when we considered intact biological cells as a model of the grains (Hoyle et al., 1984a). A grain model that is spectroscopically indistinguishable from biomaterial (including a type of algae known as diatoms and bacteria) produced an exceedingly close fit to data over the entire 7–40 μm waveband as shown in Figure 4.

With our successes over the 7–40μm waveband in relation to organic grain models we next sought to find more direct evidence of CH stretching absorp-

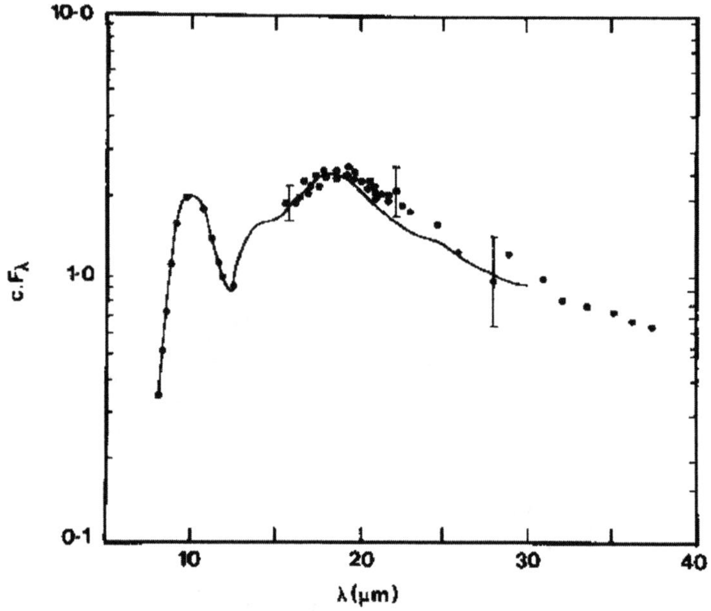

Figure 4. Fit of the Trapezium flux data to a mixed culture of diatoms heated to 175 K.

tions and emissions at 3.4μm. Such a spectral feature would provide unequivocal evidence of organics in the grains. The data was slow to come in a decisive way for the reason that the mass absorption coefficient of organic solid material (e.g. bacteria) near 3.4μm was in the range 500–1000 cm^2g^{-1}, requiring pathlengths through interstellar matter of some 10 kpc to produce measurable absorptions of a few tenths of a magnitude. The best chance of detecting organic dust in the ISM through such a signal was to be in sources of infrared radiation located near the galactic centre ~10 kpc away. Observations with poor spectral resolution that were already available before 1978 provided tentative evidence of the kind we required, but decisive data became available only in 1981. In that year D.A. Allen and D.T. Wickramasinghe used infrared equipment at the AAT to obtain a spectrum showing an unambiguous absorption feature in the galactic centre infrared source GC-IRS7 with a characteristic shape indicative of an assemblage of complex organic material in the grains (Allen and Wickramasinghe, 1981). The extinction over a 10 kpc path length at the centre of the 3.4μm band was as high as 0.3 mag, and there were no signs of any ice band to be seen.

Several months before this observation was made my graduate student S. Al-Mufti had made the remarkable discovery that the 3.3–3.5μm absorption profile of desiccated biological cells was substantially invariant, both with respect to cell type (prokaryotes or eukaryotes) and to ambient physical conditions, including temperatures up to ~700 K (Hoyle et al., 1982a,c; Al-Mufti, 1984). The prediction then was that if grains were biological GC-IRS7must not only possess absorption in this

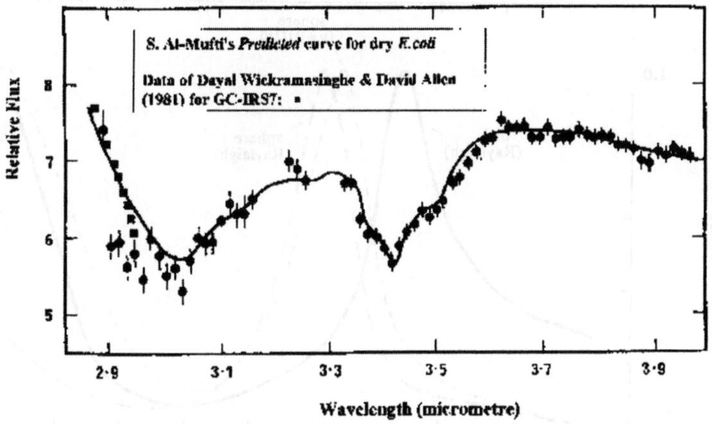

Figure 5. The first detailed observations (points) of the galactic centre infrared source GC-IRS7 (Allen and Wickramasinghe, 1981), compared with earlier laboratory sprectral data for dehydrated bacteria (curve).

waveband, but it must also show absorption with an optical depth that was a rather precisely determined function of wavelength. This is exactly what was found when the first high resolution spectral data for GC-IRS7 was compared with the bacterial model. The modelling of the data was to all intents and purposes unique, and the fit shown in Figure 5 implies that approximately 30 percent of all the available carbon in the interstellar matter is tied up in the form of material that is indistinguishable from bacteria.

This gave further confirmation of a grain model based on dehydrated bacteria that was shown earlier to match data for the interstellar extinction over the wavelength range 0.3 to 3μm without requiring any artificial parameter fitting as was needed in abiotic models (Hoyle and Wickramasinghe, 1979).

5. Disillusionment with Graphite, and Aromatic Grains

Throughout the 1970's Hoyle and I began to have serious misgivings about our earlier assignment of the 2175A extinction peak to graphite. The problem for us was the firm requirement for graphite spheres of radius precisely 0.02 μm. In the real world graphite is a highly anisotropic crystal and condenses as either whiskers or flakes. Spherical particles of a specific radius represented a travesty of common sense. A calculation carried out by Wickramasinghe et al. (1992) illustrating this point is shown in Figure 6. This indicated to us most clearly that graphite grains from carbon stars cannot contribute significantly to the overall composition of grains.

Disillusioned with the general class of graphite-based models Hoyle and I began to explore an alternative explanation of the 2175A extinction feature in terms

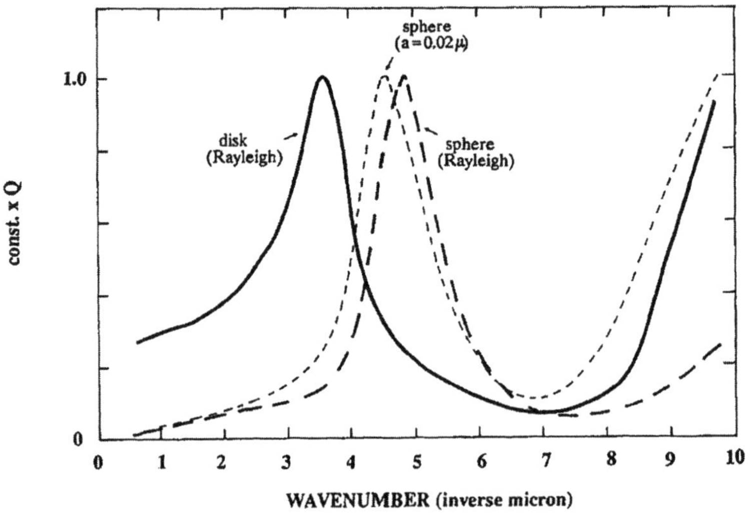

Figure 6. Calculation of extinction curve for graphite in the form of Rayleigh scattering discs and spheres, compared with that for a sphere of radius 0.02 μm (Wickramasinghe et al., 1992).

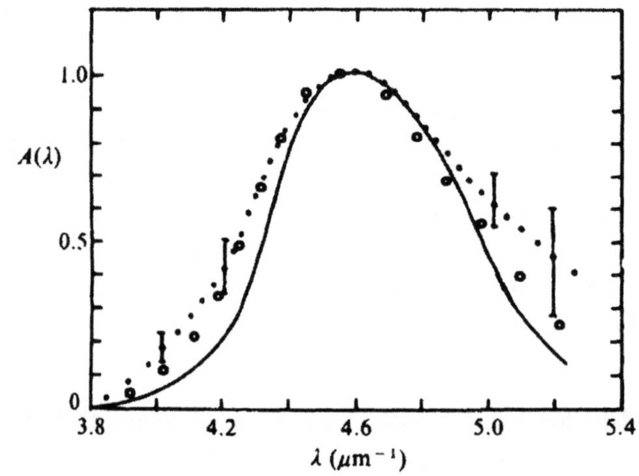

Figure 7. Normalised molar absorptivity of $C_8H_6N_2$ isomers of a bicyclic aromatic hydrocarbon compared with astronomical extinction data (Hoyle and Wickramasinghe, 1977b).

of bicyclic aromatic compounds in the mid 1970's (Hoyle and Wickramasinghe, 1977b). The fit of the astronomical data to our polyaromatic hydrocarbon model based on $C_8H_6N_2$ isomers is shown in Figure 7, the model requiring a large fraction of the carbon in the interstellar medium to be tied up in the form of such molecules. It should be stated that this was the first suggestion in the literature of polyaromatic hydrocarbons (PAHs) in interstellar space, a model that has come increasingly into vogue in recent years. Such molecules are unlikely to form *in situ* from the gas

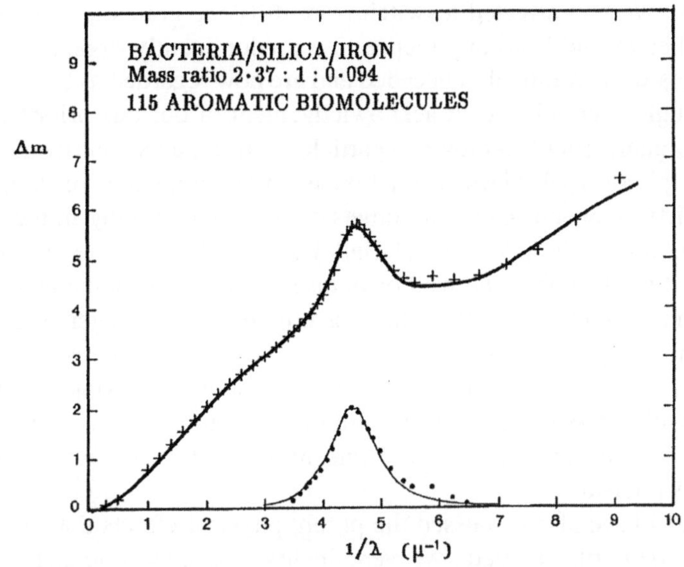

Figure 8. The theoretical extinction curve for a mixture of dehydrated bacteria and an ensemble of biological aromatic molecules. A small admixture of silica and iron is included for fine-tuning, but the fit is not sensitive to this assumption (See Hoyle and Wickramasinghe, 1991 for details).

phase under interstellar conditions, so we considered it likely that they are the result of degradation of more complex organic structures that condensed elsewhere.

After we had considered other evidence for bacterial particles in space we thought it most likely that the aromatic molecules giving rise to this feature are the result of degradation of biomaterial under interstellar conditions.

The curve in Figure 8 shows the extinction behaviour for freeze-dried bacteria (with trace quantities of silicon and iron for fine-tuning) and including 115 biological aromatic compounds resulting from bacterial degradation. The points (crosses) are the astronomical observations for the total extinction of starlight. The curve inset shows the contribution of pure absorption by the aromatics compared to the 2175A absorption excess (Wickramasinghe et al., 1989, 1990; Hoyle and Wickramasinghe, 1991). The data points in Figure 8 are a compilation of astronomical extinction data by Sapar and Kuusik (1978).

The biological ensemble of aromatic molecules responsible for ultraviolet extinction near 2175A also possesses characteristic infrared absorptions that have been observed in many astronomical sources (Wickramasinghe et al., 1990, 2002b).

6. Dust in the Solar System

In earlier sections I showed how my collaborations with Fred Hoyle led in rapid succession to two major paradigm shifts in relation to the composition of interstellar grains. We first effected a switch from dirty ice grains to refractory grains (graphite, silicates and iron) and thence to a predominantly organic composition. These changes were eventually conceded and are now regarded as being well established, although often with scant acknowledgement of our due priorities. The last move from organic grains to biogenic particles still remains controversial. Parallel changes of opinion took place in relation to the composition of cometary dust. Whipple's dirty snowball model of comets came under scrutiny in the 1970's and following the observations of Comet Halley in 1986 an organic composition for cometary dust seems now to be well recognised (Wickramasinghe et al., 1988a,b). The origin and dispersal of life in the cosmos through the agency of comets is one of the most important scientific applications of the role of cosmic dust (Hoyle and Wickramasinghe, 1981). Our original suggestion that comets delivered the chemical building blocks of life to the Earth tends now to be accepted, although the concept of fully fledged bacteria being introduced in this way is still looked upon with scepticism.

Hoyle and I have also discussed the purely physical effects that cometary dust may have on the Earth's climate. We were the first to connect cometary interactions with mass extinctions of species such as happened at the end of the Cretaceous period some 65 million years ago (Hoyle and Wickramasinghe, 1978). We have linked comets to the onset and terminations of ice ages, and also to more gentle modulations of terrestrial climate (Hoyle and Wickramasinghe, 1978, 2001; Hoyle et al., 1989).

7. Dust as Thermalisers of Galactic Radiation

The role of dust in thermalising stellar radiation at visual and ultraviolet wavelengths was also recognised at an early date (Wickramasinghe, 1970; Okuda and Wickramasinghe, 1970; Rees et al., 1969). Such processes were important in the context of new data from infrared astronomy that was becoming available in the 1970's. Infrared radiation observed from entire galaxies such as the Seyfrert galaxies and Starburst galaxies all had natural explanations in terms of dust models.

A black body with unit absorption and re-emission efficiencies at all wavelengths when placed at a typical location (in the solar vicinity) of intersellar space takes up a temperature close to 3 K. For particles of radii $\sim$0.1 μm, however, the average efficiency factor for absorption of optical and UV radiation, defined by $Q_{abs}(\lambda_{visual})$, is of order unity for most grain materials of interest, whereas the corresponding absorption (and emitting) efficiency $Q_{abs}(\lambda_{infrared})$, at an effective re-emission wavelength, is in general very much less than unity. Accordingly for a standard model of an interstellar grain of submicron size the temperature rises due 'greenhouse heating' by a factor of the general order $[Q_{abs}(\lambda_{visual})/\ Q_{abs}(\lambda_{infrared})^{1/4}]$, which leads to temperatures in the range $\sim 15 - 30$ K.

TABLE I
Energy density of some astrophysical processes

Energy density (erg cm^{-3})	Phenomenon
4×10^{-13}	Microwave background (Universe)
6×10^{-13}	H → He conversion (Universe)
$\sim 8 \times 10^{-13}$	Starlight (Galaxy)
$\sim 3 \times 10^{-12}$	Cosmic Rays (Galaxy)

Ever since the discovery of the cosmic microwave background by Penzias and Wilson (1965) Fred Hoyle became intrigued by the coincidences of energy density that existed between several astrophysical phenomena. These coincidences are illustrated in Table I.

Each of these energy density values corresponds to a black-body temperature close to 3K. The possibility that all these seemingly distinct sets of phenomena are inter-linked is attractive for the reason that the cosmic microwave background would then be connected with ongoing astrophysical processes. The crucial question, however, related to how the optical energy output of stars powered through H → He conversion could be degraded to microwaves and millimetre wave radiation. We have already noted that ordinary interstellar grains take up a much higher equilibrium temperature than 3K under interstellar and intergalactic conditions. Over the period 1968–1980 we made several attempts to invoke grain models that were somehow contrived to possess ratios $Q_{abs}(\lambda_{visual})/ Q_{abs}(\lambda_{infrared})$ of the order unity, or exceptionally high emissivities over a sub-millimetre waveband. These included considerations of the possible effects of weakly bound impurities in grains (Hoyle and Wickramasinghe, 1967), of solid hydrogen grains (Hoyle et al., 1968) and graphite needles (Hoyle et al., 1984). But on the whole these initial attempts met with little success in explaining the isotropic 2.7 K cosmic microwave background radiation. Fred Hoyle's confidence that Nature will always find a way to produce thermalisation of high grade radiation led us eventually to consider the role of cosmic iron whiskers (Hoyle and Wickramasinghe, 1988). Slender iron whiskers of cross-sectional radius 0.01μm and length ~ 1 mm, when their cryogenic optical properties were correctly calculated, yielded high opacities at the required wavelengths, without excessively high opacities in the optical spectrum (Wickramasinghe and Hoyle, 1994). We had argued that iron whiskers must form naturally in ejecta of supernovae following explosive nucleosynthesis resulting in the synthesis of iron. Such grain models were shown to explain the cosmic microwave background without requiring a contribution from a hot initial phase of a Big Bang Universe (Arp et al., 1990).

To conclude we note that grains have acquired an increasingly important role in astronomy over the past 40 years. They were elevated from their original rather negative role as mere extinguishers or starlight to many other crucial roles relating to the formation of molecules, stars, planets and even life. As we have just noted they may also be responsible for producing the cosmic microwave background, in which case their cosmological relevance cannot be ignored.

References

Allen, D.A. and Wickramasinghe, D.T.: 1981, *Nature,* **294**, 239.

Al-Mufti, S.: 1994, *PhD thesis*, University College, Cardiff.

Cooke, A. and Wickramasinghe, N.C.: 1977, *Astrophys. Sp. Sci.* **50**, 43.

Danielson, R.E., Woolf, N.J. and Gaustad, J.E.: 1965, *Astrophys. J.* **141**, 116.

Forrest, W.J., Gillett, F.C. and Stein, W.A.: 1975, *Astrophys. J.* **192**, 351.

Forrest, W.J., Houck, J.R. and Reed, R.A.: 1976, *Astrophys. J.* **208**, L133.

Hoyle, F. and Wickramasinghe, N.C.: 1962, *MNRAS* **124**, 417.

Hoyle, F. and Wickramasinghe, N.C.: 1967, *Nature* **214**, 969.

Hoyle, F. and Wickramasinghe, N.C.: 1968, *Nature* **217**, 415.

Hoyle, F. and Wickramasinghe, N.C.: 1969, *Nature* **223**, 459.

Hoyle, F. and Wickramasinghe, N.C.: 1970, *Nature* **226**, 62.

Hoyle, F. and Wickramasinghe, N.C.: 1976, *Nature* **264**, 45.

Hoyle, F. and Wickramasinghe, N.C.: 1977a, *Nature* **268**, 610.

Hoyle, F. and Wickramasinghe, N.C.: 1977b, *Nature* **270**, 323.

Hoyle, F. and Wickramasinghe, N.C., 1978, *Astrophys. Sp. Sci.* **53**, 523.

Hoyle, F. and Wickramasinghe, N.C., 1979, *Astrophys. Sp. Sci.* **66**, 77.

Hoyle, F. and Wickramasinghe, N.C.: 1980a, *Astrophys. Sp. Sci.* **68**, 499.

Hoyle, F. and Wickramasinghe, N.C.: 1980b, *Astrophys. Sp. Sci.* **69**, 511.

Hoyle, F. and Wickramasinghe, N.C.: 1980c, *Astrophys. Sp. Sci.* **72**, 183.

Hoyle, F. and Wickramasinghe, N.C., 1981, *Proofs the Life is Cosmic*, Inst. Fund Studies, Sri Lanka, Mem. 1.

Hoyle, F. and Wickramasinghe, N.C.: 1983, *Nature* **305**, 161.

Hoyle, F. and Wickramasinghe, N.C.: 1984, *From Grains to Bacteria*, Universtiy College Cardiff Press.

Hoyle, F. and Wickramasinghe, N.C., 1988, *Astrophys. Sp. Sci.* **147**, 245.

Hoyle, F. and Wickramasinghe, N.C., 1989, *Astrophys. Sp. Sci.* **154**, 143.

Hoyle, F. and Wickramasinghe, N.C.: 1991, *The Theory of Cosmic Grains*, Kluwer Academic Publishers.

Hoyle, F. and Wickramasinghe, N.C.: 1996, *Astrophys. Sp. Sci.* **235**, 343.

Hoyle, F. and Wickramasinghe, N.C.: 2001, *Astrophys. Sp. Sci.* **275**, 367.

Hoyle, F. and Wickramasinghe, N.C. and Reddish, V.C.: 1968, *Nature* **218**, 1124.

Hoyle, F., Wickramasinghe, N.C. and Al-Mufti, S.: 1982a, *Astrophys. Sp. Sci.* **86**, 63.

Hoyle, F., Wickramasinghe, N.C., Al-Mufti, S., Olavesen, A.H. and Wickramasinghe, D.T.: 1982b, *Astrophys. Sp. Sci.* **83**, 405.

Hoyle, F., Wickramasinghe, N.C., Al-Mufti, S. and Olavesen, A.H.: 1982c, *Astrophys. Sp. Sci.* **81**, 489.

Hoyle, F., Wickramasinghe, N.C. and Jabir, N.: 1983, *Astrophys. Sp. Sci.* **92**, 439.

Hoyle, F., Narlikar, J.V and Wickramasinghe, N.C.: 1984, *Astrophys. Sp. Sci.* **103**, 371.

Hulst, van de, H.C.: 1946–49, *Rech. Astron. Obs. Utrecht* **XI**, Parts I and II.

Knacke, R.F., Cudaback, D.D. and Gaustad, J.E.: 1969, *Astrophys. J.* **158**, 151.

Nandy, K.: 1964, *Publ. Roy. Obs. Edin* **3**, 142.

Okuda, H. and Wicckramasinghe, N.C: 1970, *Nature* **226**, 134.

Oort, J.H. and van de Hulst, H.C., 1946, *BAN* **376**.
Penzias, A.A. and Wilson, A.W.: 1965, *Astrophys. J.* **142**, 419.
Rees, M.J., Silk, J.I., Werner, M.W. and Wickramasinghe, N.C..: 1969, *Nature* **223**, 778.
Sapar, A. and Kuusik, I.: 1978, *Publ. Tartu Astr. Obs.* **46**, 717.
Stecher, T.P.: 1965, *Astrophys. J.* **142**, 1683.
Stecher, T.P. and Donn, B.D.: 1965, *Astrophys. J.* **142**, 1681.
Wickramasinghe, N.C.: 1967, *Interstellar Grains*, Chapman & Hall, London.
Wickramasinghe, N.C.: 1970, *Nature Phys. Sci.* **197**, 230.
Wickramasinghe, N.C.: 1974a, *Nature* **252**, 462.
Wickramasinghe, N.C.: 1974b, *MNRAS* **170**, 11P.
Wickramasinghe, N.C. and Guillaume, C.: 1965, *Nature* **207**, 366.
Wickramasinghe, N.C. and Hoyle, F.: 1994, *Astrophys. Sp. Sci.* **213,** 143.
Wickramasinghe, N.C., Hoyle, F., Brooks, J. and Shaw, G.: YEAR?, *Nature*, **269**, 674.
Wickramasinghe, N.C., Hoyle, F., Wallis, M.K. and Al-Mufti, S: 1988a, *Earth, Moon and Planets* **40**, 101.
Wickramasinghe, N.C., Wallis, M.K and Hoyle, F.: 1988b, *Earth, Moon and Planets* **43**, 145.
Wickramasinghe, N.C., Hoyle, F. and Al-Jabori, T.: 1989, *Astrophys. Sp. Sci.* **158,** 135.
Wickramasinghe, N.C., Hoyle, F. and Al-Jabori, T.: 1990, *Astrophys. Sp. Sci.* **166**, 333.
Wickramasinghe, N.C., Wickramasinghe, A.N. and Hoyle, F.: 1992, *Astrophys. Sp. Sci.* **196**, 167.
Wickramasinghe, N.C., Lloyd, D. and Wickramasinghe, J.T.: 2002, *Proc. SPIE* **4495**, 255.
Woolf, N.J. and Ney, E.P.: 1969, *Astrophys J. Letters* **155**, L181.

PART V: COMETS

GIANT COMETS AND HUMAN CULTURE

W.M. NAPIER
Armagh Observatory, College Hill, Armagh BT61 9DG, Northern Ireland
University of Cardiff

Abstract. Our understanding of Earth history on geological timescales is in a state of flux due to the realisation in recent years that the Earth is a bombarded planet. However recent findings suggest that a celestial input may also be important on much shorter timescales. Apart from the current hazard, the night sky in prehistoric times appears to have been dominated by a giant, short-period comet and its debris, and interactions with this material may have had a strong influence on human cultural development.

1. Introduction

The perception of the Earth as a bombarded planet is a very old one, and hints of past cometary and Tunguska-like events can be seen even in the earliest cosmogonic myths and artefacts in Europe and the Middle East. These describe a world in which gods strode the heavens and sometimes brought destruction to Earth below by hurling thunderbolts or burning stars to the ground, knocking down forests and setting the land alight. Impact motifs are clearly seen, for example, in Hesiod's Theogony (800 BC), and appear to be associated with a comet visible in the sky. Around 600 BC a change of perception seems to have taken place amongst European philosophers (van der Waerden, 1974; Clube and Napier, 1990). Plato (c.600 BC), in his Timaeus, could describe periodic destructions of the Earth by the return at long intervals of fire from the sky, but to his pupil Aristotle, comets were no more than sub-lunar, 'windy' exhalations. Even so, hints of a much older sky, inhabited by rampaging gods, continue to be seen in (for example) the astral prophecies of Revelation, the Sybilline Oracles, the writings of Seneca, Nonnos (4th century AD) and so on. These are derivative writings based on much older sources. Whatever problems some modern astronomers might have in interpreting these texts as literal descriptions of celestial disturbance, there was no doubt in the minds of the Romantic and other artists who depicted them (see also Malina, 1995): engravings, woodcuts and paintings from the 15th and 16th centuries routinely show the opening of the Fifth and Sixth Seals, or the falling of the burning star wormwood, as falling bolides or meteor storms in a comet-dominated sky.

However by mediaeval times these were miracles: the Aristotelean perception of the celestial vault as unchanging had become dominant. Thus the astronomical writings of Omar Khayyam in Arabia and Dante in Europe describe a sky whose planetary cycles are known with meticulous precision, but which is merely an unchanging backdrop to affairs on Earth. This uniformitarian concept lasted 2000

Astrophysics and Space Science **285:** 513–520, 2003.

years, from Aristotle to the supernova of 1570 in Cassiopeia. The appearance of that supernova, and the arrival of the Great Comet of 1577, along with Tycho's demonstration that the latter was moving a path lying beyond the moon, threw doubt on the traditional cosmology. And with the Newtonian revolution, it became possible for Whiston, Halley and other scholars throughout the 17th and 18th centuries, to contemplate once again the hugely destructive effects of a great comet colliding with the Earth as a scientific proposition.

By the mid-19th century, however, scientific trends were making a catastrophist worldview once again untenable. The geologists were finding that the Earth was shaped, not by great catastrophes, but by gradual, longacting forces. Gradual evolution was the great idea of the Victorian period (although as early as 1860 the Oxford geologist Philips drew attention to major biotic crises – mass extinctions – recorded in the fossil record). And the astronomers were finding that comets are small and insignificant bodies which, when they disintegrate, yield nothing more disturbing than meteor showers. This uniformitarian perspective lasted well into the 20th century, and only within the last 30 years has a reappraisal been under way.

To be sure, suggestions were made from time to time throughout the 20th century that 'giant meteorites' might have had damaging effects on Earth and even been responsible for the great mass extinctions which were seen in the fossil record. Lacking knowledge of impact rates, however, the stray impact hypothesis had to take its place beside many other speculative scenarios involving celestial hazards: solar flares, supernovae and passage through dense nebulae had all been speculatively associated with mass extinctions or the onset of ice ages.

A new perspective began to emerge in the 1980s, having its roots in discoveries made in the previous decade. Helin and Shoemaker (1979) undertook a research programme to discover small, Earth-crossing asteroids and found that they exist in abundance. Grieve and Dence (1979) derived terrestrial impact rates from the increasing number of impact craters which were then being found, partly through satellite imagery. The radio astronomers discovered and mapped out the molecular cloud system of the Galaxy. The significance of the latter is that a single close encounter with a giant molecular cloud exerts sufficient gravitational stress on the Oort comet cloud to send a shower of comets towards the inner planetary system and a bombardment episode on the Earth.

These new findings led Napier and Clube (1979) to propose that mass extinctions of life, and other geological phenomena, might be due to episodes of bombardment by comets and their degassed remnants, these episodes having a periodic nature due to the Sun's motion around the Galaxy. Emphasis was put on large comets as the major destroyers, because of their mass. Independently, Hoyle and Wickramasinghe (1978) proposed that if the Earth were to pass through the coma of a large comet, the stratosphere would be blanketed by cometary dust, cutting off sunlight, collapsing food chains and so causing a mass extinction of life.

From the 1980s onwards, it is probably fair to say that a revolution has been under way in the Earth sciences. The discovery by Alvarez et al. (1980) of iridium, of probable extraterrestrial origin, at the CretaceousTertiary boundary was interpreted by them in terms of a 10 km asteroid impact. Dust thrown upwards from the impact was supposed to cut out sunlight and reduce photosynthesis, as originally suggested by Napier and Clube (1979) for multiple impacts, and Hoyle and Wickramasinghe (1978) for close encounter with a large comet. Although the 10 km impact is a wonderful 'Hollywood spectacular', much favoured by science journalists and derivative writers, the evidence seems to be favouring more prolonged disturbance (Keller 2000, 2002), probably induced by cometary dusting and multiple impacts along the lines originally suggested by Napier and Clube (1979). However the causes of the great mass extinctions, and the nature and role of celestial inputs, remain controversial issues, as is the question of whether there is indeed a Galactic modulation of impact rates.

But what of events on human rather than geological timescales? Could we, for example, regard the Holocene as a microcosm of geological time writ small, in which impacts and cometary dustings still have dramatic effects? (See the Table.) Is there, even, physical evidence that the ancient 'star wars' recorded in myth and artefact are not simply artistic fancy but do indeed include depictions of real events? This is an area where angels fear to tread, in part due to the legacy left by Velikovsky (1950). Nevertheless the questions are legitimate, and attempts to address them are now being made by workers in various disciplines (Peiser et al., 1998).

A 'Guinness Book of Records' for the near-Earth environment.

Most massive discrete bodies
Rare, giant comets. Their mass influx dominates all other sources. Archetype, Chiron, is 180 km across. cf the simultaneous disintegration of 8,000 Halley comets!
Most massive single entity
The zodiacad cloud: a flattish system of dust and boulders. Current mass $\sim 10^{19}$ g.
Largest target area
Short lived dust trails. Associated with active, short period comets. Intersection timescales about a century, duration about an hour. Meteor storms, often seen to signal 'the end of the world'.
Greatest global impact hazard
About 1200 near-Earth asteroids one to 10 km across. Recurrence time between impacts about 300,000 years (to within a factor of two).

2. A Changing Sky over Holocene Timescales

As recently as 1980 it was widely held that one Tunguska-like impact, having only local effects, would take place on Earth every 2000 years or so, leading Brandt and Chapman (1982) to state that 'no one should lose much sleep over this situation even if agitated by uninformed doomsayers'. However it was all also possible for two 'uninformed doomsayers' (Clube and Napier, 1980) to state: 'a few dozen sporadic impacts in the tens of megatons, and a few in the 100 to 1000 megatons range, must have occurred within the past 5000 years'. The higher rates were claimed by Napier and Clube (1979) by the crudest of arguments: the size distribution of large terrestrial impact craters was extrapolated down to small sizes, and a simple energy-diameter relation was applied. Nevertheless these rates have been broadly confirmed over the subsequent 20 years by telescopic searches, counts of small lunar craters, military listening devices and so on. They do, however, lead to a paradox: where are these Tunguska and super-Tunguska impacts in the historical record?

This situation becomes even more paradoxical when one looks at the near Earth environment in more detail. The only active comet known to occupy a typical Apollo asteroid-like orbit is the faint telescopic comet studied by Encke in 1818. It has the shortest known period of any active comet (3.3 years), a low inclination (12 degrees) and a high eccentricity (0.847). Since its discovery two centuries ago the comet's period has been shortening by 0.1 days per revolution, due to asymmetric outgassing. Encke's comet is embedded in a complex of four meteor streams. An exceptionally large dust trail in a similar orbit has been detected in the infrared, while several of the larger known Apollo asteroids are also in orbits closely resembling that of the comet. These asteroids, the comet and the Taurid meteor streams are all embedded in a broad tube of meteoroidal material from which about half the sporadic meteors derive. This stream appears to be a bridge between the Taurids and the zodiacal cloud. The present-day comet is much too small to be the progenitor of this complex of bodies. Probably, it is simply one of a co-moving group of asteroids which has reactivated within the last two centuries.

The Taurid complex is most easily understood as the disintegration product of a very large comet. It would seem that at least half the mass of the zodiacal cloud up to 200 micron particles derives from an erstwhile giant object. Since, without replenishment, mutual collisions and radiative effects would remove the cloud within 20,000 years, it is likely that the night sky has for (say) 20,000 years or more been dominated by this giant progenitor and its debris and associated meteor streams.

3. Fluctuations in the Impact Rates

Numerical simulations by Hahn and Bailey (1992) indicate that Chiron, a comet 180 km across, is currently in an unstable orbit between Saturn and Uranus. It has probably been in a short period, Earth-crossing orbit several times in the last million years, each such epoch lasting for a few tens of thousands of years. If broken into pieces 10 km across, Chiron would yield 8000 Halley-sized comets hitting the Earth at mean intervals of half a million years, each impact being at the level associated with the great mass extinctions. If broken into pieces 50 m across, Tunguska-like fragments would strike the Earth every two or three weeks! These figures illustrate that surges of strong bombardment may occur when a large disintegrating comet enters the environment of the Earth. Orbital precession would ensure that, at about 2500 year intervals, the Earth runs into epochs of hugely enhanced fireball flux and dust inputs, of characteristic duration a century or two. Impact rates could then reach $\sim 10^3$ times the mean long-term value without violating lunar cratering data.

The importance of such multiple impacts in the interplanetary system is illustrated by the Kreutz sun-grazing comet group, which derives from the breakup of a giant comet perhaps 100 km across, 20,000 years ago. Observations by the SOHO satellite show that these comet fragments hit the Sun about once a week. This may be compared with the theoretically estimated impact rate, based on a random influx of comets from the Oort cloud, of one such impact per century. Likewise when comet Shoemaker-Levy 9 passed close to Jupiter in 1992, it splits into 20 pieces which then went on to collide with Jupiter over a period of about a week. As pointed out by Steel, such a high impact rate would never result from random impacts. Thus structure in the form of streams or concentrations of material, deriving from the breakup of a large comet, leads to a qualitatively different perspective on the nature of the impact hazard.

If Chiron or one of its offshoots was indeed visible in the prehistoric night sky, then that sky contained one or more bright, visible comets regularly seen. The major bodies would be associated with annual meteor and fireball storms of great intensity, and with occasional Tunguska impacts.

The Taurid meteor stream is broad, suggesting an old stream. Differential precession of Taurid meteors would yield a dispersion $\delta\varpi$ $\sim 90°$ in the longitude of perihelion ϖ of the Taurid meteors within 20,000 years (Steel and Asher, 1996), whereas the current two to three-week duration of the Beta Taurids suggests a major episode of meteor formation as recently as 1000–2600 BC. This takes us into the epoch of pyramid and megalith building, and prehistoric myth-making. It is scarcely credible that a disintegration yielding a major meteor shower from a comet with a period of $\sim 3.5 \pm 0.2$ yr could have taken place with the comet itself invisible. In addition to a highly active and alarming night sky, in which the gods tracked along the Zodiac, the zodiacal light itself would have been substantially brighter.

It is likely that the current Taurids were formed from a recently created fragment of the initial giant body. Orbital backtracking of known Taurid complex asteroids by Steel and Asher (1996) indicates that their observed dispersion is consistent

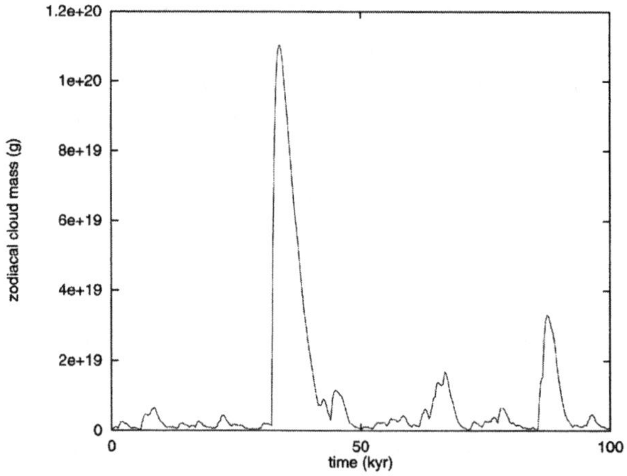

Figure 1. Mass of zodiacal cloud over 100,000 yr for a stochastic simulation in which comets arrive randomly in Encke-like orbits and disintegrate under the influence of sunlight, with the dust released undergoing collisional grinding with other zodiacal cloud material, Poynting-Robertson decay into the Sun or radiative ejection from the planetary system. Details of the model are in Napier (2001). Current mass is $\sim 10^{19}$ g for particles up to 100 g.

with the disintegration of a common giant progenitor 20,000–30,000 years ago, in the depths of the last ice age, and also within the time of existence of homo sapiens.

4. Climatic Downturns and Cometary Dustings

Going from timescales of 5 or 10,000 years to ones of 50 or 100,000 years, one finds the mass of the zodiacal cloud to be highly variable. This is illustrated in Figure 1, showing a simulation wherein comets are randomly injected into the inner planetary system and disintegrate, their dust being removed by mutual collisions and radiative effects (Napier, 2001a). Such trials have revealed that surges of dust influx to the Earth, two or more powers of ten higher than the mean background, may occur. These surges are of millennial duration or longer and may result in significant optical depths being built up in the atmosphere (Clube and Napier, 1984; Napier and Clube, 1997; Napier, 2001a,b). In that situation exhaustion of the oceanic heat buffer reduces the convective transfer of heat to the stratosphere (through dumping latent heat of evaporation of water) to the point where permanent ice crystals form at 25 km altitude and above. Thus when the comet dust disappears, the Earth is enveloped in a permanent blanket of scattering ice and the Earth is plunged into an ice age (Clube et al., 1996; Hoyle and Wickramasinghe, 2001). Arguments have been presented to suggest that, plausibly, the most recent ice age was caused by the last incursion of Chiron into the near-Earth environment (loc. cit.).

Less extreme but more frequent sharp cooling events have been recorded throughout the Holocene and earlier. According to Baillie (1999), tree-ring records reveal that global coolings have taken place on five occasions in the last 5,000 yr, namely 2354–2345 BC, 1628–1623 BC, 1159–1141 BC, 208–204 BC and 536–545 AD. The latter event overlaps with historical records and can be attributed to a worldwide dry fog of about 18 months' duration. Several of these events have been linked by various authors to influxes of cometary debris (Peiser et al., 1998). Baillie (1999), for example, has argued from the contemporaneous records that the ~540 AD dusting event, and possibly several others, had a cosmogenic rather than volcanic origin. A sudden cooling event of this magnitude, taking place in modern times, would have a calamitous effect on food production worldwide.

5. Conclusions

Fred Hoyle, with his colleagues, recognised that rare, giant comets are prolific dust machines. He and Wickramasinghe were the first to realise that over geological timescales, high-altitude dusting from such comets may have catastrophic biological effects; he saw such cometary dustings as possible triggers of ice ages; and he amongst others proposed that a recent giant, Earth-crossing comet may have played a major role in human cultural development (Hoyle, 1993).

These fields of study are only just beginning to open up. In all of them, Hoyle made distinctive and important contributions.

References

Alvarez, L.W., Alvarez, W., Asaro, F. and Michel, H.V.: 1980, *Science* **208,** 1095.

Baillie, M.: 1993, *From Exodus to Arthur,* Batsford, London.

Brandt, J.C. and Chapman, R.D.: 1982, *Introduction to Comets,* Cambridge University Press.

Clube, S.V.M. and Napier, W.M.: 1980, *The Cosmic Serpent*, Faber, London.

Clube, S.V.M. and Napier, W.M.: 1984, *MNRAS* **211,** 953.

Clube, S.V.M. and Napier, W.M.: 1990, *The Cosmic Winter*, Basil Blackwell, Oxford.

Grieve, R.A.F. and Dence, M.R.: 1979, *Icarus* **38,** 230.

Hahn, G. and Bailey, M.E.: 1990, *Nature* **348,** 132.

Helin, E.F. and Shoemaker, E.M.: 1979, *Icarus* **40,** 321.

Hoyle, F.: 1993, *The Origin of the Universe and the Origin of Religion*, Moyer Bell, New York.

Hoyle, F. and Wickramasinghe, N.C.: 1978, *Astrophys. Space Sci.* **53,** 523.

Hoyle, F. and Wickramasinghe, N.C.: 2001, *Astrophys. Space Sci.* **275,** 367.

Keller, G.: 2000, *Planet. Space Sci.* **49,** 817.

Keller, G.: 2002, *GSA Special Paper* **356**, in press.

Malina, B.J.: 1995, *On the Genre and Message of Revelation*, Hendrickson, Peabody, Massacussetts.

Napier, W.M.: 2001a, *MNRAS* **321,** 463.

Napier, W.M.: 2001b, in:B. Peucker-Ehrenbrink and B. Schmitz (eds.), *Accretion of Extraterrestrial Matter Throughout Earth's History*, Kluwer Academic, Dordrecht.

Napier, W.M. and Clube, S.V.M.: 1979, *Nature* **282,** 455.

Napier, W.M. and Clube, S.V.M.: 1997, *Rept. Prog. Physics* **60**, 293.
Peiser, B.J., Palmer, T. and Bailey, M.E. (eds.): 1998, *Natural Catastrophes During Bronze Age Civilisations*, BAR International Series No. 728.
Steel, D.I. and Asher, D.J.: 1996, *MNRAS* **280,** 806.
van der Waerden, B.L.: 1974, *Science Awakening II: the Birth of Astronomy*, Noordhof, Leiden.
Velikovsky, I.: 1950, *Worlds in Collision*, Victor Gollancz, London.

AN EXCEPTIONAL COSMIC INFLUENCE AND ITS BEARING ON THE EVOLUTION OF HUMAN CULTURE AS EVIDENT IN THE APPARENT EARLY DEVELOPMENT OF MATHEMATICS AND ASTRONOMY

S.V.M. CLUBE

Armagh Observatory, Northern Ireland

Abstract. The proposal that cometary dust particles play a significant role in the emergence and evolution of both life and disease on suitable planets was first made some 25 years ago by Hoyle and Wickramasinghe. Fundamental to this proposal was a process of punctuated seeding by particular (bio)chemical species believed to originate naturally and predominantly in larger comets, say those with diameters greater than about 100 kilometres. Rather less well known is a parallel proposal likewise favoured by Hoyle that a particular giant comet, the most recent to settle in cis-Jovian space, accounting for the latest significant phase of evolution on Earth, also had a significant part to play in the cultivation by homo sapiens of its civilization and culture. Such proposals may be seen by many as examples of excessively lateral thinking but they by no means lack independent support and have important implications for the otherwise uncertain origin of the latest ice-age (basic to climatology) and for the otherwise uncertain generation of early calendars (basic to the management of society). Aspects of these proposals are considered here in relation to a much respected supposedly Chaldean calendar probably passed down by the dynastic Isins (also the Essenes?) which evidently bears witness to known early mathematical and astronomical skills but which largely ceased to be available to subsequent scholarship beyond the Early Christians (ca. 100 CE) pending its (recent) recovery through the medium of Dead Sea Scrolls.

Keywords: giant comets, climatic cycles, ancient calendars, civilization

1. Introduction

It has long been recognised that too few ordinary comets arrive in the inner Solar System to account for the observed zodiacal dust (Whipple, 1967). It is reasonable therefore to seek to explain this dust in terms of the collisional fragmentation and erosion of sub-kilometre meteoroids in orbits close to the ecliptic. An explanation of this kind does not exclude objects in the asteroid belt (as well as those displaced as meteorites to the Earth) and some scientists have therefore considered these to be the likely primary source of zodiacal dust. On the other hand, it is more generally supposed that the so-called sporadic meteor flux is representative of this primary source and it is now well known that this flux is dominated by a prominent elliptical stream in cis-Jovian space close to the ecliptic whose helion (HE) flow intercepted by the Earth, as opposed to its antihelion (AH) flow, is observed to be coming broadly from the directions of the Gemini-Taurus-Aries constellations, that coming from the two leading constellations being for the most part less strongly comminuted (Stohl, 1983). Taken as a whole, these overall features of arguably the most momentous of the night sky's most fleeting aspect may now be regarded

Astrophysics and Space Science **285:** 521–532, 2003.

as bearing witness to an originally massive Piscean HE meteor flow close to the ecliptic whose correspondingly massive cometary source having undergone some kind of relatively modest diversion and wholesale fragmentation some five thousand years ago has subsequently fanned out under the influence of (differential) precession so as to produce the remaining so-called Taurid-Arietid complex. Since 1975, there has also been growing evidence for a massive meteoroidal swarm at the heart of the Taurid-Arietid complex which is in the 7:2 Jovian mean motion resonance; whence it is likely that the original cometary source, a probable giant comet, was itself 'captured' into cis-Jovian space under the control of the 7:2 Jovian mean motion resonance.

Accordingly we may envisage a recent cis-Jovian giant comet whose initial orbital parameters (a_0, e_0; t_0) caused its line of apsides to precess some 180° around the ecliptic until a modest planetary encounter involving tidal disruption (attributed to Mercury: see below) caused its evolved orbital parameters (a_1, e_1; t_1) to undergo perturbation within the resonance (Δa, Δe), thereafter many of its fragmentation products undergoing precession outside the resonance at a divergent rate. Assuming $t_0 = 70$ ka BP and $t_1 = 5.15$ ka BP, it follows that the inferred giant comet along with its comminution products including the zodiacal dust have a physico-dynamical timescale comparable to that of the Upper Pleistocene within that of homo sapiens (cf. Wenke, 1980). Such events as terminate the middle and upper paleolithic ($\sim$40 ka BP and $\sim$10 ka BP respectively), within the duration of the latest ice-age ($\sim$70–10 ka BP), and such as terminate the mesolithic and neolithic prior to the historical period of human civilization and culture, within the duration of the Holocene ($\sim$10–0 ka BP), are all recognised therefore as occurring within the rapid evolutionary lifespan of a specific giant comet whose remnants (both large and small) are essentially still with us.

A giant comet which is no longer visible because, like our remote ancestors, it has ceased to function as such and has largely disintegrated into dust is of course as speculative as its implied (but measured) correlations within the realms of astrophysics, geophysics, (bio)physics, (bio)chemistry, neurology and sociology – to name but a few of the primary and derivative scientific disciplines involved. Nevertheless the essential astrophysical evidence has never been seriously challenged by comet experts while at least one with recognized expertise in the formulation of holistic models of the universe (Hoyle, 1993) has neatly encapsulated some of the essential logic in readily comprehensible terms, noting that 'what Clube and Napier [have proposed (1990)] is that some [70,000] years ago the process happened for a comet that was also unusual in being far more massive than Halley's, thus adding a third exceptional circumstance to the other two. My first impression on encountering this proposal was that a third exceptional circumstance was one too many... Then I saw that the answer to this question lies in what is now called the anthropic principle, which says that the fact of our existence can be used to discount all improbabilities necessary for our existence. If history and civilization were caused by the arrival of a periodic giant comet all accident is removed from our association with such a comet'. This indeed is precisely the underlying theme here and it behoves us therefore to understand in some detail how humanity has responded and how civilization and culture have emerged in this comet's presence.

2. The History of 'Proto-Encke'

In accordance with this proposed history, it appears that Comet 2P/Encke (first detected in 1786 and since shown to be a member of the Taurid-Arietid stream) ultimately derives from a massive progenitor in a closely related orbit which (prior to its disruption at $t = t_1$) is likely (i) to have originated in cis-Jovian space at $t = t_0$ with somewhat higher eccentricity ($e \sim 0.92$, say) as a result of Jovian 'capture' during a favourable apsidal alignment of the inner and outer planets (whence $\tilde{\omega} - \tilde{\omega}_J \sim 0$ and $\tilde{\omega} - \tilde{\omega}_E \sim 0$); and (ii) to have remained in coupled Jovian-terrestrial resonance (i.e. integer p, q) based on a modulated harmonic P_0 such that

$$pP_0 = qP_E$$

$$P_0 = \{(2/7\ P_J)^{-1} + P(\omega)^{-1} - P(\tilde{\omega})^{-1}\}^{-1}$$

thereby settling in an essentially Earth-avoiding configuration (cf Asher & Clube, 1993, 1998) which was exceptionally stable against further perturbation by the inner planets. One such coupled resonance (which remains to be more fully explored) involves a specific half-period of nodal procession ($1/2\ P(\omega) = 2520$ years) and its sub-harmonic ($1/24\ P(\omega) = 210$ years) for which (p, q) = (62, 210); and a specific half- period of libration ($1/2\ P_L = 7/8.pP_0$) along with alternating nodal intersections of the Earth's orbit at intevals of 210 and 2310 years through which the Earth-Moon system and the progenitor would have been in regular proximity thereby possibly sustaining the year, month, and day in Jovian resonance as well. Assuming such resonances for the orbital evolution of the proposed giant comet, it follows that the Earth will have experienced sequences of close encounters with dense regions of the meteor stream in the vicinity of the giant comet itself **and** in the vicinity of its resonant swarm of debris, respectively during one or other nodal intersection at regular intervals of $1/2P(\omega) = 2520$ years **and** during successive near approaches to the Earth's orbit by the libration centre at intervals in the region of $\{1/2P_L^{-1} - (pP_0)^{-1}\}^{-1} = 1470$ years. The Earth will evidently undergo massive inputs of cometary debris and dust at the frequencies of these intervals resulting in corresponding patterns of climatological change during the Upper Pleistocene and Holocene (cf. Bond et al., 1993, O'Brien et al., 1995, Bond et al., 1997). In effect it is already known that the Earth's history of global cooling - global warming during the period in question is strongly paced by these frequencies; and furthermore it is already known that these frequencies have continued to beat together to produce major ice-sheet growth – iceberg calving episodes (= Heinrich events) in possible harmony with an expected 7056 year periodicity. Indeed these frequencies in the recent pattern of global cooling-global warming have also achieved considerable prominence on account of their having so far defied any other kind of explanation; whence the conclusion may be forced upon us that the last ice-age and its subsequent interglacial can only have been induced by regular surges of cometary dust to Earth originating from the latest giant comet. We name this comet 'proto-Encke' in accordance with its most recent evolution and history during the Upper Holocene.

The inferred dynamics of proto-Encke permit a schematic formulation of history for the current glacial – interglacial, in particular a reverse sequence of paired nodal surges of cometary dust to Earth (ie first and second comings) at $T_n + 140$ and $T_n - 70$ for successive values of $T_n = 2520n$ BCE where $n = 0,1,2\ldots n_f$ ($n_f \sim 24$). These surges reflecting the evolutionary state of both proto-Encke and its massive meteor stream were apparently in serious decline by the end of the Upper Pleistocene corresponding to the Older and Younger Dryas cooling events ($n = 4,3$ respectively) albeit the remaining core of proto-Encke (still avoiding Earth and Venus) appears to have survived another surge ($n = 2$) before its probable close encounter with Mercury caused its tidal disruption and produced the embryonic Taurid-Arietid stream with its resultant broad correlation of a and e. Assuming this event occurred ca 3150 BCE, a reverse sequence of successively weakened surges at T_n (e.g. $n = \ldots -1,0,1$) due to the original remnant stream is still possible along with a new set of weakening bombardments (of debris) and surges (of dust) at $T_{n,s} = T_s + 2520n$ BCE, $n = \ldots -1,0,1$ due to various Taurid-Arietid substreams here designated s (both inside and outside the original resonance) which continue to diverge and fragment. While this model of proto-Encke's history must of course be regarded as schematic and essentially provisional at the present time, it should be noted that a proto-Encke trail within the 7:2 Jovian mean motion resonance (as observed) is not excluded nor is the supposed random encounter activating and releasing Comet 2P/Encke therefrom ca 1786. The new substreams created ca 3150 BCE along with the pre-existing remnant stream, when taken together as a whole, evidently give rise to a new pattern of apocalyptic encounters with our planet occurring first during 3150 BCE $< T <$ 630 BCE and then again as a similar but declining pattern of apocalypses during subsequent periods of 2520 years. History for example reveals similar junctures of global chaos and collapse at the ends of the Sumerian ($\sim$ 2000 BCE) and Roman ($\sim$500 CE) empires some 2520 years apart, these apparently corresponding to successive pairs of nodal intersections with the densest substream, namely the proto-Encke trail. History also reverberates with prophecied first and second comings which climax at the junctures between successive eras e.g. BCE, CE. There is scope for intellectual confusion here as between the original apocalyptic eras and a new creation, the displaced recurring pattern of apocalypse. Indeed by the middle of the first millennium BCE, the comminution of proto-Encke debris was already seriously advanced and the previously self-evident notion that history was unbounded and predictably cyclic was rapidly giving way to the more fanciful notion that history (including the material universe) was not only unrelentingly uniformitarian but bounded by symbolic 'creation' and 'endtime' events. This new but misguided level of sophistication in astronomical conceptualization (still by no means eliminated), rendering cometary debris both vacuous and harmless, marked a serious and fundamental break with earlier (more realistic) traditions.

The earliest Venusian form to be observed at the horizon during the hours of dusk and dawn was apparently not the planet so-called but an upright anthropomorphic body of altogether different appearance, almost certainly the original much intensified zodiacal cloud of the current glacial-interglacial. Presumably therefore the planet so-called would have been a post-glacial novelty which emerged alongside growing awareness of the Earth's cosmic environment as enhancements during the glacial of the zodiacal cloud **and** of the Earth's exospheric dust both substantially declined. The earliest post-glacial skywatchers seeking to provide a rational explanation of eclipses were bound then to have seen Venus and the Moon in basically similar roles ie as circulating attendants of the Sun and the Earth respectively – albeit without any secure basis for establishing the preferred centre, a situation no different to that confronting late medieval Europeans. The intervening situation was different however due to the evident anti-heliocentric tendencies of Hellenistic imperialism and its break with tradition which caused corporealism for the supposedly bounded and created cis-lunar region of space to be endorsed along with incorporealism for the supposedly unbounded and uncreated region beyond. Associated with Aristotle and the neoplatonists in the first instance, later with Philo and the philistines, and yet later again with such doctrinal heavyweights as Maimonides, Averroes and Aquinas, this prioritization of the immediate cosmic locality came to pervade a vast intellectual empire in the West during modern times, providing an essentially unscientific foundation for what now may be perceived as the unacceptable face of Indo-European elitism. Thus Aristotelianism is now well known to have invoked a seemingly logical but hopelessly unscientific theory of comets which initially left open the question whether these unique bodies were corporeal, incorporeal or subject to transubstantiation at the cis-lunar interface but which very soon was settled in favour of incorporealism. In particular, fundamental significance was conferred upon the one such body clearly known to have previously and conspicuously undergone heirarchical splitting and then to have intersected the cis-lunar interface, the effects being codified as 'emanations' displaying tripartite unity. These meteoroidal effects which Greek sophists took over as primary Assyrian and Egyptian inheritances meant corporealism was infested by incorporealism thus permitting an unconventional category of magical or occult agencies responsive to supplication and propitiation to enter secular thought as myth. Such thought, by now virtually endemic despite the late medieval re-emergence of science, has of course been entirely self-serving so far as the authorities to whom it most appealed were concerned. Thus a magical spirit (possibly governed by divine law) has been supposed to infuse the universe and manifest itself principally as a dominant adjunct of the 'fittest', hence as a prerogative not only of kings and cults but also of publically sanctioned individuals and institutions not to mention the supposedly directed evolution of biological species. More recently of course the possible existence of identifiable biological factors randomly underpinning such dominance has been widely contemplated but the false intellectual edifice based on neoplatonism has by no means been removed (cf Feibleman 1959), a principal victim still being the scientific understanding of comets (which has been notably slow to develop).

3. Civilization, Astronomy and Mathematics: The Early Calendar

Since the discovery of Comet 2P/Encke it has taken mankind all of two centuries to work out its possible implications and realise that Comet proto-Encke was probably once a dominant global force. Its dominance was such that the Sun and the Moon can hardly have been recognized other than as mere but worthy celestial partners while the planets would have been perceived as having considerably less significance than the frequently recurring dense meteor storms. These relationships can be expected to have remained essentially in place until the startling and awesome break-up of proto-Encke ca 3150 BCE. Only then would the Sun have begun to be perceived as the likely dominant celestial body and only somewhat later is it likely that advanced civilizations would have regularly observed the planets. Accordingly it was those peoples with imperial ambitions who first anticipated the likely new stature of the Sun (e.g. as Shamash in Mesopotamia and as Ra in Egypt) who successfully enforced their new intellectual dominance. Indeed the Mesopotamian civilization which came to be centred on Baghdad is not without considerable significance since its Sumero-Akkadian-Seleucid precursors were the evident bearers of Chaldean (later Sabian) wisdom at its heart which, apparently long protected by the Isin dynasty, eventually found its way (via Constantinople) to medieval Northern Europe under the banner of 'Hermes Trismegistus' (Scott, 1992). This wisdom not only appears to have connected planet Mercury with the three principal entities (Anu, Enlil and Ea) heading the Mesopotamian pantheon – otherwise to be perceived as the initial Piscean-Arietid-Taurid HE flux of the third millennium BCE (see above) – but also appears to have arrived at heliocentrism through the perception first (à la Tycho) of two prominent celestial spheres, cis-lunar and cis-Venusian (the terrestrial and solar domains of Sin and Ishtar respectively), and the perception second (à la Kepler) of particular physico-numerology (Kepler's third law) which simultaneously endowed the Sun and mathematical physics with the kind of pre-eminence Newton (claiming to recover ancient knowledge) was very soon to fulfil. Newton though was also an anti-trinitarian and was markedly averse to any kind of magical interaction involving emanations ie. the Mesopotamian pantheon. This meant the European Enlightenment essentially began by discounting the effects of the recent giant comet which have now been introduced.

The European Enlightenment was not of course the first occasion when our giant comet proto-Encke and its disintegration hierarchy regarded as a divine pantheon comprising (e.g.) archangels, angels, fallen angels (demons) etc. was considered to be non-physical. Thus the fourth century CE so-called 'Arian heresy' already bore witness to an earlier emergent Roman tradition (> 70 CE) which followed the unrestrained excesses of the Augustan dynasty as it continued to suppress social attitudes based on predestinarianism (which the Chaldeans, Essenes, Pythagoreans and Early Christians for example developed), their last well known Roman exponent being the notably wise but failed philosopher-statesman Seneca. The distinctly eschatological period 200 BCE–100 CE was of course characterised as much by a disturbing increase to Earth in the flux of meteoroidal debris as by that of perceived demons and it is by no means without significance that Seneca (Corcoran, 1971) gave the cometary events immediately preceding 140 BCE and 70 CE particularly close attention.

However the effects of proto-Encke can leave us in no doubt that our ancestors were well accustomed to apocalypse and the precarious nature of their long term existence. The recurrence of bombardment catastrophes and climatic downturns would have led to a natural interest in integer harmonics such as qP_E as well as the likes of P_L, 1/24 $P(\omega)$, 1/2 $P(\omega)$ etc. It is to be expected that paleolithic, mesolithic and neolithic societies would all have been profoundly impressed by integer diurnal counts relating to recurrent solar, lunar **and** visible proto-Encke cycles. Never a casual business, such mighty rhythms would have been the principal basis of that early impression of incessant, lasting order imposed by large scale celestial processes in which mankind necessarily participated: thus arose that deep and awesome sense of 'cosmos' which later founders of western civilization such as Pythagoras and Plato would naturally insist on as the very foundation of human conduct. It can hardly be doubted that this awesome sense of cosmos overriding all human affairs, the very pounding of nature, was the main justification for harmonising human behaviour in all situations, not least agriculture **and** apocalypse.

These then are the likely circumstances under which ancient elites would have sought to control essentially predestinarian societies in the past and under which appropriate calendars and their rituals would have been established as a necessary reflection of the will of civilization. We can reasonably expect that the prevailing intellectual outlook would have been one of establishing harmony within nature (deus **in** machina) such as arose again briefly during the European Reformation: subsequently the outlook would be rejected and replaced by a desire to be free of nature's restraints (deus **ex** machina) such as arose again during the Counter Reformation and Enlightenment. Thus the reversion to little ice-age and unsettled conditions in medieval Europe for several centuries meant a redoubling of efforts to search out the biblical timescale and comprehend nature's physical principles so as to adapt more precisely to the adverse environment. Newton's success however – on the brink of a climatic upturn – became a general signal to exploit these principles to other ends and to lose sight of their distant environmental roots in Babylonia (cf. Keynes, 1947). These possible analogies between the emergence of early ancient civilization in Mesopotamia and early modern civilization in Western Europe are perhaps helpful when we consider the early calendar.

The so-called Metonic cycle of fifth century BCE Greece (19 tropical years = 235 synodic months) is often considered to be the first calendrical cycle of any consequence to Western civilization. However it has long been known (de Cheseaux, 1754) that the apparent mean orbits of the Sun and Moon are tied to a much longer diurnal resonance (multiples of 315 years = $3/2 \cdot pP_0$) which is an order of magnitude more precise than the Metonic cycle. Thus the attention of experts has frequently been drawn to the very ancient chronologies of Mesopotamia which are the basis of the so-called Daniel cycle (1260 years) found in the Books of Enoch and Jubilees (e.g. Guinness, 1897). Aspects of these chronologies were preserved within the biblical canon and subsequent biblical scholarship was much concerned with evaluating history in terms of an apparent longterm (Day, Year) = 360 (civil day, civil year) record which evidently discounted conventional leaps and particularly noted a $3^1/_2$ Year period. The idea that the Daniel cycle has an observational basis has not been accepted in modern times but the circumstances relating to this ill-defined judgement are now materially altered by (i) discoveries in the Taurid-Arietid stream and their related findings above; (ii) the discovery of a very accurate calendric basis for long term cycles as revealed by the Dead Sea Scrolls; and (iii) the discovery that so-called 'artificial memory systems' (numerical counters or so called 'tokens') with likely calendrical associations were already in widespread use (by temple elites) throughout Mesopotamia by 10 ka BP some 5 ka before the invention of writing (e.g. Rudgeley, 1998; Schmandt-Besserat, 1996).

Recent investigations into the origins of Second Temple Judaism and Early Christianity based on findings at Qumran and elsewhere (e.g. Davies, 1987, 1996; VanderKam, 1998) suggest the Essene movement was instrumental in maintaining an accurate 'perpetual' calendar with an apocalyptic cycle of $2520 = 7 \times 360 = 50 \times 49 + 70$ years. Perpetual calendars are not widely used these days and their general purpose is often no longer immediately appreciated: this was to introduce regular leap weeks or months so as to harmonise the civil and tropical years between regularly spaced apocalypses whilst also ensuring that each day of the civil year always arises on the same day of the week. Such calendars have their attractions, especially for tidy minds and highly ordered societies. For the Essenes, it seems that the impending apocalypse of the time had more to do with the Flood and its symbolic date $\sim$ 2520 BCE along with the Book of Enoch (implied creation date $\sim$ 3150 BCE) than it had to do with the Exodus and its symbolic date $\sim$ 1260 BCE along with the Book of Jubilees (implied creation date $\sim$ 3780 BCE). Indeed the perpetual calendar was evidently regularised over periods of 36 jubilees (1764 years; see later) and there appears to have been widespread awareness especially in Babylonia of the countdown to apocalypse from about (2520–1764 =) 756 BCE onwards throughout the cradle of western civilization. Nowadays there is growing recognition that the Book of Enoch is more closely linked with the original Hebrew (OT) Bible (and the Synoptic (NT) Gospels) whereas the Book of Jubilees along with the Greek Septuagint are later reconstructions of the biblical record by mainstream scholars of Second Temple Judea who were apparently intent on placing apocalyptic history and creation more firmly in the past. The Essene luni-solar calendar can thus be accepted as the likely original one from Babylonia presumably passed down through representatives of the Isin dynasty and Chaldean magi. The reconstruction as such would then have been the work of Second Temple leaders and scholars upholding supposedly prior Sabian wisdom who reverted to a purely lunar calendar and thereby sought to abandon the apocalyptic associations (first and second comings cf. Section 2) of the Essene calendar. If so, it is not inconceivable that the failure of an anticipated apocalypse based upon the date of creation rather than that of the Flood would have served to undermine the public authority of the magi during the period 630–560 BCE. There is no agreed explanation for the massacre of the magi around this time (e.g. van der Waerden, 1974) but it is possible that more sophisticated survivors amongst the magi adapted to the requirements of society, modified the history of apocalyse and successfully pushed back the date of creation to an earlier time $\sim$ 3780 BCE. Political correctness is by no means unknown amongst politically motivated scholars and it is not unreasonable to suppose that mainstream society in Second Temple Judea would have dissembled in this manner. Nevertheless there was a skeleton in the cupboard, namely the Essenes and their vital luni-solar calendar and it was by no means suppressed as the CE began.

The Essene calendar (e.g. VanderKam, loc. cit.) is remarkable for its precision and for its combination of 3 perpetual cycles: solar, lunar and apocalyptic. The solar cycle evidently comprises a basic (civil) year of $364 = 52 \times 7$ days which is then probably adjusted by interpolating regular leap weeks at intervals of 6, 98 and 1764 years such that the average tropical year becomes:

$$\begin{aligned} Y(d) &= 364 + 7\ \{1/6 + 3/6.49 + 1/6^2.49\} \\ &= 365.2421 \text{ days} \end{aligned}$$

The lunar cycle running in parallel with the solar cycle has a basic year of $354 = 12 \times 29^1/_2$ days subject to the interpolation of 30 day leap months every 3 years. There is then a residual lunar cycle count (1 day every 6 years) added to the solar cycle leaps which has an accumulated value over 2520 years of

$$3550 = 31080\varepsilon + n_0 M(d) \text{ days}$$

in which the average synodic month $M(d) = 29^1/_2 + \varepsilon$, the necessarily integer n_0 and the implicit addition of 88 months alternating with 29 and 30 days are such that

$$M(d) = 29.530608 \text{ days}$$

closely in accordance with the Metonic cycle.

The tally of 'long periods' each comprising five jubilees (245 years) and a decade of year-weeks (70 years) characterised by 2 interpolated months per jubilee and 1 interpolated month per decade of year-weeks is then potentially interesting because it represents a possible passage of 'signs' such that a completed 'week' of 'long periods' would mark the start of a final 'long period' completing 88 leap months as well as containing 2 apocalypses. Such completed 'weeks' of 7 × 315 years may well be significant in fact since there is a near-coincidence between the start of that for T_n ($n = 0$) and the occasion of known calendrical reform (the so-called Canopus Decree) which no longer appears to be altogether coincidental! The Essene passage of 'signs' has not yet been fully evaluated but the remarkable accuracy of Y(d) and M(d) coupled with the known 2520 year count and pattern of enforced leap weeks already suggests we should have considerable regard for the mathematicians and astronomers who had apparently invented this scheme at least two and a half millennia previously. Indeed the pattern itself is doubly interesting on account of its successive denominator values which were probably first introduced by way of successive empirical improvements to the length of the civil year. If so more primitive formulations might have preceded the replacement of leap months by leap weeks such that (say)

$$Y'(d) = 360 + 30\ \{1/6 + 3/6.60 - 6/60.360\}$$
$$= 365.2417 \text{ days}$$

Thus it is by no means inconceivable that the inventors of the first successful luni-solar calendar would have adopted 360 day years and 30 day months along with intercalation every 6 years thereby having each year acquire 5 extra or epagomenal days and producing a drift of 120 years prior to the insertion of a further leap month – possibly consistent with the early calendars of the Persians, the Egyptians and the Dogons. Such a perpetual calendar is not particularly attractive however since it remains significantly out of step with the seasons for rather long periods and the next stage of interpolation requires the added complication of *omitted* leap months. These disadvantages might well provoke reform and it is certainly interesting that the improved formulation of Y(d) over Y'(d) is founded on a modification of certain fractions (1/6, 1/60, 1/360) which were critical to Babylonian enumeration and arithmetic from their inception along with their peculiar (but as yet unexplained) alternating base, as follows:

.... (1/3600, 1/360), (1/60, 1/6), (1,10), (60,600), (3600,36000) ...

Thus it is entirely possible that the earliest known arithmetic had its origins in luni-solar calendrical activities which required regular leaps every 6 or (3 + 3) years – precisely in accordance with the known apparent evolution of this arithmetic (e.g. Schmandt-Besserat, loc. cit.). Also implied is its likely Afroasiatic provenance throughout Persia, Mesopotamia, Arabia and Northern Africa, well before the invention of writing (cf. Bernal, 1987). The ramifications of this relatively straightforward historical scheme and the supposedly later, more sophisticated, luni-solar calendar apparently maintained by the Isins and their possible successors the Essenes are even more far-reaching since we might well suppose that writing itself does not have a primarily economic raison d'etre (the currently preferred view) but simply arose on account of the massive workload encountered by the ancient astronomer-priest elite when it was suddenly confronted (in circumstances of perceived danger) with the need to observe a multitude of cometary fragments ca 3150 BCE following a close planetary encounter (à la Shoemaker-Levy). History after all tells us that it was the gods who taught mankind to write!

4. Conclusion

This paper highlights a supposed academic scandal par excellence (Baigent and Leigh 1991). On the one hand there has long been widespread public interest in the Dead Sea Scrolls and what they tell us about a very critical period in the early development of those modern religions (Judaism, Christianity, Islam) which continue to have the most pervasive influence on the development of western civilization. On the other hand there is a widespread but dedicated group of experts exercising the most exacting methods of modern scholarship to penetrate these Scrolls who have now repeatedly alerted us to the over-arching significance of a previously lost but once highly valued ancient calendar. And yet the worldwide community of scholars best suited to cast judgement on the technical insights that emerge (not least the astronomical and cosmological elite) does not wish to say anything at all. Never before, so it would seem, have the interests of the public been so comprehensively flouted by the academic mainstream.

This paper also ventures to get to the root of the scandal. It recognizes a three thousand year old disposition on the part of scholars to run away from perpetual, predestinarian calendars which tend to shackle society in chains. Instead of nature's recurrent apocalypse we have always been given a much narrower 'once and for all' creation-endtime scenario briefly confining us within a necessarily finite cosmos which is nevertheless pervaded by an infinite domain of free spirits with which we can associate in supposedly eternal bliss. Intellectual 'creationism' as the antidote to periodic apocalypticism and intellectual 'dualism' as the provider of (good and evil but entirely vacuous) spirits are of course part and parcel of modern fundamentalist thinking whether it be in the scientific mainstream or on the scientific fringe. While the Dead Sea Scrolls bear witness to the foundations of this emergent intellectualism (e.g. Collins, 1997 Ch 3), the so-called big bang and universal expansion, Hoyle's infamous bête noir, are but the latest scientific concoctions to satisfy these stupefying requirements. Hoyle was never simply negative however and recognized well enough that today's cosmology is basically yesterday's theology with an account that has departed even further from the reality on which it was originally based. The solution in principle is simple since past reality in principle can be reconstructed and is always within the purview of science. The question raised here therefore is whether proto-Encke should now be in the public domain.

References

Asher, D.J. and Clube, S.V.M.: 1993, An Extraterrestrial Influence During the Current Glacial-Interglacial, *Q.J.R. astr Soc* **34**, 481–511.

Asher, D.J. and Clube, S.V.M.: 1998, Towards a dynamical history of 'proto-Encke', *Cel. Mech. and Dyn. Astr.* **69**, 149–170.

Baigent, Michael and Leigh, Richard: 1991, *The Dead Sea Scrolls Deception*, Jonathan Cape.

Bernal, M.: 1991, *Black Athena; The Afroasiatic Roots of Classical Civilization, Vol I: The Fabrication of Ancient Greece 1785–1985*, Vintage, London.

Bond, G., Broecker, W., Johnsen, S., McManus, J., Labeyrie, L., Jonzel, J. and Bonani, G.: 1993, Correlations between climate records from North Atlantic sediments and Greenland ice, *Nature* **365**, 143.

Bond, G., Showers, W., Cheseby, M., Lotti, R., Almasi, P., Menocal, P. de, Priore, P., Cullen, H., Hajdas, I. and Bonani, G.: 1997, A pervasive millennial-scale cycle in North Atlantic Holocene and Glacial climates, *Science* **278**, 1257–1266.

Cheseaux, M. de: 1754, *Mémoires posthumes: 'Remarques historiques, chronologiques et astronomiques, sur quelques endroits du livre de Daniel'*.

Clube, Victor and Napier, Bill: 1990, *The Cosmic Winter*, Basil Blackwell, Oxford.

Collins, J.J.: 1997, *Apocalypticism in the Dead Sea Scrolls,* Routledge, London and New York.

Corcoran, T.H.: 1971, *Seneca X, Naturales Questiones II, Book VII Comets*, Heinemann, London.

Davies, P.R.: 1987, *Behind the Essenes: History and Ideology in the Dead Sea Scrolls*, Scholars Press, Atlanta.

Davies, P.R.: 1996, *Sects and Scrolls: Essays on Qumran and Related Topics*, Scholars Press, Atlanta.

Feibleman, J.K.: 1959, *Religious Platonism*, George Allen & Unwin, London.

Guiness, H.G.: 1897, *The Approaching End of the Age*, Hodder & Stoughton, London.

Hoyle, Sir Fred: 1993, *The Origin of the Universe and the Origin of Religion*, Moyer Bell, Rhode Island and London.

Keynes, Lord: 1947, *Newton, the Man* in *Newton Tercentenary Celebrations*, Royal Society, Cambridge University Press.

O'Brien, S.R., Mayewski, P.A., Meeker, L.D., Meese, D.A., Twickler, M.S. and Whitlow, S.I.: 1995, Complexity of Holocene Climate as Reconstructed from a Greenland Ice Core, *Science* **270**, 1962–1967.

Rudgeley, Richard: 1998, *Lost Civilizations of the Stone Age*, Century, London.

Schmandt-Besserat, Denise: 1996, *How Writing Came About*, University of Texas Press, Austin.

Scott, Walter: 1992, *Heretica* (1924 edition), Solos Press.

Stohl, J.: 1983, On the distribution of sporadic meteor orbits, *Asteroids, Comets, Meteors*, Uppsala University Press.

VanderKam, James C.: 1998, *Calendars in the Dead Sea Scrolls: Measuring Time*, Routledge, London and New York.

Van der Waerden, Bartel L.: 1974, *Science Awakening II; The Birth of Astronomy*, Oxford University Press, New York.

Wenke, Robert J.: 1980, *Patterns in Prehistory*, Oxford University Press, Oxford and New York.

Whipple, F.L.: 1967, On maintaining the meteoritic complex, *NASA SP* **150**, 409.

PART VI: PANSPERMIA

PANSPERMIA ACCORDING TO HOYLE

CHANDRA WICKRAMASINGHE
Cardiff Centre for Astrobiology, 2 North Road. Cardiff, CF10 3DY, UK

Abstract. Fred Hoyle's involvement in panspermia, recasting an ancient idea in a modern scientific framework marks an important turning point in the fortunes of this theory. Panspermia is discussed nowadays as a serious alternative to a purely terrestrial origin of life.

In the mid-1970's panspermia theories were thought by many to be on the fringes of science. The idea that life-seeds are distributed widely in the cosmos has ancient roots. It was discussed in some form in the writings of the Greek philosopher Anaxoragas in the 5th Century BC. It was also a widely held belief amongst Vedhic and Buddhist philosophers of the Orient at even earlier times. The revival of panspermia in a purely scientific context came about in the early 20th Century with the writings of Svante Arrhenius (1903). Arrhenius' ideas soon fell into disfavour, but for reasons that later turned out to be wrong. Becquerel (1924) and others used laboratory experiments to argue that bacteria would not survive the radiation environment of space, particulary ultraviolet radiation. It was later shown that microorganisms are not too easily destroyed by UV, but are mostly rendered inactive and that this process can be relatively easily reversed. Hoyle and I also argued much later that for bacteria in interstellar molecular clouds ultraviolet radiation poses no problem. Only a thin coating of condensed carbonaceous material around a bacterium would shield it from potentially damaging radiation. So Becquerel's original objections to panspermia turned out to be flawed.

Moreover, it has been amply demonstrated that that bacteria are resistant to a wide range of extreme conditions that would prevail in space, including their ability to withstand high doses of ionising radiation and extremely low temperatures. Despite all these developments the stigma against panspermia continued to linger well into the 1970's, the time when Fred and I began to work on these ideas.

It has often been remarked that Fred Hoyle rashly embarked on a campaign to re-instate panspermia. But the facts reveal a different story altogether. Fred and I had worked together on models of interstellar grains throughout the 1960s (Hoyle and Wickramasinghe, 1962, 1969).

By the late 1960's we had found that purely inorganic models of interstellar dust were difficult to reconcile with the observational data that was available. After Fred's resignation from Cambridge in 1972, his involvement with the AAT and his frequent trips to the USA led to a temporary cessation of our collaboration. When I had published a series of papers arguing for the presence of vast quantities of complex organic polymers in space I became a little uneasy about the profundity of the implications that might follow. Frightened perhaps by the prospect, I wrote Fred a series of letters with the seemingly outrageous proposition that the grains may be

Astrophysics and Space Science **285:** 535–538, 2003.

real live bacteria. After a long silence he responded from Cornell in what appeared to me to be a disappointingly negative way. He was exceedingly cautious in the way he approached this question, more cautiously than his critics have thought.

In the autumn of 1976 Fred was happy with the idea that interstellar dust contained the molecular building blocks of life that were subsequently mopped up by comets. This is in fact the limited version of panspermia that nowadays has come to be accepted almost without dissent – in striking contrast to the situation that prevailed in the late 1970's. In this respect at least Fred was 30 years ahead of his time. For a few years Fred held back from the proposition that interstellar dust (or a fraction of it) had a biological connotation. The first tentative steps towards biochemistry were thought to take place in interstellar clouds, but assembly of monomers into primitive living cells was considered to occur either within the interiors of hundred of billions of comets in the outer regions of the solar system, or even perhaps on the Earth (Hoyle and Wickramasinghe, 1976a,b)

It should be stated at this point that Fred's interest in the outer solar system as a nursery of life goes back to the 1940's when he embarked on a correspondence with biologist J.B.S. Haldane, the co-inventor of the primordial soup theory of the origin of life (Hoyle, 1994). In his classic book *Frontiers of Astronomy* (1953) Fred Hoyle already asserts a preference to the outer solar system for an origin of life on the grounds that there was vastly more carbonaceous matter available there.

The transition from prebioltic organics in the interstellar medium to bacterial grains came to us in slow stages. Firstly we argued that there is no logic to demand an origin of life on the Earth, and that the odds against such a purely terrestrial origin are superastronomical (Hoyle and Wickramasinghe, 1981). Secondly interstellar grains continued to exhibit properties that defied a proper explanation in terms of inorganic dust. Finally, through refinements of astronomical observations that continued though the 1970's the dust was found to possess properties that were indistinguishable from particles of bacterial composition (Hoyle et al., 1982; Hoyle, 1982). The infrared and ultraviolet properties of dust (Hoyle and Wickramasinghe, 1992) imply that about one third of all the interstellar carbon is tied up in the form of hollow organic particles with an uncanny resemblance to freeze-dried bacteria. Alternative non-biological explanations of this data seemed to Fred (and to me) to be highly contrived, and moreover, any such explanation was hard put to explain the astounding efficiency of conversion of inorganic matter to particles invariantly of bacterial size, shape and composition.

Since biological replication could not of course operate in the interstellar medium, we were led to suggest that amplification followed the cyclic processing of interstellar matter into stars, comets and planetary systems. The average timescale for this cyclic process is a few billion years. The overall logic of the scheme is illustrated in Figure 1.

Comets are thus regarded as the amplifying sites of cosmic microbial life. In the early days of the solar system comets contained radioactive materials that would have heated their cores and preserved a warm liquid watery environment

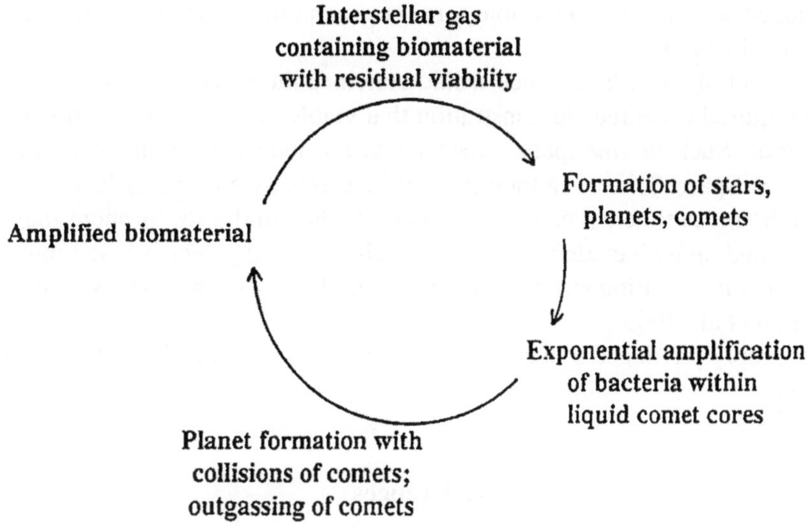

Figure 1. Cosmic life cycle.

for millions of years. Exponential amplification of microbes in such environments ensures that even the minutest surviving fraction, $<10^{-20}$ reaching a new comet forming stage in this cycle is thus vastly amplified. The logic of panspermia would then be inescapable.

In the years from the 1980's to the present various lines of evidence from a variety of disciplines have added greatly to the plausibility of panspermia. The antiquity of microbial life on the Earth has been pushed back to a time before 3.8 Gy (Mojzsis et al., 1996) when the Earth was being severely bombarded by comets and asteroids. In recent years the limits of microbial life on our planet have expanded to encompass an extraordinarily wide range of habitats. Microbes are found in geothermal vents, the ocean floor, in radioactive dumps and in Antarctic soil. Microorganisms have been recovered from depths of 8km beneath the Earth's crust, and laboratory studies have shown that bacteria can survive pressures at ocean depths of thousands of kilometres or more. The long-term survivability of bacteria has also been extended from 25 to 40 million years (Cano and Borucki, 1995) to a quarter of a billion years in the case of a bacterium entrapped in a salt crystal (Vreeland et al., 2001). Direct proof of the survival of bacteria exposed to radiation environments in the near Earth environment has also been demonstrated using NASA's Long Exposure facility.

The Hoyle-Wickramasinghe panspermia theory requires life to have been introduced to Earth for the first time by comets some 4 billion years ago, with an ongoing incidence of microorganisms continuing to the present day. Recent discoveries of organic molecules and fragile structures within the Mars meteorite ALH84001 have gone in the direction of supporting the idea that microbial life could indeed be transferred in viable form between objects within the solar system (McKay et al., 1996).

No amount of indirect evidence would convince the hardened sceptic, however. What is required is a direct demonstration that viable microbes exist within cometary material. Such *in situ* space experiments are indeed in train as a long-term objective of Space Science, although definitive results may be at least a decade away. In the mean time collections of cometary dust in the stratosphere may have already turned up with evidence that comes close to being decisive, so vindicating one of Fred's most daring and provocative scientific assertions (Harris et al., 2002; Wainwright et al., 2003).

I feel deeply privileged to have been Fred Hoyle's friend and accomplice in this endeavour.

References

Arrhenius, S.: 1903, *Die Umschau* **7**, 481–485.
Becquerel, P.: 1924, *Bull. Soc. Astron.* **38**, 393.
Cano, R.J. and Borucki, M.: 1995, *Science* **268**, 1060–1064.
Harris, M.J. et al., 2002, *Proc. SPIE Conf.* **4495**, 192–198.
Hoyle, F.: 1953, *Frontiers of Astronomy*, Heinemann.
Hoyle, F.: 1994, *Home is Where the Wind Blows*, University Science Books, California, USA.
Hoyle, F. and Wickramasinghe, N.C.: 1962, *MNRAS* **124**, 412.
Hoyle, F. and Wickramasinghe, N.C.: 1969, *Nature* **223**, 459.
Hoyle, F. and Wickramasinghe, N.C.: 1976a, *Nature* **264**, 45–46.
Hoyle, F. and Wickramasinghe, N.C.: 1976b, *Nature* **268**, 610–612.
Hoyle, F. and Wickramasinghe, N.C.: 1981, in: C. Ponnamperuma (ed.), *Comets and the Origin of Life*, D. Reidel.
Hoyle, F. and Wickramasinghe, N.C.: 1982, *Astrophys. Sp. Sci.* **86**, 321–329.
Hoyle, F. and Wickramasinghe, N.C.: 1993, *The Theory of Cosmic Grains*, Kluwer Academic Publishers.
Hoyle, F. et al.: 1982, *Astrophys. Sp. Sci.* **83**, 405–409.
McKay, D.S. et al.: 1996, *Science* **273**, 924.
Mojzsis, S.J. et al.: 1996, *Nature* **384**, 55–59.

ASTRONOMY OR BIOLOGY? *

FRED HOYLE† and CHANDRA WICKRAMASINGHE
Cardiff University, Cardiff Centre for Astrobiology, 2 North Road, CF10 3DY, Cardiff, UK
E-mail: Wickramasinghe@cardiff.ac.uk

Abstract. The role of biology in astronomical phenomena and processes was first discussed extensively by us in the period from 1979-1982. The two sections reproduced below are the concluding chapters of 'Space Travellers' which we published in 1981. The ideas discussed here have turned out to be forerunners to several recent developments in astrobiology.

1. Bacteria on Planets

In this section we deal with two particular issues. Our picture has been of an immensely powerful universal biology that comes to be overlaid from outside on a planet such as our own. Wherever the broad range of the external system contains a life form that matches some local planetary habitat, the form in question succeeds in establishing itself. The external system contains a wide spectrum of choices which the local environment proceeds to select according to the particular niches that the local system provides. In our view the whole spectrum of life, ranging from the humblest single-celled life forms to the higher mammals (and beyond) must be overlaid on a planet from outside.

Bacteria exist everywhere on the Earth and in astonishing profusion. A handful of garden soil contains typically about a billion bacteria, and so do certain samples of deposits taken from the floors of the oceans. The seemingly hostile continent of Antarctica teems with bacteria – they are to be found in the soils and inside the rocks of dry valleys. and they exist and multiply towards the bottom of vast glaciers. In total, there are something of the order of 10^{27} bacteria on the Earth, of which only a minute proportion are pathogenic to man.

It is essential to all life-forms that they have access to energy. Photosynthesis, whereby sunlight provides the energy supply, is of course well-known, with chlorophyll playing a crucial role in the production of sugars from carbon dioxide and water. There is a widespread but erroneous popular view that photosynthesis is the basis of all life. Rather is it the case that the majority of bacterial forms derive their energy quite otherwise. Wherever an energy yielding chemical reaction exists, and which proceeds only very slowly under inorganic conditions, some bacterium will usually be found to be living on it, speeding the reaction catalytically. There are bacteria that live by oxidizing sulphites to sulphates, which one might perhaps

* Reprinted from 'Space Travellers' published in 1981
† Deceased 2001

Astrophysics and Space Science **285:** 539–554, 2003.

have expected on an oxidizing planet like the Earth. More unexpected are bacteria that use free hydrogen, an entirely evanescent substance on the Earth, to reduce sulphates to free sulphur. Nor would one expect to find the bacteria that use only free hydrogen and carbon dioxide to produce methane water.

It is generally recognized that bacteria requiring free hydrogen cannot have evolved on the Earth over the last three billion years, and so they are supposed to be survivors from an earlier epoch in which free hydrogen is assumed to have been in plentiful supply. Yet the survival thread of these bacterial forms is so thin that it would surely be snapped in only a few centuries, or at most a few millennia. The clear implication is that the Earth receives a steady, continuing supply of sulphur and methane-producing bacteria, and that whenever unusual local conditions happen to set up an environment in which such bacteria can prosper they do so temporarily until the conditions change.

Modern industry, with its dumps, tips, and tailings has created exceptional niches for bacteria which are always filled, even though those conditions never existed before on the Earth, except conceivably as a short-lived fluke. There is almost nothing that, one can do in daily life which does not create a niche for some form of bacterium or protozoa. The farmer's bales of hay create a niche, and so does a bird's nest.

It is usual in biology to argue that bacteria and protozoa evolve genetically with great speed and so manage to adapt themselves to whatever possibilities the environment may offer. This view persists in spite of there being a considerable weight of evidence against it. Experiments in the laboratory over many generations have shown that bacteria do not evolve. Bacteria (and viruses) are incorrigibly stable. Experimenters can do three things to bacteria. Where varieties are already present initially, one of them can be encouraged to multiply preferentially with respect to the others. Another thing that can he done to bacteria and viruses is to ruin them. The third alternative consists in dividing up what is already there and reassembling the bits in a different order, as for instance if one were to take selected bits from two different kinds of bacteria and reassemble them in an attempt to form a new bacterium. What experiments in the laboratory cannot do, however, is to make effective new bits. This is to be expected if the bits owe their origin to a galaxy-wide evolution which gave rise to a much higher order and more subtle system than is usually supposed.

True terrestrial evolution has consisted in the fitting of bits into aggregates that were optimal to the environment, as was demonstrated for plants and higher animals by classical biology. The bits have not been evolved to fit the environment, however. They are simply the inflexible cosmic bits that could manage to survive in the environment. The distinction shows clearly in the phenomenon of adaptation. Birds, rodents, fish, and insects are clearly well-adapted to the environment in their different ways. Bacteria, on the other hand, are not closely adapted. Bacteria found in the ocean depths are not particularly adapted to those depths – they are simply bacteria that happen to function in a manner that is insensitive to high pressure.

Many forms of bacteria are found surviving under temperatures that are far from optimal. Many found in warm soils multiply better at lower temperatures than are ever found in the tropics. Others multiply best at temperatures much higher than exist on the Earth at all, above 75 °C. There is indeed a very considerable measure of *disadaptation,* which shows itself in bacteria possessing properties that are irrelevant to survival on the Earth. Resistance to enormous doses of X-rays, ultraviolet light, and resistance to great cold were examples mentioned elsewhere in our writings. The cell walls of bacteria are stronger than reinforced concrete, and this is another example. Biologists are constantly being surprised by new unearthly properties, and the erection of almost any really novel environment reveals some such property. The canning of food provided a new environment. There was a practical need to find a cheap and convenient means of sterilizing food after its sealing inside a metal can. Flooding by X-rays was tried, and it was then discovered that *Micrococcus radiophilus* was discovered. Heating was also tried and some bacteria were found to be capable of surviving temperatures above 100 °C. Water in the open expands slightly as it is heated. Water in a closed can of fixed shape cannot expand, however. Instead a high pressure is created, which can prevent the formation of bubbles of steam even above 100 °C. Likely enough it was the prevention of steam bubbles which permitted bacteria to survive in these experiments. But how could such a property ever have evolved on the Earth?

Manifestly, bacteria are a life-form that has simply been plastered over the Earth, to survive wherever it is possible to do so. The properties of bacteria have galaxy-wide quality, much wider than is needed for survival on the Earth alone. Indeed, because of the great breadth of their survival characteristics, the potentiality of bacteria to establish themselves on planets must be much greater than one would permit oneself to suppose within the narrow confines of an Earth-bound theory of the origin of life. Just because the other planets of our solar system are different from the Earth is no reason at all for thinking that life in our system must he unique to the Earth. The spread of the survival characteristics of bacteria are so great that they may well include the ability to fit environmental possibilities on some other planet or planets. In this article we shall find suggestive evidence that this is actually so for Venus, Jupiter and Saturn, and that it is possibly so for Mars, Uranus and Neptune.

We take as an indispensable requirement for the survival of bacteria that water can condense on them, and can pass to their interiors and be held there in a liquid form. This condition rules out Mercury and the Moon. The lifeless states of Mercury and the Moon show immediately in their drab lack of colour. As a valid general correlation, life and colour may be associated together, although perhaps not in what mathematicians call a one-to-one correspondence. A stricter indication of life comes when particle sizes close to 1 μm turn out to be dominant, in which connection we recall the size distribution of spore-forming bacteria given in Figure 1 taken from data given in *Bergey's Determinative Bacteriology* (Buchanan and Gibbons, 1974).

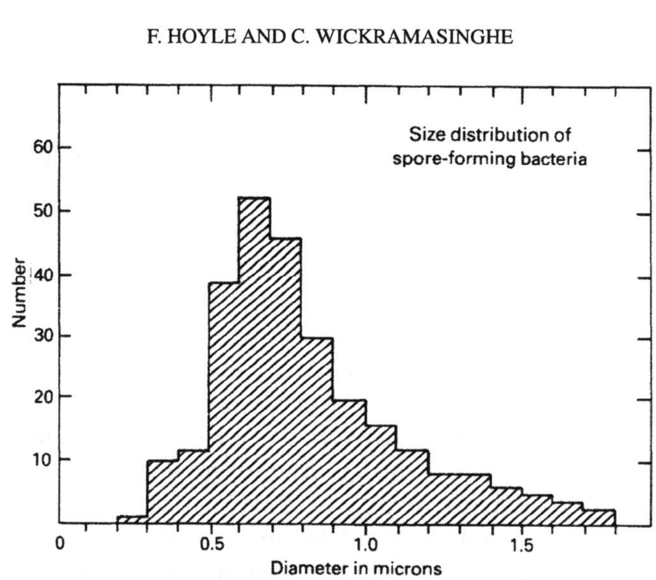

Figure 1. The size distribution of species of known spore-forming bacteria.

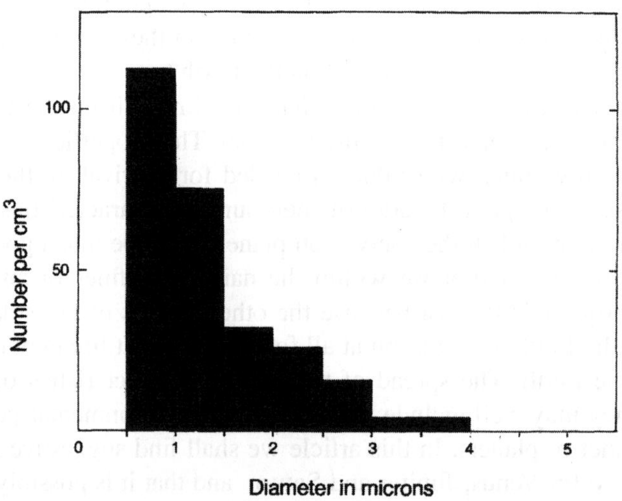

Figure 2. Distribution of particle sizes in the upper clouds of Venus (approximately 55–65 km in altitude) measured by Pioneer Venus.

The upper clouds of Venus produce a rainbow, indicating that the cloud particles are spherical and that they have sizes in the region of 1 μm (Coffeen and Hansen, 1974). The distribution of particle sizes actually measured in the upper clouds by Pioneer Venus is shown in Figure 2 (Knollenberg and Hunten, 1979). Although a considerable amount of further argument is needed before one can be satisfied about details, the correspondence of the observed particle sizes with those of bacteria tells the main story immediately.

There are two ways in which bacteria can be spherical in shape, one through forming spores and the other through restriction to the class of micrococci. Of these, spores appear the more likely possibility for the reason that the clouds of Venus are in convective motion, extending from an upper level at an altitude of 65–70 km down to an altitude of 45 km.

The temperature at 45 km is about 75 °C and at the top is about –25 °C. While survival over this range is easily possible for bacteria, the repeated variations of temperature caused by a circulating cloud system would be better resisted by bacteria capable of forming spores which are still more hardy than the bacteria giving rise to them. Thus bacteria could be rod-shaped in the lower warmer regions, but giving place to spherical spores in the cold upper regions of the clouds. In this way there is no requirement for the bacteria to be exclusively micrococci.

A quantity called the 'refractive index' of a particle is relevant to its light scattering properties. The refractive index of water is about 1.33 and that of biological material about 1.5. Bacterial spores contain about 25% water by weight, which causes them to behave like a uniform particle with refractive index 1.44, which is exactly what is known by observations on the degree of polarization of the upper clouds of Venus.

The refractive index of sulphuric acid is 1.48, and sulphuric acid droplets with 25% water would also behave like uniform particles with refractive index 1.44. But this explanation of the upper clouds of Venus fails on a number of important points. There is no reason for such droplets to have sizes of 0.6–1.2 μm (Figure 2). Sulphuric acid droplets could, and very likely would, have a much broader size distribution than Figure 2. The 25% water concentration is an arbitrary choice, whereas for bacterial spores there is no choice, about 25% is what the water concentration must be. The chemical sampler carried by Pioneer Venus (1978) measured vapour pressures of oxygen and sulphur dioxide in the cloud regions that were thousands of times lower than one can easily have in the laboratory, and yet, even at the much higher laboratory concentrations, oxygen and sulphur dioxide do not go easily to sulphur trioxide, and thence by the addition of water to sulphuric acid. Commercial sulphuric acid is produced in two ways. The old way by a complex of reactions involving nitric acid, and the modern way by catalytic processes involving either platinum metal or vanadium oxide. Inorganic catalysts like platinum tend to become 'poisoned' under natural conditions, however – they take part in other reactions which change them. The highly effective catalysts found in nature are bacteria, which not only maintain themselves (which is all that inorganic catalysts can do) but actually increase in number through the chemical reactions which they promote. Hence to produce some sulphuric acid, which is apparently required to explain certain details of the infrared radiation emitted by the clouds of Venus, it is to bacteria that we should look. Sulphur bacteria are yellow in colour, and it is to their presence (among other non-yellow bacteria) that we attribute the pale yellow colour of the light reflected by Venus. Sulphuric acid is colourless, on the other hand, and droplets of it would not produce any such colouring.

Water is not a particularly abundant constituent of the atmosphere of Venus. The second atmospheric sample taken by Pioneer Venus at an altitude of about 46 km near the base of the lower clouds gave about 0.5% water. With this concentration, and choosing –10 °C as the lowest temperature at which the bacteria replicate, the relative humidity can be calculated to be about 85%, which is adequate for bacteria with an appreciable internal content of dissolved salts. It is important here that the problem of maintaining liquid water inside a bacterium is not simply evaporation from a free water surface but of evaporation from the outer surface of the bacterial membrane, which is markedly water-attractive. At 85% relative humidity only a quite small extra bit of holding power against the evaporation of water molecules is sufficient to stop a bacterium from drying out. However, under exceedingly dry conditions, near zero per cent humidity, bacteria must largely dry out. As bacteria circulated through the middle and lower cloud regions of Venus, they would be subject to relative humidity values that ranged from near zero in the hotter, lower regions to about 85% in the cooler, upper regions. This would provide a natural water pump with alternating phases of filling with water and of subsequent evaporation, and with each water-filling episode giving a fresh supply of nutrients to the interior of the cell.

Since Venus is exceedingly hot at ground-level (about 450 °C) it is a matter of some surprise to find evidence of life existing there. The circumstance which makes life possible is the dual circulatory pattern of the Venusian atmosphere. From the temperatures and pressures measured by Pioneer Venus one can infer the presence of a lower convective zone from ground level to a height of about 30 km. There is then still atmosphere up to about 45 km, where the second convection zone begins. It is the non-moving in-between region from 30 km to 45 km which protects life in the higher zone from being quickly swirled down to the impossibly hot conditions near the ground.

Although bacteria are small and are able to ride easily with the atmospheric motions, there must nevertheless be occasional situations where a bacterium carried down to the base of the higher convective zone fails to find an up-current on which it can ascend again. Inexorable gravity will then cause the bacterium to fall slowly down through the in-between zone until it reaches the lower convective zone, where it will quickly be snatched downward and destroyed by the heat. It is therefore to be expected that the in-between zone will contain a thin haze of slowly falling doomed bacteria, and the existence of a thin haze between altitudes of 30 km and 45 km was indeed found by Pioneer Venus.

From the data obtained by Pioneer Venus one can infer that the fall-out through the in-between zone would begin to denude a population of high-level bacteria in about 30 000 years, if the bacteria did not renew themselves. For renewal a supply of nutrients is required. Water, nitrogen. and carbon dioxide are amply available, but in addition to these main ingredients other elements – sodium, magnesium, phosphorus, sulphur, chlorine, potassium, calcium, manganese, iron, cobalt, copper, zinc, and molybdenum – are needed in smaller proportions. It can be calculated that the fall-out through the in-between zone would carry about 10 000 tonnes of these essential ingredients each year from the higher convection zone to the lower zone. How, one can ask, may such a loss from the higher zone be compensated? The answer appears to be by the infall of meteoric material from space, rather than by gas diffusion upward from the lower atmosphere. The meteoric supply to the Earth is known to be about 10 000 tonnes per year, and the supply to Venus – a very similar planet should be about the same. The problem for gas diffusion upward lies in the lack of volatility of the compounds of many of the needed elements. There is no such problem for meteors in the size range from about 0.1 mm to 1 cm, which are gasified as shooting stars on plunging at great speed into the high atmospheres of the Earth and of Venus.

The size distribution of particles measured for the upper clouds of Venus is essentially maintained in the middle and lower clouds and in the haze zone of the in-between region. In the middle and lower clouds, however, there is also a considerably less numerous population of much larger particles. These we attribute to the tendency of bacteria to aggregate into colonies, a property that may well be helpful in preventing too much evaporation of water in the dry, hot conditions of the lower clouds.

The correspondence between the rate of supply of meteoric material and the rate of loss of essential nutrients down through the in-between region, about 10,000 tonnes per year in each case, suggests that the quantity of bacteria in the clouds is nutrient-limited. The clouds have built up until the drop-out at the bottom equals the supply from outside. If the supply from outside were to cease then the bacterial population would decline steadily, until after a few tens of millennia it would be possible to see down to the ground level of Venus. Such a situation may well have occurred in past times, only then there were no humans to observe the lower excessively heated surface zone of the planet.

Humans have for long looked at the red colour of Mars and seen evidence there for the existence of life. The argument was that the red colour implies a highly oxidized condition, a conclusion that a recent Mariner landing on Mars has shown to be correct. The supply of oxygen needed to produce this condition might have come from photosynthetic organisms, and this further aspect of the old argument may also be correct. But if so the organisms must have existed in the remote past when liquid water was present at the surface of the planet. The Martian surface is cut by many sinuous channels, of which that in Figure 3 is an example. It is generally agreed that these channels were made by a liquid much less viscous than molten lava, and water is a likely possibility. But there is no liquid water nowadays at the surface of Mars, and consequently (if we take the need for liquid water to be essential) there can he no active surface life. It is surprising therefore that so great an effort was made in the recent Mariner landing to look for evidence of life. It is true that bacteria in a dormant condition might survive for long periods in the Martian surface dust, but such bacteria might well need unusual conditions for growth, which could not be anticipated in a terrestrially designed experiment.

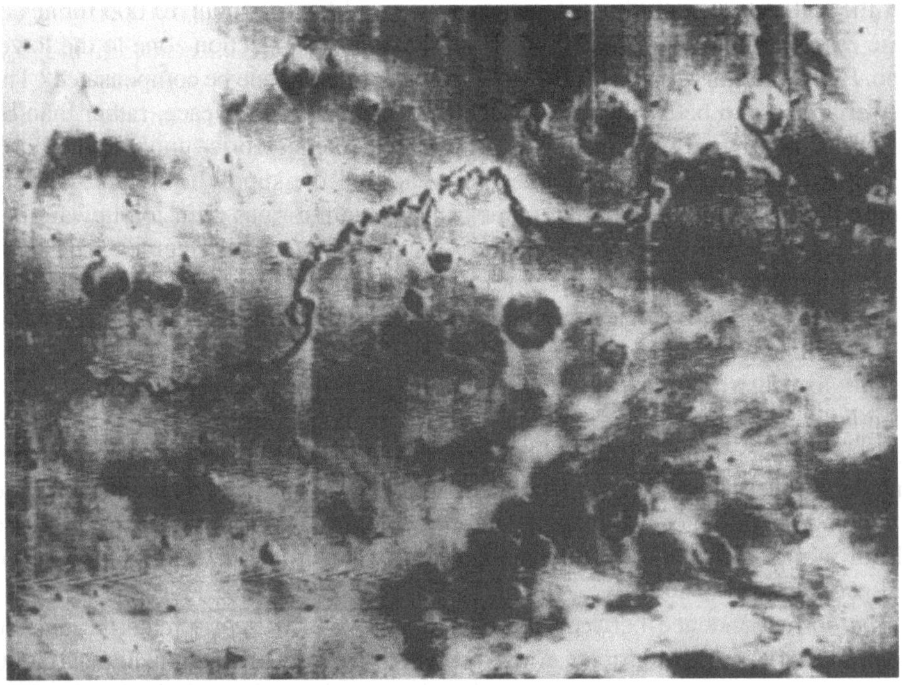

Figure 3. A sinuous channel on Mars, probably produced by the flow of liquid water (Courtesy of NASA).

If one argues analogously to the Antarctic, the best chance for life to be active on Mars would be deep inside glaciers where the temperature might rise sufficiently for water to become liquid. There would still be problems of nutrient supply, but if the glaciers themselves turn over, top to bottom from time to time, such problems would be capable of solution. The bacteria would need to live on some energy-producing chemical reaction, and, if the reaction had a gaseous product (such as carbon dioxide or methane), the possibility would exist for the building up of subsurface pockets of gas, which might explode sporadically to the surface unleashing quantities of bacteria, spores, and inorganic dust into the Martian atmosphere. In this connection we recall the vast atmospheric dust storm which greeted the arrival of a Mariner vehicle in 1971. This storm has been attributed to high winds generated by the normal Martian meteorology, but if so one might wonder why such winds are not a seasonal phenomenon. The outward explosion of microbiologically-generated gas could prove to be a better explanation.

From an analysis of reflected sunlight it has been estimated that the clouds of Jupiter consist of particles with diameters close to 0.5 μm and with refractive index 1.38. This diameter value is somewhat less than the main peak of the distribution of Figure 1 at about 0.7 μm. The calculations for Jupiter were made, however, on the assumption of spherical particles. Since Jupiter does not show a rainbow effect like Venus, it is to be doubted that this assumption of sphericity is correct, and for rod-shaped particles the calculation would be somewhat changed. The difference between the calculated 0.5 μm and the 0.7 μm of Figure 1 does not seem, therefore, to be a barrier to the particles of the Jovian clouds being interpreted as bacteria. Unlike bacterial spores, which contain about 25% water, rod-shaped bacteria, when active, contain about 70% water by weight. Estimating the equivalent refractive index of a uniform particle along the same lines as before, we obtain 1.37, which is close to the value 1.38 estimated from the reflecting properties of the clouds of Jupiter.

Water is not a problem on Jupiter. Much of Jupiter's oxygen is combined with hydrogen into water, and in the lower atmosphere below the visible upper clouds liquid water drops probably exist in profusion

We have emphasized already that bacteria are able to operate in almost any sense in which energy can be released by chemical reactions. Their many varieties simply take what is available. But if Jupiter is considered as a closed chemical system there can be very little that is available. In the absence of photosynthesis, which seems unlikely to take place low down in the clouds (except possibly under an exceptional condition to be considered later) degradable materials are not produced continuously within the system, and any degradable materials there may have been in the beginning, would long ago have been used up. So Jupiter can operate as a biological system only if degradable materials are supplied from without. Injected into the atmosphere of Jupiter, most meteoric material would, after vaporization, be degradable in this sense. Metallic iron is oxidizable in the presence of water, and there are bacteria which operate precisely on this transformation. As we have already remarked, sulphates are reducible to sulphur, and carbon dioxide is reducible to methane.

The flux of meteoric material in the neighbourhood of Jupiter has been measured to be about 100 times greater than it is near the Earth, due to a focusing effect produced by Jupiter's gravitational field. Moreover, the surface area of Jupiter is about 100 times greater than that of Earth. It follows therefore that the incidence of meteoric material onto Jupiter must be some 10,000 times greater than the incidence onto the Earth. Since the latter may be as high as ten thousand tonnes per year, the incidence onto Jupiter may well be 100 million tonnes per year. The energy produced by the chemical degradation of such a quantity of material is comparable with, but rather less than, the energy to be obtained from the burning of a similar quantity of wood. While very considerable, this amount of energy is smaller than the energy involved in photosynthetic processes on the Earth (the amount of wood grown annually is much more than 100 million tons). It must be remembered, however, that terrestrial biology operates in an oxidizing environment, which forces a high metabolic rate, otherwise the biological system would soon burn up. No such tendency to burn up exists on Jupiter. The great quantity of free hydrogen in the Jovian atmosphere removes all oxygen, and bacteria could survive there in a dormant condition for a very long time. Destruction rates would be low and populations could gradually be built up over extended periods. The ochre, red, and brown belts and spots of Jupiter can be identified with sulphur and iron bacteria.

It is one of the remarkable features of meteoric material that all size ranges contribute appreciable quantities of mass. This is true even for large bodies several kilometres in radius. It has been estimated that two or three such bodies hit the Earth in a million years, and for Jupiter there would correspondingly be two or three large bodies in a century, each contributing about 100 million tons of nutrient material in one sudden burst. A kilometre-size object hitting Jupiter's atmosphere at high speed would be disintegrated into heated gas that would be sprayed out over a considerable area, but still over an area that was small compared to the vast total area of Jupiter. It would form a spot, a spot in which the supply of nutrients for biological activity was exceptionally high, so that a large bacterial population would be built up in the resulting area. It is possible in this way to understand the origin of the spots of Jupiter, of which the Great Red Spot is the best-known example.

The possibility exists for a feedback interaction to be set up between the properties of a localized bacterial population (for example, the infrared absorption and emission properties of the bacteria) and the general meteorology of Jupiter's atmosphere. If bacterial populations on Jupiter have become adapted to the meteorology there, it is possible that evolution has produced a situation in which populations are able to prevent supplies of nutrient materials from being swept away from them by atmospheric motions. In a measure they may have become able to control the meteorology, and thus to hold together spot concentrations of nutrient materials. This may well be the explanation of the persistence of spots, and in particular the remarkable persistence of the Great Red Spot.

Let us now turn to meteoric materials in the form of tiny submicron particles. These too must make an appreciable contribution to the supply of nutrient material. Such particles are always electrically charged by sunlight, and being of small mass they are subject to deflection by the strong magnetic field of Jupiter, which causes them to rain down on the polar regions of the planet rather than on the equatorial zone. Bacterial activity resulting from this specifically polar accretion of fine meteoric particles may well be responsible for the remarkable dappled appearance of Jupiter at its poles.

At first sight one might think that photosynthesis would he impossible on Jupiter. At the upper cloud levels where sunlight is available the temperature is probably too low to permit bacterial activity, and at lower levels where the temperature rises to appropriate higher values there can be little penetration of sunlight. In the neighbourhood of spots the clouds are by no means smoothly layered, however. If sunlight can ever penetrate far enough to reach places where water is liquid, photo effects become possible. We remarked above that without sunlight, and in the presence of free hydrogen, bacteria can reduce sulphates in the presence of water. With sunlight, however, both reduction and oxidation become possible. A

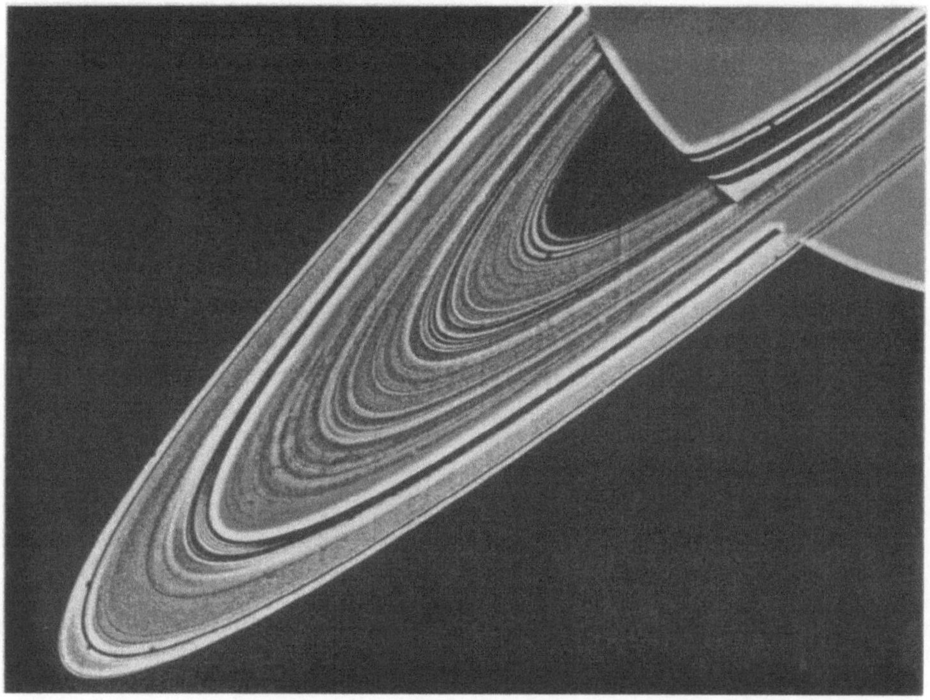

Figure 4. The Saturnian ring system photographed by Voyager 1 (Courtesy of NASA).

wider range of bacterial colours then arises, with the addition of purple, blue, and green to the yellows, reds, and ochres of the reducing bacteria. Such a situation may not be inconsistent, with recent observations from Voyager I. Occasional blue colours are seen on Jupiter, suggesting that the special conjunction of sunlight and liquid water may sometimes arise there.

Astronomical observations, particularly the most recent Voyager 1 data, show a generally similar situation with respect to colour on the main body of Saturn, including the presence of spots although with more yellow and green than Jupiter. The well-developed polar coloration and the absence of exceptionally large spots suggests that meteoric additions to Saturn may well be much more in the form of a rain of tiny particles than as large spot-forming bodies.

The rings of Saturn seem also to contain large quantities of bacteria-sized particles. Spectacular pictures of the ring system relayed from cameras on Voyager 1 in November 1980 revealed many surprising features. The so-called Cassini division between the two outer bright rings, the A- and B-rings, was seen to be populated by many rings of fine particulate material. The Saturnian ring system, which has been likened to the grooves on a gramophone record, is shown in Figure 4.

Two aspects of these recent observations are particularly noteworthy. Several spoke-like structures have been found to appear in the B-ring – Saturn's brightest

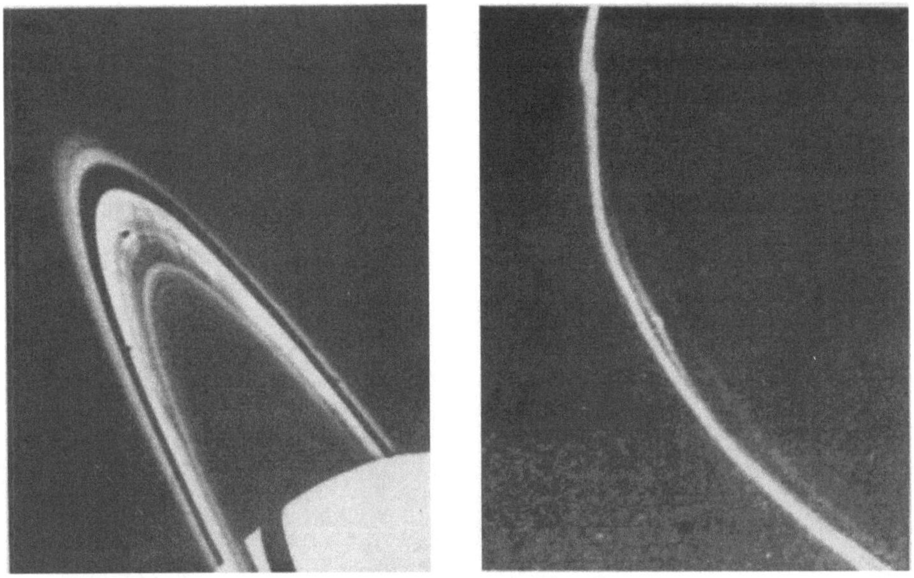

Figure 5. (Left) Rings of Saturn showing radial togues in the B-ring. The touges are comprised of bacteria sized particles (Courtesy of NASA). (Right) The outer F-ring comprised of bacteria-sized particles has a braided structure (Courtesy of NASA).

ring seen from the Earth (Figure 5 Left). These spokes appeared as dark tongues in images that Voyager 1 relayed whilst being above the rings on their sunward side but as the spacecraft went below the rings the same structures appeared as bright tongues. From this single observation we can infer that the particles in the tongues have the dimensions of bacteria. Such particles scatter sunlight mainly in the forward direction, so that they appear bright when the Sun is behind them, but dark when the Sun is in front. Another baffling feature of the new data concerns the braided and twisted appearance of the outer F-ring, which is also made up of bacterial-size particles (Figure5 Right). Since two small icy satellites have been discovered in orbits that straddle the F-ring, a possible resolution of this peculiar ring structure is that its particles are currently being spewed out from the two icy satellites designated S13 and S14. These satellites, which have sizes appropriate to large comets, may have become trapped in the gravitational field of Saturn and been melted in their interiors due to tidal effects, or due to the effect of collisions with smaller objects. Such satellites would act like biological pressure cookers, releasing jets of microorganisms, which had been produced in their melted interiors. We envisage a situation where the F-ring is woven *in situ* from jets of material thus released from S13 and S14. Magnetic effects, which involve the coupling of electrically charged bacteria with the planetary magnetic field, could play a role in maintaining both the twisted structure of the F-ring and the structure of the radial tongues seen in the B-ring.

No observations are yet available to determine the sizes of the particles which constitute the clouds of Uranus and Neptune. The atmospheres of these two planets are known, however, to contain vast quantities of methane, as do Saturn to a lesser extent, and Jupiter to a still smaller extent. The existence of methane may also be indicative of bacterial activity. Astronomers have been content to accept the methane observed in the atmospheres of the four large outer planets of the solar system as the outcome of a thermodynamic trend of carbon compounds to go to methane at low temperature and in the presence of an ample supply of free hydrogen. But if one takes a vessel containing a mixture of free hydrogen and carbon monoxide or carbon dioxide the separate gases persist unchanged for an eternity. The 'trend' is so slow as to be essentially zero. It is in such circumstances that catalysts are used in the laboratory and in industry. Of all catalysts, bacteria are the most efficient. Indeed the methame reproducing bacteria have evolved precisely to speed the conversion of carbon dioxide and hydrogen into methane and water. If the conversion happened at all readily under inorganic conditions there would be no niche for these bacteria. The natural explanation for the quantities of methane present in the atmospheres of Jupiter, Saturn, Uranus, and Neptune is that methanogenic bacteria have been active there in the reduction of carbon monoxide and carbon dioxide. The scope for such bacteria on the outer planets is clearly enormously greater than it is on the Earth, where this class of bacteria has been able to establish only a small toe-hold. Such bacteria could be responsible for the white zones of Jupiter, and for the generally uncoloured appearances of Uranus and Neptune.

2. Biology or Astronomy in Control?

For an externally impressed biological system to gain a toe-hold in a new environment, as for instance bacteria which are incident from space upon some new planet, it is essential that the external system be already equipped for survival in the new environment. Once a toe-hold becomes established, however, a two-way adaptation can develop. The basic components (genes) of the biological system could become selected to match the environment, or the cosmically-imposed components could be built (without being much changed individually) into structures better suited to the environment. In both of these cases it is the biological system that makes changes in response to the restrictions imposed by the physical aspects of its surroundings. This is one direction in which adaptation can take place. The other direction is through the biological system altering the environment itself. It is this second way with which we shall be concerned in this final section.

Terrestrial biology has changed the composition of our atmosphere, it affects the break-up of rocks into soil, it affects the rates of run-off of surface waters and of evaporation from the land, and it even influences in some degree the Earth's climatic zones. We similarly expect biology elsewhere to influence local conditions. The meteorology of the atmosphere of Jupiter was mentioned in the preceding section as a possible example, and perhaps the meteorology of Venus may also be affected. These are minor effects, however, compared to what biology may be capable of achieving on an astronomical scale.

It is interesting that issues that would influence the well-being of life on a cosmic scale turn out to involve problems which have been under active consideration in astronomy for more than a generation, and for which no satisfactory inorganic solutions have been found. These issues involve the star formation process and its relation to the interstellar grains. The unresolved issues are as follows:

What decides the rate at which stars form from the interstellar gas?

When they are formed what decides the mass distribution of stars?

What decides the rotations of the stars. and whether they are formed with planetary systems?

What decides how star formation is correlated throughout a whole galaxy, leading to the production of both grains and stars on a far-flung basis, often with the appearance of a new set of spiral arms for a galaxy?

The answers to these questions are almost certainly connected with the existence of a magnetic field everywhere throughout our galaxy. But the nature and origin of the galactic magnetic field is a further unresolved problem, and so the additional question must be added:

How did the magnetic field of our galaxy come into being?

This further question has proved so baffling that many astronomers have given up hope of answering it, by claiming the magnetic field to be truly primordial, it being imposed on the Universe at the moment of its origin. The magnetic field is what it is, because it was what it was, right back to the first page of Genesis. If this view, is correct the situation is crude, uncontrolled, and unsatisfactory.

To understand the difficulty of the problem, suppose we were to try to generate the magnetic field of our galaxy with the aid of an enormous electric battery, with one terminal connected to the centre and the other terminal to the outside of the galaxy. With the battery switched on, an electric current would begin to flow through the interstellar gas. Owing to a phenomenon known as 'inductance', the electric current would at first be exceedingly small, but as time went on the current would become stronger, and as it did so the magnetic field associated with the current would increase in its intensity. Suppose we allow the current to grow for the whole age of our galaxy, about 10,000 million years. What voltage do we need for the battery so that after 10,000 million years the resulting magnetic field will be as strong as the galactic magnetic field is actually observed to be? The answer is about 10,000,000 million (10^{13}) Volts. What process, we may ask, could have produced a battery of such enormous voltage that could operate for a time as long as 10,000 million years? A conceivable answer is a stream of electrically charged interstellar grains projected at high speed, 100 km s^{-1} or more, into an electrically neutral gas, a process in some respects similar to that which drives a terrestrial thunderstorm. The required projection speeds of $\sim$100 km s^{-1} are attainable by radiation pressure. The remaining problems are first to attain systematic directivity for a stream of grains, and second to maintain electrical neutrality in the gas through which the stream passes (when neutrality fails in a terrestrial thunderstorm there is a lightning flash, and the electric battery is instantly dissipated). With inorganic grains it is difficult, if not indeed impossible, to resolve these issues. Bacteria, on the other hand, have far more complex properties than inorganic grains, and may be able to exert a control both on stream directions and on the neutrality of the interstellar gas. The issue is not proved, but it is conceivable, and if it were to happen, bacteria would be well-placed to control the whole process of star formation.

The nutrient supply for a population of interstellar bacteria comes from mass flows out of the large galactic population of old stars (100,000 million of them), which may well have had an inorganic origin. 'Giants' arising in the evolution of such stars experience a phenomenon in which material containing nitrogen, carbon monoxide, water, hydrogen, helium, some refractory solid particles and supplies of trace elements flows continuously outward into space. In total from all giant stars, a mass about equal to the Sun is expelled each year to join the interstellar gas. This is the nutrient supply.

The problem for interstellar bacteria is that the nutrient supply cannot be converted immediately into an increase of the bacterial population, because of the need for liquid water, which cannot exist at the low pressures of interstellar space. Water in interstellar space exists either as vapour or as solid ice, depending on its temperature. Only through star formation. leading to associated planets and smaller bodies, can there be access to liquid water. Conditions suited to the presence of liquid water can exist over long periods of time on planets like the Earth. Liquid water need not exist, however, for long periods of time, since bacteria can multiply so extremely rapidly given suitable conditions. Shorter periods could exist on bodies much smaller than planets and in the early high-luminosity phase of newly formed stars the bodies could lie far out from the stars, at much greater distances than the Earth is from the Sun. In the case of our own solar system, liquid water could quite well have existed in the early days far out towards the periphery, and it could have existed at the surface of bodies of lunar size or inside still smaller bodies. The nutrients present in the outer regions of the solar system must have exceeded by many millions the amount at the Earth's surface. Hence the short-lived conditions associated with star formation must be of far greater importance to the population of interstellar bacteria than the long-lived, more or less permanent environment provided by planets like the Earth.

It has long been clear that the detailed properties of our own solar system are not at all what would be expected for a blob of interstellar gas condensing in a more or less random way. Only by a very strict control of the rotation of various parts of the system could such an arrangement as ours have come into being. The key to maintaining control over rotation would seem to lie once again in a magnetic field, as indeed does the whole phenomenon of star formation. The surest way for interstellar bacteria to prosper in their numbers would be through maintaining a firm grip on all aspects of the interstellar magnetic field. By so doing they would control not only the rate of star formation but also the kinds of star systems that were produced.

The multiplying capacity of bacteria is enormous, as we pointed out in earlier chapters. To go from an individual bacterium to the number of all the interstellar grains requires about 170 doublings. When conditions are optimal a bacterial population can double in a few hours, so 170 doublings take less than a month. Of course such a prodigious explosion in number would never literally be achieved because of practical limitations occurring in the availability of nutrients. Nevertheless, there is evidence that whole galaxies are overwhelmed from time to time by comparatively rapid and very large scale episodes of grain formation. as for instance the galaxy M82. This example is far from unique. There are many cases of galaxies embedded in a vast cloud of particles. On a lesser scale, there is a similar distinction between two appendages to our own galaxy, the Large and Small Magellanic Clouds.

For a generation or more astronomers have been accustomed to thinking of star-forming episodes accompanied by the production of vast clouds of interstellar grains. The episodes are sometimes local but they are often galaxy-wide. They are thought to be triggered by some large-scale event, the after effects of which linger on for some considerable time, several hundred million years. The condensation of the exceptionally bright stars which delineate the spiral structures of galaxies has often been associated with these episodes. From our argument it seems then that even the origin of the spiral structures of galaxies may well be biological in its nature.

The potential of bacteria to increase vastly in their number is enormous. It should occasion no surprise, therefore, that bacteria are widespread throughout astronomy. Rather would it be astonishing if biological evolution had been achieved on the Earth alone, without the explosive consequences of such a miracle ever being permitted to emerge into the Universe at large. How could the Universe ever be protected from such a devastating development? This indeed would be a double miracle, first of origin. and second of terrestrial containment.

Some biologists have probably found themselves in opposition to our arguments for the proprietary reason that it seemed as if an attempt were being made to swallow up biology into astronomy. Their ranks may now be joined by those astronomers who see from these last developments that a more realistic threat is to swallow up astronomy into biology.

References

Buchanan, R.E. and Gibbons, N.E. (eds.): 1974, *Bergey's Manual of Determinative Bacteriology*, The Williams and Wilkins Co., Baltimore.

Coffeen. D.L. and Hansen, J.E.: 1974, in: T. Gehrels (ed.), *Planets, Stars and Nebulae, Studied with Photopolarimetrv,* University of Arizona Press.

Hoyle, F. and Wickramasinghe, C.: 1981, *Space Travellers: The Bringers of Life*, University College, Cardiff Press.

Knollenberg, R.G. and Hunten, D.M.: 1979, *Science* **203**, 792.

A BALLOON EXPERIMENT TO DETECT MICROORGANISMS IN THE OUTER SPACE

J.V. NARLIKAR
Inter-University Centre for Astronomy and Astrophysics, Post Bag 4, Ganeshkhind, Pune 411 007, India
DAVID LLOYD
Cardiff School of Biological Sciences, Cardiff University, P.O. Box 915, Cardiff CF10 3TL, Wales, UK
N.C. WICKRAMASINGHE, MELANIE J. HARRIS, MICHAEL P. TURNER, S. AL-MUFTI and MAX K. WALLIS
Cardiff Centre for Astrobiology, Cardiff University, 2 North Road, Cardiff, Wales, UK
M. WAINWRIGHT
Department of Molecular Biology and Biotechnology, University of Sheffield, Sheffield, S10 2TN, UK
P. RAJARATNAM
Indian Space Research Organization, Antariksh Bhavan, New BEL Road, Bangalore 560 094, India
S. SHIVAJI and G.S.N. REDDY
Centre for Cellular and Molecular Biology, Uppal Road, Hyderabad 500 007, India
S. RAMADURAI
Tata Institute of Fundamental Research, Homi Bhabha Road, Mumbai 400 005, India
F. HOYLE
Cardiff Centre for Astrobiology, Cardiff University, 2 North Road, Cardiff, Wales, UK
*(*since deceased)*

Abstract. The results of biological studies of a cryosampler flown with a balloon, in which air samples were collected at altitudes ranging from 20 to 41 km, well above the Tropopause over Hyderabad, are described. In the analysis carried out in Cardiff, voltage-sensitive dyes that could detect the presence of viable cells were used on these air-samples. Clumps of viable cells were found to be present in samples collected at all the altitudes. The images obtained from electron microscopy are consistent with the above finding. Reference is also made to another paper presented at this conference describing the identification of bacterial species in the sample carried out in Sheffield. Counter arguments are discussed against the criticism that the detected cells and microorganisms (in the samples collected above the local tropopause at 16 km) are due to terrestrial contamination.

1. Introduction

Though the phenomenon of life was speculated to be of universal origin since very early times, scientific investigations of producing life from non-life, starting from a constituent mixture of inorganic gases were attempted in the early part of the last century by Oparin (1953) and Haldane (1929). The Urey-Miller experiments

Astrophysics and Space Science **285:** 555–562, 2003.

of the mid-1950s showed how amino acids and nucleotides might form from a mixture of inorganic gases (Miller and Urey, 1959). Since then several experiments have demonstrated the production of hexaglycine under conditions similar to those prevailing in terrestrial hot springs (Imai et al., 1999) and the production of biologically relevant molecules like alcohols, quinones and ether from UV irradiation of polyaromatic hydrocarbons in water-ice (Bernstein et al., 1999). But what is important to recognise is the fact that while the production of chemical building blocks is a necessary condition, it is not sufficient. What is relevant is their involvement in the highly specific processes for the production of nucleic acids, enzymes and other proteins, membranes and organelles that are involved in life.

In the mid-1970s, Fred Hoyle and Chandra Wickramasinghe advocated the alternative scenario in which microbial life is considered to be of universal origin and brought to the Earth as panspermia from comets which visit the Earth's neighbourhood from time to time. They argued that life on Earth began with this input. Hoyle and Wickramasinghe (1980) have given theoretical arguments that life cannot originate in a small terrestrial pre-biotic pool; a cosmic origin is instead suggested. See also the article by Wickramasinghe (2002) in this volume.

If life is truly cosmic, then several questions arise: i) Is there evidence in favour of *'panspermia'* ? ii) Can microorganisms survive in the harsh extraterrestrial environments? iii) What mechanisms are available for transport of such microrganisms to the Earth? We consider these briefly.

Microorganisms expelled from any source into unshielded regions of interstellar space will firstly become deactivated and subsequently degraded by exposure to cosmic rays, UV and other intense electromagnetic radiation. This will lead to the production of free organic molecules and polymers. An impressive array of such molecules have been detected. The production of such a variety of interstellar organic molecules from wholly non-biological sources is highly unlikely, if not totally impossible.

Over the last decade it has been demonstrated by several laboratory experiments that the microorganisms can survive extreme conditions of temperature, pressure and even radiation (Hoyle and Wickramasinghe 2000). Further a carbonaceous coating of even a few microns thick provides essentially total shielding against UV radiation (Secker et al., 1994). So far as their transport to the Earth is concerned Hoyle and Wickramasinghe (2000) have been advocating a cometary transport of these microorganisms for more than two decades, adducing several lines of evidence in favour of this hypothesis. We will not go into details here but simply state that this hypothesis was one of the motivating factors for the present experiment.

2. ISRO Cryosampler Experiment

Though some attempts to detect direct evidence for extraterrestrial life-forms entering our upper atmosphere were made in the 1960s and 1970s by mainly NASA supported Balloon Programmes (see Bruch, 1967 for a summary) and a Soviet Rocket Experiment (Lysenko, 1979), no definite conclusions could be drawn due to the primitive nature of the sterilization procedures that were used. Actually some indications of extraterrestrially-derived microorganisms were claimed, but in view of the lack of sound techniques to conduct the experiments aseptically, it was impossible to rule out terrestrial contamination. In view of these difficulties, very stringent procedures to completely exclude terrestrial contamination had to be evolved. Such techniques became available in the late 1990s (Shyamlal et al., 1996). Biochemical, chemical and molecular biological studies to identify the collected microorganisms were also being developed (Smibert and Kreig, 1994). Furthermore extremely sensitive new dye-based detection methods for living organisms were developed in Cardiff (Lloyd and Hayes, 1995; Lopez-Amoros et al., 1995). So an experiment was proposed to collect direct evidence for extraterrestrial life in stratospheric balloon flights using an aseptic cryosampler and highly sensitive voltage-sensitive dyes (Narlikar et al., 1998).

2.1. Collection of Air Samples Using the Cryosampler

The cryogenic sampler instrumentation comprised a 16-probe assembly. Each probe had a volume of 0.35 L and was made of high vacuum grade stainless steel. It was capable of holding a vacuum of 10^{-6} mb and pressure of 600 b. The temperature cycling ability of the probes was tested between −246°C and 140°C. To minimize contamination, the probes were machined from the above stainless steel stock, only the minimum required electron-beam welds were made, and the interior was electropolished. Just before the experiment, the probes and their manifold were cleaned with acetone and then four times with demineralised water. The assembly was then steam-baked and finally heated with infrared lamps to temperatures of 140°C. To prevent collection of any outgassed substances from the gondola, an intake tube of 2 m length formed a part of the payload ensemble and was sterilised as above. The probe mouth consisted of a metallic (Nupro) valve which was motor driven to open/close at a given altitude through ground command, using a telecommand-transmitter-receiver-decoder cryocontrol unit chain. During the flight, the probes remained immersed in liquid neon to create the cryopumping effect that allowed the ambient air samples to be collected on ground command.

2.2. First Experiment

In order to test the feasibility of collection of air samples aseptically and to test the rDNA sequencing and other procedures for identification of the microorganisms, a preliminary analysis was conducted by S. Shivaji and G.S.N. Reddy at CCMB and by P.M. Bhargava at Anveshna, Hyderabad, of one of the probes having an air sample collected from the altitude range of 10–36 km. This sample was collected in a balloon flight launched from Hyderabad on April 29, 1999. All procedures were carried out under aseptic conditions using stringently sterilised equipment.

The sample air was passed first through a 0.45 micron and then through a 0.22 micron pore filter. The exiting air was passed through a calibrated flowmeter. Each filter was placed in a nutrient agar plate; no growth occurred at 25°C in seven days. The filter was then transferred to a blood agar plate and incubated at 25°C for 11 days, when six distinct colonies had grown from the 0.45 micron filter; none were obtained from the 0.22 micron filter. These colonies were sub-cultured and maintained on nutrient agar. The cultures have been deposited in the MTCC Type Culture Collection of the Institute of Microbial Technology, Chandigarh, India.

Based on the morphological, physiological and biochemical characteristics and additionally on the basis of the 16S rDNA sequencing the isolates were identified as *Pseudomonas stutzeri*. But, these isolates were distinct from all the earlier described strains of *P. Stutzeri* with respect to i) ability to grow well on lysine, ii) high percentage of C15:0 and C18:0 fatty acids, iii) presence of carotenoid pigments, and iv) the yellow pigment which was produced only at room temperature (25°C), and then only in the stationary phase. In addition it is noteworthy that *P Stutzeri* has so far not been described as an airborne organism, its natural habitat being soil and water. So it is tempting to speculate that this is evidence for an extraterrestrial bacterium. But the fact that the air samples were collected as low as 10 km altitude, where even debris from jet planes can provide terrestrial contamination, puts a question mark over the extraterrestrial nature of the detected organism. At the same time, the unique properties of the detected colonies on comparison with terrestrially known strains of *P Stutzeri* are remarkable. While the experiment demonstrated the capability of the techniques for collection of air samples and detection of microorganisms, it also showed the need for a more careful experiment, where samples are collected at heights well above the ones where normal terrestrial contamination is even remotely possible.

2.3. AIR SAMPLES COLLECTED BY THE CRYOSAMPLER BALLOON FLIGHT ON 21 JAN. 2001

The balloon carrying the cryosampler payload was launched on 21 January 2001 from Hyderabad. Air samples were collected at different heights above the local tropopause at 16 km. The lowest height was 19.8 Km and the highest at 41.06 km. Probes in this flight were in duplicate sets at each height. One set is being analysed in Cardiff and the other one at CCMB, Hyderabad. In this meeting the preliminary results from the probes being analysed in Cardiff are given, whose details are given in Table I.

We now describe the results of biological studies carried out in Cardiff.

TABLE I
Details of probes being analysed in Cardiff

Probe	Collection height range	Collected NTP volume (litres)
A	19.80–20.32 km	81
B	24.36–27.97 km	70.5
C	28.47–39.05 km	38.4
D	39.75–41.06 km	18.5

3. Detection of Viable Cells

All recovery procedures were conducted in a sterile system in a laminar-flow chamber. From each of the 4 probes, air was passed first through a 0.45 micron filter and then through 0.22 micron filter. Eight filters were thus derived. The probes were stored at −70°C before sample preparation, and likewise the derived membrane filters were stored at this temperature before isolates were obtained and tested.

Since only the 0.45 micron filters are expected to have trapped microbial sized particles, these have been analysed. Approximately 4 mm^2 squares were aseptically cut from the filters and treated with fluorescent membrane potential sensitive dyes; either a cationic carbocyanine or an anionic oxonol dye was employed. Cationic dyes penetrate the cell membranes of viable cells, but not of dead cells, whereas anionic dyes penetrate the membranes only of non-viable cells. Any viable living cell present in a sample would therefore be expected to give rise to a fluorescent image when excited by light and could be located using an epifluorescence microscope. Light from laser Argon at the wavelength of 488 nm was used for this purpose. Depending on the optics employed for measurement, the images represent a single organism or a cluster. For details of the method see Lloyd and Hayes (1995).

Isolates treated with cyanine dye showed fluorescent images in the form of clumps of 0.3–1 micron sized organisms, the clumps themselves measuring 5–15 microns across. Higher resolution images will require deployment of a confocal microscope, but already the detection of viable cells by this technique (not found in the sterile controls) is beyond doubt.The use of anionic dyes revealed a comparable detection rate of dead or non-viable cells. In this report attention is focused only on the take-up of a cationic carbocyanine dye as an idicator of viable cells in the stratosphere.

Electron microscopy of aseptically isolated squares of membrane filters was performed next. The squares were mounted on 12 mm diameter sticky carbon tabs which in turn were mounted on 10 mm diameter aluminium tabs. The samples treated in this way were gold sputter-coated and imaged in a JOEL 5200 LV scanning electron microscope under a vacuum of 7 nanobar. The procedure adopted is suited to imaging bacteria because bacterial cell walls do not collapse or explode under the conditions of observations. The structures were similar to the ones revealed by the epifluorescence microscope. Further analysis on the nature of the microorganism is in progress.

On each of the micropore filter isolates examined so far, measuring approximately 2 mm × 2 mm, a number N in the range of 1–3 microbial clumps were found. Since the air volume passing through each filter in A and B was about 80 litres at NTP (Table I) and the area of the entire filter is about 2000 mm^2, with N=3, the density of microbial clumps at 25 km is estimated as $[(3 \times 2000)/4]/80 = 18.75$ per litre at NTP. With an atmospheric pressure at 25 km of 0.025 bar, the estimated number density of microbial clumps at 25 km is 0.47 per litre. At a height of 40 km, the average number of clumps per 4 mm^2 of membrane filter may provisionally be taken as N=1 subject to further analysis. The NTP equivalent volume of air that passed through 2000mm^2 of this filter being 18.5 litres (Table I) and the ambient air pressure being 0.0025 bar, the density of microbial clumps is about 0.068 per litre of air at 40 Km. With a clump comprised of about 100 bacterial cells the number density of viable cells at 40 Km, corresponding to this would be about 7 per litre.

Kasten (1968) had shown that if there is a steady infall of matter from outside the atmosphere, the number density $\mathcal{N}(h)$ of matter particles of any given kind would be inversely proportional to the terminal velocity $w(h)$ at that height. Using this formula, one gets an exponential drop in the number density with height:

$$\mathcal{N} \propto \exp(-ah), \tag{1}$$

where a can be determined from atmospheric parameters. Using the values available, one estimates that the number density of viable cells in steady state should follow a similar distribution and that the number density at height 41 km should be approximately one tenth of the number density at 25 km. The above calculation, given the experimental errors, does appear consistent with this hypothesis.

3.1. Terrestrial Contamination

Considering the small number of about 7 living cells per litre of air at 41 km, as determined above, the main concern will be whether what has been found is of terrestrial contamination. As noted earlier, the possibility of contamination due to the instrumental collection process is completely ruled out by the stringent sterilization methods adopted. Hence the only possibility is the presence of terrestrial cells being carried aloft to such great heights by some extraordinarily rare events or spacecraft debris.

Under normal circumstances atmospheric mixing takes place only up to the tropopause, that acts as a barrier against the transport of waves responsible for mixing the terrestrial material with the upper atmosphere. However during some extraordinary events such as a very powerful volcanic eruption, it is possible for the material to reach large vertical heights. This was demonstrated by the volcanic eruption of Mt Pinatubo in Phillippines on June 15, 1991, which was estimated to have injected nearly 20 million tonnes of SO_2 upto a maximum vertical height of 32 km (Grant et al., 1994 and references therein). The stratospheric mixing has been almost continuously monitored using several balloon flights (Deshler et al., 1992). It was shown that the gravitational settling to lower altitudes is rapid and after a few months, no significant mixing even in the lower stratosphere was taking place. Considering this direct evidence of no contamination of even as powerful a volcanic eruption as that of Mt Pinatubo, we can immediately rule out terrestrial contamination as an explanation of our results, as no volcanic eruption or any other extraordinary event took place even months prior to our experiment.

On the positive side, as mentioned above, the observation of the depth profile of the organisms matches that predicted for extraterrestrial cells injected at the top of our atmosphere, using the method of Kasten (1968). Furthermore, the same calculation yields the clearance of any transient injection (e.g. of space debris) through gravitational settling in a time of order of months. Hence it stretches credulity to maintain that the cells detected in our experiment are of terrestrial origin.

Milton Wainwright at Sheffield university (who is reporting his findings in another paper in this meeting) has claimed to have isolated two species of bacteria, *Bacillus simplex* and *Staphylococcus albus* from the samples at 41 km. Details of this finding may be seen in his paper (Wainwright, 2002) in this volume.

4. Concluding Remarks

Viable living cells have been detected using the cationic cyanine dyes at all heights ranging from 21 km to 41 km. Terrestrial contamination is ruled out because of the sample collection at altitudes well above the tropopause during a time when there were no extraordinary terrestrial events like volcanic eruption etc. Furthermore the use of stringent procedures for sterilization rules out the contamination due to the instruments or balloon. With an average falling speed for 3 micron sized clumps at 40 km of about 0.3 cm/s (Kasten, 1968), the infall rate of clumps with a number density of 0.068/litre over the entire Earth would be

$$(0.068 \times 10^{-3}) \times (0.3) \times (5 \times 10^{18}) \text{ per second}$$

Assuming an average of 100 individual bacterial cells each of mass 3×10^{-14} g in a clump, a daily mass input of about a third of a tonne of biomaterial is deduced. Although these estimates are very tentative, they serve the purpose of illustrating the amount of infall matter involved, if the results of this investigation are accepted, and a *prima facie* case for a space incidence of bacteria into the Earth is seen to be established.

Acknowledgement

We are grateful to Professor K. Kasturirangan, Chairman, Indian Space Research Organisation (ISRO), for making available ISRO's state of the art cryogenic sampler payload together with avionics and special sterilisation systems specifically developed by ISRO for this experiment, as also for his sustained support, guidance and encouragement.

References

Bernstein, M.P., Stanford, S.A., Allamandola, L.J., Gilette, J.S., Clement, S.J. and Zare, R.N.: 1999, *Science* **283**, 1135.

Bruch, C.W.: 1967, in: P.H. Gregory and J.L. Montieth (eds.), *Airborne Microbes Symposium of the Society for Microbiology*, **17**, p. 345, Cambridge University Press).

Deshler, T., Hofmann, D.J., Johnson, B.J. and Rozier, W.R.: 1992, *Geophys. Res. Lett.* **19**, 199.

Grant, W.B. et al.: 1994, *J. Geophys. Res.* **99**, 8197.

Haldane, J.B.S.: 1929, *The Origin of Life*, Chatto and Windys.

Hoyle, F. and Wickramasinghe, N.C.: 1980, *Evolution from Space*, J.M. Dent & Sons.

Hoyle, F. and Wickramasinghe, N.C.: 2000, *Astrophys. Sp. Sci.* **268**, 1.

Imai, E., Honda, H., Hatori, K., Brack, A. and Matsumo, K.: 1999, *Science* **283**, 831.

Kasten, F.: 1968, *J. Appl. Meteorology* **7**, 944.

Lloyd, D. and Hayes, A.J.: 1995, *FEMS Microbiology Lett.* **133**, 1.

Lopez-Amoros, R., Mason, D.J. and Lloyd, D.: 1995, *J. Microbiol. Meth.* **22**, 165.

Lysenko, S.V.: 1979, *Mikrobiologia* **48**, 1066.

Miller, S.L. and Urey, H.C.: 1959, *Science* **130**, 245.

Narlikar, J.V. et al.: 1998, *Proc. SPIE Conf. On Instruments, Methods and Mission for Astrobiology* **3441**, 301.

Oparin, A.I.: 1953, *The Origin of Life*, translated by S.Margulis, Dover.

Secker, J., Wesson, P.S. and Lepock, J.R.: 1994, *Astrophys. Sp. Sci.* **329**, 1.

Shyamlal et al.: 1996, *Ind. J. Radio Sp. Phys.* **25**, 1.

Smibert, R.M. and Kreis, N.R.: 1994, in: P. Gerherdt, R.G.E. Murray, W.R. Wood and N.R. Kreig (eds.), *Methods for General and Molecular Bacteriology*, Am. Soc. Microbiol., p. 603.

Wainwright, M.: 2003, Reference to article in this volume.

Wickramasinghe, N.C.: 2003, Reference to article in this volume.

A MICROBIOLOGIST LOOKS AT PANSPERMIA

MILTON WAINWRIGHT
Department of Molecular Biology and Biotechnology, University of Sheffield, Sheffield, S10 2TN, UK

Originally, the term 'panspermia' was used by microbiologists (of the late Victorian period, e.g. Roberts, 1874) to refer to the observation that terrestrial air is full of microorganisms. Panspermia, was later used to cover the view that life on Earth originated from space, while more recently, it has been extended to describe the hypothesis that life continues to rain down to Earth from space. In order to avoid confusion, here I will use the term panspermia in its original astrobiological sense, while adopting the term neopanspermia to refer to the hypothesis that life continues to arrive to Earth from space.

Although belief in both the panspermia and neopanspermia is becoming somewhat fashionable, many authors habitually fail to make reference to the originators of the modern view of these hypotheses, namely Fred Hoyle and Chandra Wickramasinghe (Wainwright, 2001). Some authors instead date the resurrection of panspermia *sensu lattu* to the far more speculative, and essentially unprovable, views of Francis Crick on so-called directed panspermia (i.e. the view that life on Earth was seeded by some unknown, cosmic civilisation) (Crick, 1981). There is no doubt however, that priority of the modern view of neopanspermia belongs to Hoyle and Wickramasinghe (Hoyle and Wickramasinge, 2000). Hoyle and Wickramasinghe (1979) have also championed the view that pathogens arrive to Earth from space; an idea ('pathospermia') that has come in for even more derision than their views on neopanspermia.

Here I intend to discuss both panspermia and neopanspermia from a microbiologist's viewpoint, taking the opportunity to point out the problems inherent in trying to demonstrate that Earth is being continually seeded with space-derived microorganisms.

Survival of Space-borne Micro Organisms

It goes without saying that any microorganism reaching the Earth from space must have survived the extreme rigours of the space environment. Although microorganisms need not grow and reproduce in space, they obviously must survive in a form that can then reproduce on arrival on Earth. The ability of microorganisms to withstand environmental extremes is being increasingly widely recognised. Hoyle and Wickramasinghe (Wickramasinghe, 2001) pointed out that Earth microorganisms (geomicrobes) possess all the characteristics necessary to allow them to survive in space and, by reversing this argument, suggested that the possession of such

Astrophysics and Space Science **285:** 563–570, 2003.

characteristics suggests that microorganisms originated in a more demanding environment, i.e. space. This viewpoint has however, been criticised by, amongst others, Battista et al. (1999) who suggests that microbial resistance to ionising radiation may involve DNA repair mechanisms that are evolutionary responses to other environmental factors, such as water shortage.

Although emphasis is currently being placed on the study of microorganisms, notably bacteria, that can live, and are specifically adapted to growth in, extreme environments (i.e. extremophiles), it is important to note that many common bacteria, inhabiting non-extreme environments, are able to survive extreme conditions. Such organisms can be termed *extremodures*. At the turn of the century for example, scientists showed that common bacteria could survive at a temperature of –252 °C (McFayden and Rowland, 1900). More recently, *Escherichia coli* has been shown to be capable of surviving exposure to high pressure (Al-Mufti et al., 1984).

Exposure to radiation, notably UV, is likely to be the major factor limiting the survival of microorganisms in space. Bacteria can however, resist UV by forming clumps of individual cells (as in *Sarcina* or *Staphylococcus*), while the spores of both bacteria (e.g. *Bacillus*) and fungi are often UV-resistant. The shading provided by adherence to particulates might also help protect microorganisms from UV, as might the formation of a UV-resistant layer resulting from the carbonisation of the outer cells of a mass of space-inhabiting bacteria (Hoyle and Wickramasinghe, 2000). Such considerations suggest that panspermic microorganisms need not be particularly unusual or especially specialised in order to survive the rigours of space.

The greatest degree of protection from the rigours of the space environment will obviously be afforded to microorganisms surviving in comets or laying dormant within masses of interstellar dust or meteorites. It is not surprising therefore to find that bacteria-like fossils have been found in the Murcheson meteorite and in the much discussed Mars meteorite, ALH 84001. There have also been a number of claims that living bacteria (species of *Bacillus* and *Staphylococcus*) have been isolated from meteorites), although these have generally been regarded as being contaminants (Orio and Tornabene, 1965). Nevertheless, claims that ancient living bacteria exist in meteorites continue to appear (Abbott, 2001).

Contamination and the Problem of Common Versus Unusual Microorganisms

The theory of neopanspermia predicts that microorganisms are continually raining down upon our planet and that by definition most, if not all Earth microorganisms, are derived from space. Earth organisms are therefore space organisms. The hypothesis states that common Earth organisms such as species of *Pseudomonas, Staphylococcus* and *Bacillus* originate from space. The fact that common geomicrobes can survive under extreme conditions explains how this might be possible. Unfortunately a problem arises with human perception that makes this suggestion unpalatable. Most microbiologists would assume that space-derived microorganisms would *a priori* be a) novel, b) primitive, and c) possess unusual, marker-physiologies. The demonstration that common Earth organisms occur in space would doubtless induce the knee-jerk response of most microbiologists that such isolates were contaminants, or at best originated from Earth. The isolation of space-derived microorganisms would probably only be generally accepted if they proved to be novel or unusual, despite the fact that this is not a requirement of the neopanspermia hypothesis.

Since microorganisms exist everywhere on Earth, any attempt to isolate microorganisms from space-derived samples is obviously bedevilled by problems of contamination. Any claim that geomicrobes have been isolated from space can be readily dismissed by invoking contamination, even where highly rigorous sterile techniques have been employed to exclude this possibility. Although contamination is a real problem it is often too readily used to dismiss all claims for the isolation of ancient, or space-inhabiting, microorganisms.

Recent Studies of the Isolation of Microorganisms from the Stratosphere

Scientists in India recently collected samples at 41 km above the Earth's surface. Every effort was made to avoid contamination of these samples, portions of which were sent to Cardiff University and transferred to filter membranes for microbial analysis. Here, viable, but none culturable bacteria were isolated. The bacterial masses isolates consisted of coccoid clusters with the occasional rod (Harris et al., 2001). Samples of the filters were forwarded to this laboratory where attempts were made to grow bacteria and any other microbial isolates present on the filters. The membrane filters, through which washout from the 41 km was passed, yielded two bacteria and one fungal isolate (Wainwright et al., 2002) The bacterial isolates were independently identified, using 16srRNA analysis, as *Bacillus simplex* (100% similarity) and *Staphylococcus pasteuri* (98% similarity, this isolate would however, have been classified as a *Micrococcus* had traditional identification techniques been employed). The fungal isolate was independently identified as *Engyodontium album.* All of these isolates are common Earth organism found in soil and vegetation.

None of the isolated organisms had ever been used in the laboratory where the work was performed. The work was conducted in a laminar airflow cabinet using standard microbial techniques. The fact that the isolated bacteria were rod and coccoid correlated with observations made at Cardiff on the uncultureable isolates, using scanning electron microscopy. The knee-jerk reaction of critics to the culturing of common bacteria from 41 km would be that they are laboratory contaminants; however, the following evidence suggests otherwise.

The filters were shaken in sterile deionised water; this was transferred to the surface a range of isolation media and incubated at 25 °C. Initially no surface growth and colony formation was observed on the media after 4 days of incubation. The surface of each medium was then removed and examined under the microscope, In the case of the PDA medium alone, two bacteria (mainly cocci with the occasional rod) were seen under the light microscope. These bacteria were then transferred to liquid LB medium and incubated at 30 °C for 4 days. A degree of serendipity was involved in the isolation of the bacteria, since the only PDA available at the time was some 20 years old. When reconstituted, an unusually soft medium resulted, which presumably because of oxidation, was browner in colour than fresh PDA. Bacteria were only isolated when this soft, aged, PDA was employed. Since airborne, or other contaminants, would readily have grown to form colonies on all the media used in this study, the implication is that the organisms isolated were not contaminants, but must have been exposed to unusual conditions and that only soft PDA supported their growth. It is possible that unknown growth promoting oxidation products may have provided the stimulus for growth, alternatively the fact that the medium was soft and wet may have been important. It is noteworthy that freeze-dried cultures are generally resuscitated using liquid media. This suggests the possibility that the isolates have been in freeze-dried state, suggesting that they had a non-terrestrial, i.e. space origin.

Problems Resulting from Attempts to Replicate these Isolations

A technician independently isolated the above named bacteria from the same membrane sample used in the initial isolation; the technician worked in a separate laboratory within this Department. The same isolate protocol was used, but the technician was unaware of any expected outcome. He was able to successfully isolate *S. pasteuri* and *B. simplex* (but not *Engyodontium album*) from the membrane. Subsequently however, workers at two other universities were unable to isolate the organisms when other membranes were sampled. The inability of independent workers, in separate University laboratories would, at first sight, appear to invalidate the original isolation work conducted in Sheffield. However, it must be borne in mind that such attempts at replication represent isolation experiments and that every membrane need not possess viable microorganisms. It may be that, while the majority of organisms present on the membranes are non-viable, viable cells were, by chance present on only one membrane.

If we assume that the isolates were not laboratory or otherwise contaminants then they must have originated at a height of 41 km above the Earth. The obvious question that arises is – how did they get to the altitude? The two obvious conclusions are that they were carried up from Earth and remained, or were falling through, the atmosphere at 41 km from where they were isolated. Alternatively, they may have arrived from space and were sampled at 41 km before falling to Earth. Proof of the latter possibility would of course validate neopanspermia.

The apparently obvious conclusion is that the bacteria and the fungal isolates were carried up from Earth and floated in space at 41 km from where they were collected. However, the only means by which bacteria can achieve a height of 41 km would appear to be via a volcanic eruption. The residence time of such particles, derived from volcanoes, would be only a matter of days. Since no volcanic eruptions occurred on Earth during the years prior to the 41 km-sampling event this would appear to rule out this possibility, although some authorities continue to believe that particles can cross the tropopause and reside at 41 km for longer periods (Gregory, 1961).

The Neopanspermia Paradox

Most microbiologists would assume that the isolates, if not contaminants, must have arrived to a height of 41 km from the Earth's surface. The isolates are regarded as common Earth bacteria. The fact that the *S. pasteuri* isolate is, unlike *S.aureus* and *S.epidermidis* capable of solubilising insoluble phosphates suggests that this isolate is not a skin contaminant, but possesses a property typical of a 'common' environment-derived, isolate. The use of the term, common in relation to microorganisms is however, questionable since microorganisms (with the exception of certain extremophiles) tend to be found ubiquitously over the Earth's surface. Even microorganisms that are regarded as being extremophiles (e.g. halophilic organisms, capable of growing in high salt concentrations) can often be readily isolated from non-saline soils. The ubiquitous distribution of microorganisms would of course support the view that such organisms are continually raining down from space, over the total surface of the Earth. Clearly the suggestion that the bacteria isolated from 41 km cannot have come from space because they are Earth organisms is in direct opposition to the neopanspermia hypothesis. If however, the Earth origin of these isolates is accepted than we are presented with a paradox since this would mean that bacteria are able to leave the Earth and infect space. Such a paradox would in fact demonstrate the correctness of the neopanspermia hypothesis and would suggest that space is contaminated by geomicrobes. Whether or not these organisms would find there way into deep space is however, debatable. In short those who believe that organisms isolated at 41 km must be from Earth are unwittingly validating the neopanspermia hypothesis.

The Problem of Phylogeny

From the biologists point of view the major limitation on the view that modern microorganisms can arrive from space relates to evolution and phylogeny. Essentially, because it should have evolved, any ancient microorganism should, in terms of its nucleotide sequence, be markedly different from modern microorganisms. If an organism has a very similar nucleotide sequence to a modern organism then it must, according to current molecular biology, be a modern organism. Similarly, an organism originating from space would be expected to possess nucleotide sequences which differ from modern organisms found on Earth, simply because one would not expect the direction and rates of evolution to be the same in space as on Earth. An example of how such considerations can lead to conflict is provided by the recent findings of bacteria in ancient salt crystals. Vreeland et al. (2000) isolated a bacterium from a 250 million year old salt crystal. They claimed that the bacterium was not a contaminant, but was the same age as the crystal. Graur and Pupko (2001) however, claim, based on the use of molecular methods, that the bacterium is modern and, despite the fact that Vreeland et al. (2000) used extremely rigorous sterile technique, must be a modern contaminant. Similar findings have been reported following the examination of Permian salt crystals by Fish et al. (2001). Clearly, if the bacterium is not a contaminant then there must be something wrong with our current understanding of molecular phylogeny and palaeontology. The general acceptance of the viewpoint that the nucleotide sequences of microorganisms must have changed over time (due to evolution) makes most microbiologists believe that any microorganism with nucleotide sequences close to that of modern microorganisms must, by definition, be a modern microorganism. Invoking Occam's Razor, most biologists would state that if an Earth-like organism is found at 41 km above the Earth's surface then it is an Earth organism; anyone who believes that the tropopause acts as a barrier to such contamination must as a result, alter their own, rather than the evolutionary paradigm. Similarly, the presence of a modern organism in an ancient salt crystal must, say the biologists, indicate only thing, contamination; it should however, be noted that the whole question of molecular phylogeny is in a sate of constant flux (Maher, 2002).

Panspermia and the Origin of Life

As was mentioned above, the term panspermia was originally used in its astrobiological connotation to refer to the view that life did not originate on Earth, but was seeded by organisms (and not just microorganisms) from space. A small number of microorganisms arriving in this way would have been capable of replicating at an incredibly rapid rate and would, in the absence of any competition from native organisms, have rapidly colonised the planet. It is noteworthy that many microorganisms have the ability to grow at very low nutrient concentrations (i.e. they are oligotrophic). Such oligotrophic growth would have been essential because nutrients to support microbial growth would have been present at very low concentrations in the un-inoculated prebiotic soup. It is generally accepted that heterotrophic life predated life based on photosynthesis, a view that ties in well with the view that heterotrophs arrived exogenously.

Wainwright and Falih (1997) showed that a fungus could grow, without added nutrients, on buckminsterfullerene. Since fullerenes have been transported to Earth in meteorites and could protect bacteria from UV, this novel carbon allotrope may have played an important role in the early origin of life on Earth.

Once the conditions on Earth were suitable life took off remarkably quickly; a fact that is in agreement with the view that it arose exogenously with Earth having been continually showered with microorganisms which lay inert until the point at which conditions were suitable to support their growth.

The fact that life got going so quickly has been used recently to suggest that the origin of life is a simple and frequent event, a viewpoint that is not however, supported by any compelling evidence. Finally, it has been suggested that if living organisms did not make the journey through space to begin life on Earth then perhaps DNA or RNA may have rained down on the prebiotic Earth and given life a 'kick-start' (Gribbin, 2001).

Conclusion

In 1928, a certain Professor F.G. Donnes had this to a say about panspermia:

> Perhaps the chief objection to the doctrine of panspermia is that it is a hopeless one. Not only does it close the door to thought and research, but also it introduces a permanent dualism into science and so prejudices important philosophical issues.

Many scientists clearly continue to express this viewpoint. It is often said for example that panspermia solves nothing simply because it does not answer the question of how life arose; if life arose in space, it is argued, then it could just as easily have arisen on Earth. Of course such arguments ignore the contrast between the finite nature of Earth and the infinity of space, in terms of both time and the limitless variety of astronomical bodies on which life could have arisen.

Donnes was clearly wrong when he said that panspermia closes the door to thought and experimentation. Such is the current level of theoretical and practical work being devoted to the question of neopanspermia that much new knowledge will be gained from its study, even if it does eventually prove to be a 'false doctrine' Neopanspermia is a scientifically valid idea simply because it can be refuted by experimentation.

Unfortunately, microbiological studies on Earth are unlikely to provide overwhelming evidence of the validity of the neopanspermia hypothesis. This is simply because cynics can dismiss any results by invoking contamination, especially because the Hypothesis predicts that microorganisms found in space will be identical to those found on Earth. As a result, the only way to convincingly demonstrate panspermia is to conduct experiments in space where the problems of contamination can be eliminated. However, in the absence of the facilities to study microorganisms *in situ* in space, evidence can only be accumulated from Earth-based experiments. If the neopanspermia is correct then a critical mass of information will accumulate that hopefully, will be sufficient to convince even the most diehard sceptic.

References

Abbot, A.: 2001, Resuscitated 'alien' microbes stir up Italian storm, *Nature* **411**, 229.

Al-Mufti, S., Hoyle, F. and Wickramasinghe, C.: 1984, The survival of *E.coli* under extremely high pressures, in: C. Wickramasinghe (ed.), *Fundamental Studies and the Future of Science*, Cardiff, University of Cardiff Press, pp. 342–352.

Battista, J.R., Earl, A.M. and Park, M.J.: 1999, Why is *Deinococcus radiodurans* so resistant to ionizing radiation? *Trends in Microbiol.* **7**, 362–365.

Crick, F.: 1981, *Life Itself*, London, Macdonald.

Donnes, F.G.: 1928, The mystery of life, *Review of Reviews* **76**, 349–354.

Fish, S.A., Shepherd, T.J., McGenity, T.J. and Grant, W.D.: 2001, Recovery of 16s ribosomal RNA gene fragments from ancient halite, *Nature* **417**, 432–436.

Graur, D. and Pupko, T.: 2001, The Permian bacterium that isn't, *Mol. Biol. and Evol.* **18**, 1143–1146.

Gregory, P.H.: 1961, *The Microbiology of the Atmosphere*, Leonard Hill, London.

Gribbin, J.: 2001, *Space*, BBC Books, London.

Harris, M.J., Wickramasinghe, N.C., Lloyd, D., Narlikar, J.V., Rajaratnam, P., Turner, M.P., Al-Mufti, S., Wallis, M.K., Ramadurai, S. and Hoyle, F.: 2001, *Proceedings SPIE Conference*, 4495, in press.

Hoyle, F. and Wickramasinghe, N.C.: 1979, *Diseases from Space*, Dent, London.

Hoyle, F. and Wickramasinghe, N.C. (eds.): 2000, *Astronomical Origins of Life-steps towards panspermia*, Kluwer, Dordrecht.

Macfayden, A. and Rowland, S.: 1900, Influence of the temperature of liquid hydrogen on bacteria, *Proc. Roy. Soc.* **66**, 488–489.

Maher, B.A.: 2002, Uprooting the tree of life, *The Scientist* **16**, 26–27.

Orio, J. and Tornabene, T.: 1965, Bacterial contamination of some carbonaceous meteorites, *Science* **150**, 1046–1048.

Roberts, W.: 1874, Studies on biogenesis, *Proc. Roy. Soc.* **22**, 289–291.

Vreeland, R.H., Rosenzweig, W.D. and Powers, D.W.: 2000, Isolation of a 250 million year old halotolerant bacterium from a primary salt crystal, *Nature* **407**, 897–900.

Wainwright, M.: 2001, In praise of Hoyle and Wickramasinghe, *Microbiology Today* **28**, 2.

Wainwright, M. and Al Falih, A.M.K.: 1997, Fungal growth on buckminsterfullerene, *Microbiology* **143**, 2097–2098.

Wainwright, M., Wickramasinghe, N.C., Narlikar, J.V. and Rajaratnam, P.: 2002, Evidence for microorganisms in stratospheric air samples collected at a height of 41 km, *Proc. SPIE Conf., Hawaii*, in press.

Wickramasinghe, C.: 2001, *Cosmic Dragons*, Souvenir Press, London.

WHAT DARWIN MISSED

ANTHONY K. CAMPBELL
Department of Medical Biochemistry and the Darwin Centre, University of Wales College of Medicine, Heath Park, Cardiff, CF14 4XN, UK; E-mail: campbellak@cf.ac.uk

Abstract. Throughout his life, Fred Hoyle had a keen interest in evolution. He argued that natural selection by small, random change, as conceived by Charles Darwin and Alfred Russel Wallace, could not explain either the origin of life or the origin of a new protein. The idea of natural selection, Hoyle told us, wasn't even Darwin's original idea in the first place. Here, in honour of Hoyle's analysis, I propose a solution to Hoyle's dilemma. **His** solution was life from space – panspermia. But the real key to understanding natural selection is 'molecular biodiversity'. This explains the things Darwin missed – the origin of species and the origin of extinction. It is also a beautiful example of the mystery disease that afflicted Darwin for over 40 years, for which we now have an answer.

Hoyle and Darwin

'The Darwinian theory is wrong because random variations tend to worsen performance'. Thus wrote Fred Hoyle in his famous book 'The intelligent universe'. Hoyle pointed out three important things in this book (Hoyle, 1983). First, that the idea of natural selection had been around for several decades before Darwin wrote The Origin. Secondly, that it was Wallace's clear letter of 1858 that really clarified Darwin's mind on the matter. Thirdly, and more important, natural selection as conceived by Darwin and Wallace just won't work mathematically (Hoyle, 1987, 1999). The odds are stacked hugely against random change producing even one new protein. What then is the solution to Hoyle's dilemma? Life from space is one. But still the numbers are so astronomical that we need something else in the equation. The solution I believe is molecular biodiversity.

The Problem with Natural Selection

In 1859 Charles Darwin published what many believe still to be the most important biological book ever written (Darwin, 1859, 1868). Ever since, there has been fierce debate about what exactly is the process of evolution, and how it occurred. Fred Hoyle, in a number of his books, pointed out that in fact Darwin was not the first to write about the process of evolution. Nor was he the first to come up with a mechanism – natural selection. In 1858 Darwin received to his horror a letter from a little known explorer writing from the East Indies and the Malay Archepelego, Alfred Russel Wallace (Wallace, 1858). Even today this letter is as good a description of the principle of natural selection as has ever been written. In it Wallace argued also about the force of the 'struggle for existence'. Darwin was

Astrophysics and Space Science **285:** 571–585, 2003.

distraught. His friends knew that he had been working on this idea for some 20 years. There were notes, even essays sent to friends and colleagues, such as Asa Gray, Professor of Natural History at Harvard. But he had never dared to publish the concept. Was this because he was afraid of the conflict with The Church, to whom his dear wife Emma was so dedicated? Or was it because he hadn't really got the idea fully formed, as Hoyle suspected?

Charles Darwin's first notes on evolution in his Red Notebook were entitled 'Zoonomia', in honour of his Grandfather's seminal classification of disease on a Linnaean framework. Yet it wasn't until quite late in life that Charles acknowledged fully the debt he owed to this outstanding polymath, whom he never met. But what is more significant is, where are the notes that describe natural selection as clearly as in Wallace's letter? The title of Wallace's 1858 letter reveals it all, 'On the tendency of varieties to depart indefinitely from the original type'. There has been much argument about possible delays in Darwin contacting Lyell about the letter, implying improper behaviour by Darwin. I reject this. A much more likely explanation was that Darwin was suffering from one his many bouts of vomiting, that so severely affected him for 40 years (Colp, 1977).

In spite of some analyst's attempts to build a wall of aggression between them, Wallace and Darwin became friends and respected colleagues. Darwin helped Wallace gain a pension in later life. And Wallace's seminal contribution cannot take away Darwin's extraordinary scientific efforts, which remain amazing. His work on barnacles remains a classic even today. In all of Darwin's books, extending to 39 volumes in my own library, the theme of evolution by natural selection can be found. But it wasn't until The Origin, which he hastily finished in less than a year after the Wallace letter, that we find Natural Selection as a real concept. Wallace respected Darwin greatly. This was why he wrote to him in the first place. There is no evidence of any animosity or competitiveness between them. Wallace even coined the term 'Darwinism' (Wallace, 1889), and wrote a book with this title. Hardly the behaviour of a man who thought his brilliant idea had been ripped off by a competitor! No, all the evidence points to the fact that Darwin had come up with the idea earlier. Both men deserve the credit of history. In the event Charles Lyell and Joseph Hooker, two friends to whom Darwin had confided his problem, agreed that the Wallace letter with some of Darwin's notes and the Asa Gray letter should be read to the Linnean Society on 1st July 1858. The President in his summing up remarked that nothing of major significance had been presented at that night's meeting. How wrong can you be? Wallace couldn't be there as he was still in Far East. But why was Darwin not there to read his own paper? Maybe his friends Hooker and Lyell suggested that he stayed at home. But, what is more likely is that he was ill, from the mystery illness that afflicted him for 40 years, and defied the twenty or more doctors he saw, including his father Robert Darwin. And yet as we shall see Darwin missed what was wrong with him, something of major significance to his concept of natural selection.

As Hoyle also pointed out in several of his books, Darwin was not the best at acknowledging earlier ideas from other scientists. He never acknowledges his polymathic, paternal grandfather until the Preface of the 6th edition of The Origin. Though Charles did celebrate his grandfather in the biography about him. And yet as I have pointed out already, his first notes on what was to become natural selection were entitled Zoonomia in honour of his grandfather's famous medical text, where Erasmus Darwin first spells out the process of what we now call evolution. Erasmus shows his insight into this process in his wonderful poem – The Temple of Nature – published in 1803, the year after Erasmus' death. There he wrote 'Organic life beneath the shoreless waves was born and nursed in oceans pearly caves. First forms minute, unseen by spheric glass, move on the mud and pierce the watery mass'. Furthermore as Hoyle points out there were several other authors who had written about 'evolution', and even natural selection, decades before The Origin. James Hutton, and then Charles Lyell, established the science of modern geology, showing that the contemporary forces of nature were the same that occurred over millions of years, and thus could explain how the landscape and seas formed. The theory was called Uniformitarianism. It was a simple intellectual step for biologists to apply this principle to the evolution of life, thereby revealing that this had taken millions of years as well, contrary to the literal description in the Bible. Lamarck, in 1802, was the first to use the word biology scientifically. He had developed a theory of 'acquired characteristics' to explain the evolution of life. Ironically Darwin rejected the idea initially, but his pangenesis of later years is remarkably similar to Lamarck's hypothesis. Applying Hutton's and Lyell's ideas to biology, Hoyle tells us that Edward Blythe, who spent many years in India, wrote two seminal papers in 1835 and 1837, which were published in a well read journal called The Magazine of Natural History. Darwin must have seen them. Blythe wrote 'may not then, a large proportion of what we considered species have descended from a common parentage?' Blythe was asking whether varieties arise in nature by random effects. Just what Darwin and Wallace were to argue some 20 years later. Chambers in 1844 published anonymously 'The Vestiges of the natural history of Creation…'. But Darwin never mentions this either. Though he was aware of it. Contrary to what some have written natural selection is not a **theory** of evolution. It is a **principle** we apply to the process of evolution, which, as we have seen, was first described by Erasmus Darwin, one of the great geniuses of the 18th century.

But it wasn't ascribing the priority of the idea of natural selection that was Hoyle's real difficulty. His mathematical mind told him it that the variety of DNA and protein molecules that contemporary molecular and cellular biology had uncovered simply cannot be formed by random processes. He likened it to forming a Jumbo jet in a scrap yard by a random process. He puts numbers on the argument, at one point telling the reader that the odds involved ten to power of forty pages of zeros! This 'astronomical' problem can be illustrated by examining a protein we have cloned, the enzyme luciferase that catalyses the light emitting reaction in the British glow-worm *Lampyris noctiluca* (Sala Newy et al., 1996). This has

DNA sequence	Base
ATG GAA GAT GCA AAA AAT ATT ATG CAC GGT CCA GCG CCA TTC TAT CCT TTG GAG GAT GGA ACT GCT GGA GAA CAA	75
TTG CAC AAA GCA ATG AAG AGG TAT GCA CAG GTT CCA GGG ACA ATT GCT TTT ACT GAT GCG CAT GCA GAG GTA AAT	150
ATT ACA TAT TCC GAA TAT TTT GAA ATG GCT TGC CGG TTA GCC GAA ACT ATG AAG AGG TAC GGA CTT GGT TTA CAA	225
CAC CAC ATT GCT GTT TGT AGC GAA AAT TCT CTT CAG TTT TTT ATG CCT GTA TGC GGG GCT CTA TTT ATT GGA GTT	300
GGA GTT GCA TCA ACA AAT GAT ATT TAC AAT GAA CGT GAA TTA TAC AAC AGT TTG TCC ATA TCA CAA CCT ACA ATA	375
GTA TCC TGT TCC AAA AGA GCG CTG CAA AAA ATC CTA GGG GTA CAA AAG AAA TTA CCT ATA ATT CAG AAA ATT GTT	450
ATT CTG GAT TCT CGA GAG GAT TAT ATG GGG AAA CAA TCT ATG TAC TCG TTC ATT GAA TCT CAT TTA CCT GCA GGT	525
TTT AAT GAA TAT GAT TAC ATA CCG GAT TCA TTT GAC CGC GAA ACA GCA ACA GCA CTT ATA ATG AAT TCA TCG GGA	600
TCT ACT GGA TTG CCC AAG GGA GTT GAG CTT ACT CAC CAA AAT GTG TGT GTT AGA TTT TCT CAC TGC AGA GAT CCT	675
GTG TTT GGT AAT CAA ATT ATT CCC GAT ACT GCG ATT TTA ACA GTT ATA CCA TTT CAT CAT GGT TTT GGA ATG TTT	750
ACA ACA CTA GGA TAT TTA ACG TGT GGA TTT CGT ATT GTG CTT ATG TAT AGA TTT GAA GAG GAA TTA TTT TTA CGA	825
TCA CTT CAA GAT TAT AAA ATT CAA AGT GCG TTG CTG GTA CCT ACT CTA TTT TCA TTC TTT GCC AAA AGC ACC TTA	900
GTC GAT AAA TAC GAT TTA TCC AAC TTA CAT GAA ATT GCG TCT GGT GGA GCT CCC CTC GCG AAA GAA GTT GGA GAA	975
GCT GTA GCA AAA CGT TTT AAG CTG CCG GGA ATA CGA CAA GGG TAT GGA CTT ACT GAA ACT ACC TCA GCT ATT ATA	1050
ATT ACA CCA GAA GGG GAT GAT AAA CCA GGA GCA TGT GGT AAA GTT GTT CCA TTC TTT TCT GCC AAA ATT GTT GAT	1125
CTG GAT ACG GGT AAA ACT TTG GGT GTT AAT CAG AGG GGG GAA TTA TGT GTG AAA GGC CCA ATG ATA ATG AAG GGT	1200
TAC GTA AAC AAC CCA GAA GCA ACA AGT GCA TTG ATA GAC AAA GAT GGT TGG TTA CAC TCT GGT GAC ATA GCT TAC	1275
TAC GAC AAA GAT GGT CAC TTC TTC ATA GTA GAT CGT TTG AAA TCG TTA ATT AAA TAC AAA GGT TAT CAG GTA CCG	1350
CCT GCC GAA TTA GAA TCG ATA TTG CTG CAA CAT CCC TTC ATA TTT GAT GCA GGT GTT GCA GGA ATT CCC GAC CCA	1425
GAT GCC GGT GAA CTT CCT GCA GCC GTT GTT GTC TTA GAG GAA GGC AAA ACG ATG ACT GAA CAA GAA GTG ATG GAT	1500
TAT GTT GCG GGA CAA GTA ACT GCT TCT AAG CGT TTA CGT GGA GGA GTT AAG TTT GTG GAC GAA GTA CCT AAA GGT	1575
CTA ACT GGA AAG ATT GAT GGA AGA AAA ATC AGG GAG ATC CTT ATG ATG GGA AAA AAA TCC AAA TTG TAA	1644

Figure 1. DNA and protein sequence of glow-worm luciferase. Virtually all proteins start with ATG = methionine, and end with a stop, in this case TAA. The coding sequence thus = 1641 base pairs.

Amino acid sequence																				No
M	E	D	A	K	N	I	M	H	G	P	A	P	F	Y	P	L	E	D	G	20
T	A	G	E	Q	L	H	K	A	M	K	R	Y	A	Q	V	P	G	T	I	40
A	F	T	D	A	H	A	E	V	N	I	T	Y	S	E	Y	F	E	M	A	60
C	R	L	A	E	T	M	K	R	Y	G	L	G	L	Q	H	H	I	A	V	80
C	S	E	N	S	L	Q	F	F	M	P	V	C	G	A	L	F	I	G	V	100
G	V	A	S	T	N	D	I	Y	N	E	R	E	L	Y	N	S	L	S	I	120
S	Q	P	T	I	V	S	C	S	K	R	A	L	Q	K	I	L	G	V	Q	140
K	K	L	P	I	I	Q	K	I	V	I	L	D	S	R	E	D	Y	M	G	160
K	Q	S	M	Y	S	F	I	E	S	H	L	P	A	G	F	N	E	Y	D	180
Y	I	P	D	S	F	D	R	E	T	A	T	A	L	I	M	N	S	S	G	200
S	T	G	L	P	K	G	V	E	L	T	H	Q	N	V	C	V	R	F	S	220
H	C	R	D	P	V	F	G	N	Q	I	I	P	D	T	A	I	L	T	V	240
I	P	F	H	H	G	F	G	M	F	T	T	L	G	Y	L	T	C	G	F	260
R	I	V	L	M	Y	R	F	E	E	E	L	F	L	R	S	L	Q	D	Y	280
K	I	Q	S	A	L	L	V	P	T	L	F	S	F	F	A	K	S	T	L	300
V	D	K	Y	D	L	S	N	L	H	E	I	A	S	G	G	A	P	L	A	320
K	E	V	G	E	A	V	A	K	R	F	K	L	P	G	I	R	Q	G	Y	340
G	L	T	E	T	T	S	A	I	I	I	T	P	E	G	D	D	K	P	G	360
A	C	G	K	V	V	P	F	F	S	A	K	I	V	D	L	D	T	G	K	380
T	L	G	V	N	Q	R	G	E	L	C	V	K	G	P	M	I	M	K	G	400
Y	V	N	N	P	E	A	T	S	A	L	I	D	K	D	G	W	L	H	S	420
G	D	I	A	Y	Y	D	K	D	G	H	F	F	I	V	D	R	L	K	S	440
L	I	K	Y	K	G	Y	Q	V	P	P	A	E	L	E	S	I	L	L	Q	460
H	P	F	I	F	D	A	G	V	A	G	I	P	D	P	D	A	G	E	L	480
P	A	A	V	V	V	L	E	E	G	K	T	M	T	E	Q	E	V	M	D	500
Y	V	A	G	Q	V	T	A	S	K	R	L	R	G	G	V	K	F	V	D	520
E	V	P	K	G	L	T	G	K	I	D	G	R	K	I	R	E	I	L	M	540
M	G	K	K	S	K	L	*													547

Figure 1. Continued. The coding DNA sequence = 1641 base pairs, equivalent to 547 amino acids. This is three shorter than that for the luciferase from the firefly *Photinus pyralis*- (Sala Newby et al., 1996). The sequence ends with SKL which targets the luciferase to the peroxisome in the live cell.

547 amino acids, formed from 1641 bases (Figure 1). Each base position has four choices from A,T,G and C. Thus the number of combinations randomly are $4^{1641} = 10^{987}$. Yet a sphere based on the Milky Way filled with hydrogen atoms would only contain 2×10^{78} atoms. These astronomical numbers highlight Hoyle's dilemma. On its own, random small change cannot account for the origin or the evolution of life. But the clue to what I believe is the most important thing Darwin missed can be found in Chapter VI of The Origin – Difficulties on Theory. Here he highlights his own difficulties that he can't explain. Darwin even wrote that bioluminescence, the phenomenon I have studied for over 30 years (Campbell, 1988), was a problem for him (Campbell, 1994). Here he wrote, 'The electric organs offer another and more serious difficulty; for they occur in about a few dozen fishes... The presence of luminous organs in a few insects offers a parallel case of difficulty.'

Origins and Rubicons

As Hoyle pointed out Darwin was firm that 'Natura non facit saltus' – Nature takes no leaps. Darwin was wrong! But the solution still cannot be found from Eldridge and Gould's ingenious theory of 'punctuated equilibrium'. The mathematical argument still remains. Natural selection is a scientific truth. It is a principle that works. Those of us who carry out genetic engineering use it every day. Bacteria transformed by the gene we are trying to isolate are selected for the antibiotic resistance in the plasmid used to transform the bacteria. But does natural selection explain the whole process of evolution? Like Fred Hoyle I think not! The one thing Darwin never really addressed was the title of his most famous book (Darwin, 1859, 1868). 'The Origin' deals with the **development** of species not their **origin**. Hoyle argued that small variations of the type Darwin insisted on would be lost in the noise, a simple principle in physics. Yet larger variations have never been found in the fossil record. Some 10,000 fossil insects have been found, and some 30,000 fossil spiders, yet there are no intermediary forms. The origin of wings in insects and birds still remains a major difficulty for Darwinian evolution. Archaeopteryx has not solved the 'missing links' problem.

I have pointed out in one of my own books – Rubicon – (Campbell, 1994) that there are in fact three categories of evolutionary processes that Darwin missed:

- Biochemical events
- Origins
- Extinctions

Rubicons, thresholds, in biochemical evolution include: the origin of chirality – left handed amino acids and right handed sugars; the origin of the four bases – A,T,G,C; ATP, and not GTP, TTP or CTP as the energy currency of life: the origin of oxygen generating systems – photosynthesis; haemoglobin, oxygen binding proteins and scavengers; and the origin of internal cell signals that switch cells on and off, including calcium, central to living systems responding to environmental stimuli (Campbell, 1994). Other key rubicons crossed over 4000 millions years of evolution, and for which there are no satisfactory mechanisms, include the origin of new phenomena such as the eye, bioluminescence, and the electric organs of some fish; the origin of new cell types and organs such as nerve, muscle, secretary organs; the origin of organelles in a eukaryotic cell – the endosymbiotic theory just doesn't hold up. For example, the DNA in a mitochondrion is just too small to have originated as an endosymbiotic bacterium. Mitochondrial DNA only codes for a few proteins. Yet a live bacterium such as *E. coli* needs at least 300 to survive. The origin of the mitochondrion is much more likely to have been an encapsulated plasmid. Could this have come from space? Schemes that show how species have evolved, or protein families all come back to one point – 'the origin'. Natural selection can explain well the development of a species or genus from this point of origin. But the problem Darwin missed was how does a living system get to that point of origin, only after which can the random process of natural selection operate?

Hoyle's solution was life from space. This is certainly an option. But it simply removes the problem out of reach. The 'astronomical' size of the possible variations in glow-worm DNA remains a difficulty even if life is coming from space. However I believe there is a solution. And it can be found by examining one of the 'difficulties' Darwin highlighted – the marvellous phenomenon of bioluminescence.

The Origin of a New Protein

Curiosity about how animals make light - bioluminescence – has transformed biology. It has lead to a technology for lighting up the chemistry of living cells (Sala Newby et al., 1999, 2000). Using the DNA from jelly fish, fireflies and glow-worms we can now watch signals such as Ca^{2+} in defined parts of an individual cell, or even an intact organisms such as mouse, a fly or a plant. We can target the proteins to different parts of the cell using genetic engineering – the nucleus, the endoplasmic reticulum, the mitochondrion, the plasma membrane and the cytosol. We can use bioluminescence genes as reporters to watch individual genes switching on and off in live cells, and beetle luciferases to measure ATP in living cells. The green fluorescent proteins (GFP) from jelly fish and other coelenterates form their fluor by cyclising three amino acids within the protein sequence. As a result GFP can be used to locate specific proteins and organelles within live cells. The movement of these can be followed.

I became particularly curious about how luminous organisms produce different colours. Bioluminescence is a major communication system in the sea. In the deep sea most animals are luminous. I found that Nature has given us a complete rainbow, from violet to deep red. A minimum of three components are required for bioluminescence – a small organic molecule known generically as the luciferin, a protein catalyst – the luciferase, and oxygen. Some organisms also require other components such as ATP in beetles, and Ca^{2+} to trigger light emission in jelly fish, but these are not the energy source. The enthalpy for light is always an oxidation reaction using oxygen or one of its peroxy metabolites. I then found that these creatures have evolved three ways to emit different colours to produce Nature's rainbow:

- The luciferin
- The protein environment around the luciferin
- Energy transfer

For example, the most common chemistry responsible for light in the sea is called coelenterazine (Campbell and Herring, 1990), after the group of organisms where it was discovered. Organisms using coelenterazine produce blue or green light. Whereas the benzothiazole in luminous beetles produces green, yellow, orange or red light. Evolution has genetically engineered the luciferases in beetles to produce subtly different electrochemical environments around the luciferin. Thus the luciferase of the British glow-worm is 80% identical to that of US firefly *Photinus pyralis*. Just one or two amino acids change the colour from green to yellow, altering either the ionisation state, 3D conformation or electrochemistry of the excited product from the luciferin, called the oxyluciferin. GFP in some luminous coelenterates accepts the energy from the excited oxyluciferin, without the direct transfer of a photon, thereby changing the colour emitted from blue to green.

I realised that by using genetic engineering we could mimic these three ways Nature has evolved to produce its rainbow. By engineering into the luminous pro-

TABLE I
An astronomical baby?

Baby day	Month	Radius	Volume (l)	Object
1	First	10 microns	4.2×10^{-12}	Cell
11		100 microns	4.2×10^{-9}	Seed
21		0.1 cm	4.2×10^{-6}	Pin head
28	Second	1 cm	4.2×10^{-3}	Walnut
38		5 cm	5.2×10^{-1}	Organ
45		15 cm	14	Normal baby
48		0.5 m	524	Adult
56	Third	3 m	113,000	Car
61		10 m	4.2×10^{6}	House
71		100 m	4.2×10^{9}	Lake
107	Fourth	500 km	5.2×10^{20}	Asteroid
115		3196 km	3.3×1^{-23}	The Earth
129	Fifth	71500 km	1.5×10^{27}	Jupiter
142		1.4×10^{6} km	1.2×10^{31}	The sun
182	Sixth	1.5×10^{10} km	1.4×10^{43}	The solar system
252	Ninth	15 cm	14	Normal baby
259		70000 light years	1.2×10^{66}	The milky way

The baby starts on day 1 as one fertilised cell. Each cell divides once every 24 hours.

teins amino acids that would react with enzymes, such as kinases or proteases, I predicted that we should be able to measure these in a live cell by a change in colour or intensity. When the reactive site reacts with the molecular target in the live cell the light emitted changes in intensity and/or colour, enabling the intracellular event to monitored and imaged (Waud et al., 1996, 2001). This is the Rainbow effect. We have now engineered three types of Rainbow protein, based on the three ways that Nature has generated colour in bioluminescence. In type 1, two bioluminescent proteins emitting two colours are engineered together. In type 2, the site is engineered within the bioluminescent protein. In type 3 the reactive site is engineered between an energy transfer donor-acceptor pair. One family of Rainbow proteins is able to monitor activation of the cell suicide pathway in living cells. Without cell death there can be no life. Without cell death we would have webbed fingers, a tadpole would not convert to a frog, we would not have a brain, and a leaf would not fall in autumn. The importance of cell death in development can be illustrated by a Fred Hoyle type calculation. If a baby begins as a single fertilised egg, and divides once every twenty four hours, a typical time for a human cell. how big will the baby be when it is born? The answer is truly astronomical (Table I). By twenty

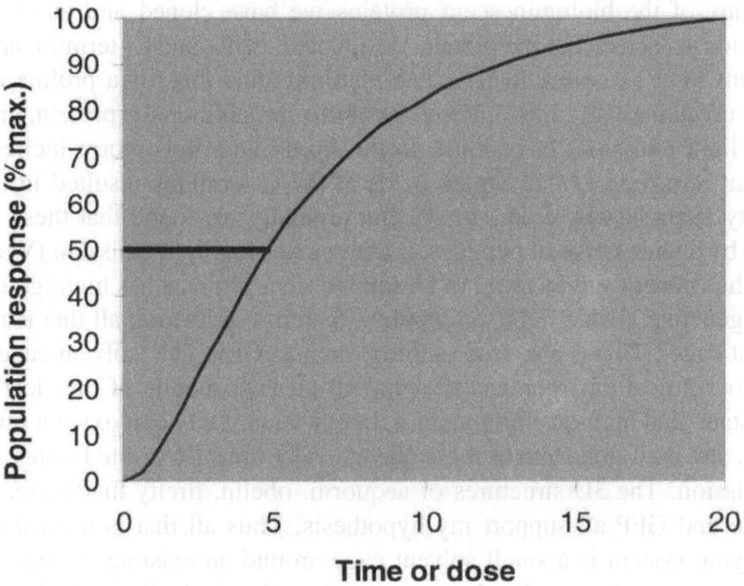

Figure 2. Rubicon. The classic Rubicon question is: When studying a cell population, as a dose response or as time course, at 50% of the maximum have all of the cells been activated to half their maximum or have only 50% of the cells been switched on.

nine weeks it will be the size of an asteroid, and by birth it will be nearly the size of the Milky Way!

The Rainbow protein technology has lead to some important discoveries about the Rubicons that determine what a cell does, and when it does it. The endoplasmic reticulum (ER) is a three dimensional spider's web that wraps itself around the nucleus and extends to the plasma membrane. It is a key decider of cell events. The ER is the major Ca^{2+} store in cells. But is also has a signalling system inside it using two proteins known as calreticulin and BiP (grp78) to communicate to the plasma membrane, nucleus, mitochondrion and cytosol. When the right Rubicon is crossed the cell decides to fire, to divide, to defend itself or to die. And this can be modelled mathematically (Baker et al., 2002). Thus when examining a dose response or time course experiment on a population of live cells, the key question Rubicon asks (Campbell, 1994) is: at 50% of the maximum population response have all of the cells been activated to half of their maxima or have only half of the cells been switched on (Figure 2). Bioluminescent and fluorescent probes have enabled us to examine signalling in single cells. This analysis has shown that in many cell populations individual cells switch on and off at different times and at different concentrations of stimulus. The Rubicon principle has major implications for understanding disease and in the development of drugs. Whilst developing this technology I realised that we might have solved Fred Hoyle's problem.

In many of the bioluminescent proteins we have cloned and engineered the amino acids at each end of the protein, designated the N- and C- termini, are critical to efficient light emission. In jelly fish photoproteins there is a proline at the C-terminus (Watkins et al., 1993). Removal of this destabilises the protein, preventing efficient light emission. In contrast, in the firefly and glow-worm luciferases we found that removal of 9–12 amino acids at the C-terminus resulted in 98% loss of activity (Sala Newby et al., 1994). But crucially we found that these could be replaced by a wide range of peptides, thereby restoring light emission (Waud et al., 1996). Thus genetic engineering of bioluminescent proteins has highlighted that in order to generate visible light, susceptible to natural selection, all that is needed is a 'solvent cage'. This is also true in fluors such as GFP. The 'solvent cage' creates an electrochemical environment allowing efficient formation of the electronically excited state, and high quantum yield light emission. Only one oxygen must be allowed in, and no water. Both of these quench electronically excited states, reducing light emission. The 3D structures of aequorin, obelin, firefly luciferase, bacterial luciferase and GFP all support my hypothesis. Thus all that is needed to start a new enzyme system is a small solvent cage around an existing molecule, that in bioluminescence when oxidised can generate an electronically excited state.

This hypothesis could be very important for Hoyle and Wickramasinghe's ideas about Panspermia, as it predicts that a large, complete protein does not have to hit the Earth for a new phenomenon to be initiated. All that is needed is a new peptide. And this could be as little as ten amino acids long. Using genetic engineering this hypothesis can now be tested.

Molecular Diodiversity

A further problem Darwin missed was the origin of mass extinction. Bioluminescence can again give us a clue. Whether the mass extinctions observed in the fossil record at the K-T boundary 65 million years ago, and at other Rubicons in evolution, were precipitated by asteroid impacts or volcanic eruptions misses the point. None of these catastrophic geological events could be physically responsible for losing some 80% of all species over a time span of several million years. Nor can a physical impact explain why some organisms survived. We wouldn't be here otherwise.

The stability and survival of ecosystems depends critically on biodiversity. Conventionally biodiversity has three components: diversity of species, genetic diversity, and diversity of habitats. But bioluminescence highlights a fourth – molecular biodiversity (Campbell, 2003). This is subtly different from genetic diversity, even though it can be ultimately dependent on inheritable DNA. In several luminous organisms multiple genes within an individual produce several different protein luciferases catalysing the same reaction. Within a luminous family there are related, but molecularly diverse, proteins producing light. Within an ecosystem there are several chemistries all producing similar coloured light, or different colours, all susceptible to the same forces of natural selection. This could therefore explain the Darwin finch problem, and Hoyle's problem. How could two finches evolve from one, whilst all the finches exist in the same environment? Not all the Galapagos finches were isolated individually on separate islands, as highlighted by Lack in his famous book, Darwin's Finches in 1948. As a population evolves it develops molecular biodiverisity in many of its biochemical processes. This leads to cellular diversity within organs, as each individual adapts to different environments. Mathematically it will be possible to show that this will lead to a bimodal distribution within the population, until some of the individuals in one generation cross the Rubicon, and can only mate with each other. Their molecular biodiversity is such that their DNA just won't mix with the rest of the population. A new species has been formed.

But as a species starts to lose its molecular biodiversity it becomes less adaptable to changes in the environment. A catastrophic event could certainly hasten the crossing of the extinction rubicon, but is not absolutely essential. Once a species has crossed the molecular biodiversity rubicon there is no return, it is on the road to extinction. This might happen within just a few thousand years, or it might take several million years, as in the case of some dinosaur and ammonite species at the K-T boundary.

A particular example of this molecular biodiversity principle is highlighted by something else that Darwin missed – the cause of the illness that afflicted him for over 40 years.

Darwin's Disease

'I have had a bad spell. Vomiting every day for eleven days, & some days after every meal.' 'The sickness starts about two hours after a meal'. Thus wrote Charles Darwin, aged 54, on 5th December 1863 to his friend Hooker. For 40 years, Darwin suffered from a range of symptoms, which devastated him for weeks on end. Explanations have been organic or psychosomatic. None have produced a satisfactory answer. On December 10th, 1831, aged 22, at Plymouth awaiting better weather for the Beagle to depart, Darwin suffered the first bout of the illness that was to plague him for the rest of his life. He had chest pain and heart palpitations, but told no one at the time, fearing it might prevent him taking his trip of a lifetime. While on the Beagle he suffered continuously from seasickness. But on land he seems to have been healthy, apart from one bad bout of fever. Returning home in 1836, he married in 1837, and set up house with Emma. Then, on 20th September 1837, the problem struck again – heart palpitations! For the rest of his life he was to be plagued by ill health. Some bouts were so bad that, to his great frustration, he had to stop work for weeks on end. He recorded his symptoms in many letters, and in his Diary of Health, which he wrote between 1st July 1849 and 16th January

1855. His worst problems were recurrent nausea, retching and vomiting, flatulence, headaches and a swimming head. He also suffered intermittently from gut pain, eczema, particularly on the face, boils, and continual fatigue. No wonder he got depressed! The diagnosis defied the twenty or more doctors who examined him, including his father.

Since his death there have been extensive attempts to explain his illness. Organic explanations include over twenty aliments, the two most convincing being Chagas disease and arsenic poisoning. Chagas' disease is endemic in South America, and can be transmitted by the famous great bug of the Pampas, *Triatoma infestans*, a known carrier of *Trypanosoma cruzi*, which infects the small intestine. Darwin describes being bitten by one. Alternatively several authors have argued strongly for a psychosomatic cause. But none explain how such severe physiological problems, and an extraordinary diverse set of symptoms, could have a wholly psychological origin. None explain why they would suddenly remit, and then inexplicably flare up again. These ups and downs did not correlate with stress events in Darwin's life, as has often been claimed. The solution can be found in his following quotes: 'The sickness starts usually about two hours after a meal'; 'At no time must I take any sugar, butter, spices, tea bacon, or anything good'; 'Thank, also, my Father for his medical advice – I have been very well since Friday, nearly as well as during the first fortnight & am in heart again about the non-sugar plan.' Darwin had the syndrome we have discovered – systemic lactose intolerance (Table II; Matthews and Campbell, 2000 a,b; Campbell and Matthews, 2001)!

Lactose, galactose β 1–4 glucose, is the sugar in mammalian milk, apart from *Pinnepedia* (sea lions and walruses). It is hydrolysed in the small intestine by the enzyme lactase (lactase-phlorizin-hyrolase, EC 3.2.1.62/108). The resulting monosaccharides are then absorbed here. However, all mammals, apart from most white Northern Europeans, and few tribes such as the Bedouins and the Fulani in Africa, lose 80-90% of their lactase after weaning. Thus all Asians, black Africans, Jews, Chinese, Japanese, and American red-Indians are potentially lactose intolerant. Too much lactose results in unpleasant gut symptoms, including pain, distension, tummy rumbling, flatulence and diarrhoea. However we have discovered that lactose can also cause systemic symptoms (Table II). The syndrome can be so severe that many, like Darwin, suffer from depression. Until we pinned it down to lactose. Removing lactose from the diet has revolutionised their lives. Nowadays removing lactose from the diet is not as easy as it would have been for Darwin. There is a huge lactose industry, adding it to foods and drinks without proper labeling. But the only source of lactose for Darwin would have been fresh milk, white sauces and custard, and cheese. Darwin ate these regularly.

We propose that systemic lactose intolerance is caused by toxins generated by hypoxic bacteria in the large intestine. It takes about two hours for food to reach here, the precise time Darwin reported it took for him to be sick after a meal. The sickness is caused by stopping of the intestine (ileus), resulting in reflux back into the stomach. No medical treatment alleviated Darwin's distress. The only time he

TABLE II
Darwin's disease – systemic lactose intolerance

Symptoms of systemic lactose intolerance	% people with systemic sugar intolerance*	Darwin's description of his symptoms	Occurrence
Gut symptoms (pain, dissention, diarrohea)	84%	Stomach ache	Common
Flatulence (farting)	84%	Flatulence (belching)	Common
Headache and light headedness	78%	Headache and swimming head	Common
Tiredness and chronic fatigue	63%	Chronic fatigue	Very common
Nausea and vomiting	61%	Vomiting	Very common
Muscle and joint pain	38%	Muscle and joint pain	Not mentioned
Allergy (eczema, hay fever, rhinitis, sinusitis)	21%	Skin rash and boils	Often
Depression	Common, but not quantified	Depression	Frequent

Campbell and Matthews, unpublished.
* represents proportion of people who have this particular symptom within 24 h of taking lactose. Occurrence is based on his notes and letters during periods of the attacks. The data for systemic lactose intolerance are based on 157 patients aged 6–76, with 108 females and 49 males, representing white Northern Europeans (Caucasians) (137), Asians (18) and black Africans (1), others (1) were given 50g oral lactose (equivalent to 1 litre of cow's milk) in 200 ml water (1 g/kg for those under 16 years). Symptoms were recorded for up to 3 days after ingesting the lactose, and breath hydrogen monitored every 30 min for at least 3 h. 84 patients exhibited at least one systemic lactose symptom after lactose. For ease of memory, we have designated this the MATHS syndrome (Matthews and Campbell, 2000, 2001a,b), standing for **M**uscle and joint pain, **A**llergy (rhinitis, eczema, hay fever, sinusitis), **T**achycardia and tiredness, **H**eadache, nausea and vomiting caused by **S**ugar overload in the intestine. The patients were divided into six groups (A-F), based on the clinical symptoms and the breath hydrogen. Conventionally a patient is diagnosed as lactose intolerant if they show an elevation in breath of 20 ppm within 2–3 of lactose ingestion. However we found that 50% of patients with systemic lactose intolerance did not have an elevation in breath hydrogen. The average number of symptoms after lactose was three. These data show that lactose intolerance must now be re-defined.

got better was after a drastic dietary change! Dr Clark took him off 'much sugar', so no custard! A notable success also was the water therapy of Dr Gully. Hydrotherapy involved not only cold water bathing, but also drinking large amounts of water, thus reducing milk intake. Darwin never suffered from his syndrome on the Beagle. With no fridges, there would rarely have been any fresh milk, and only a little cheese.

Lactose intolerance is caused by loss of the enzyme lactase. There are three mechanisms: (a) Congenital loss (rare); (b) Inherited loss after weaning (very common); (c) Reversible loss caused by gut infections or hormonal imbalance. The family history of the Darwin is consistent with inherited hypolactasia. So how did lactase evolve? Since dairying is only some 6000 years old (Dudd and Evershed, 1998); what could be the selective advantage of retaining any lactase after weaning? Darwin does not mention this in The Descent of Man, or The Origin. Darwin's disease, systemic lactose intolerance, can give us unique insights into the evolution of our own species. He missed the evolutionary significance of his own illness, the evolutionary origin of milk.

Conclusions

Like Erasmus Darwin Hoyle was a polymath. He was a distinguished scientist, inventor, writer and philosopher. He was fascinated by many aspects of science apart from astronomy. He was the first to put his finger on what Darwin missed. We can now use his inspiration and exploit modern concepts and technology to understand what perhaps was the problem above all that intrigued Fred Hoyle for most of his life – the origin of the complexity of living systems, not just on our own planet, but throughout Hoyle's Universe.

Acknowledgements

I thank my wife, Stephanie Matthews for the joint development of our new syndrome – Systemic lactose intolerance, the disease we show that Darwin had for some 40 years.

References

Baker, H., Errington, R.J. and Campbell, A.K.: 2002, Mathematical model predicts that calreticulin interacts with the endoplasmic reticulum Ca^{2+} – ATPase, *Biophys. J.* **82**, 582–590.

Blythe, E.: 1835, *The Varieties of Animals*, Magazine of Natural History.

Blythe, E.: 1837, *Distinctions between Man and Animals*, Magazine of Natural History.

Campbell, A.K.: 1994, *Rubicon: The Fifth Dimension of Biology*, Duckworth, London, pp. 304.

Campbell, A.K.: 1994, in: A.K. Campbell and L.J. Kricka (eds.), *Bioluminescence and Chemiluminescence: Fundamental and Applied Aspects, Darwin and the glow-worm*, Wiley, Chichester, pp. 392–396.

Campbell, A.K.: 1988, *Chemiluminescence: Principles and Applications in Biology and Medicine*, Horwood/VCH, Chichester and Weinheim, pp. 608.

Campbell, A.K.: 2003, Save those molecules – Molecular biodiveristy and life, *J. Appl. Ecology* **40**, 193–203.

Campbell, A.K. and Matthews, S.B.: 2001, *Lactose Intolerance and the MATHS Syndrome: What are They and how can I Cope*, ISBN 0-9540866-0-0, Welston Press, Pembrokeshire.

Campbell, A.K. and Herring, P.J.: 1990, Imidazolopyrazine bioluminescence in copepods and other marine animals, *Marine Biol.* **104**, 219–225.

Chambers, R.: 1844, 1969, *Vestiges of the Natural History of Creation*, Leicester University Press, New York.

Colp, R.: 1977, *To be an Invalid. The Illness of Charles Darwin*, University of Chicago Press, Chicago and London.

Darwin, C.R.: 1859, 1st edition, 1868, 6th edition, *The Origin of Species by Means of Natural Selection or the Preservation of Favoured Races in the Struggle for Life*, Murray, London.

Darwin, C.R.: 1858, Essay and letter of evolution, *Proc. Linn. Soc. July 1st*.

Dudd, S.N. and Evershed, R.P.: 1998, Direct demonstration of milk as an element of archeological economies, *Science* **282**, 1478–1481.

Hoyle, F.: 1983, *The Intellegent Universe. A New View of Creation and Evolution*, Joseph, London.

Hoyle, F.: 1987, 1999, *Mathematics of Evolution*, Acorn Enterprises, Tenessee.

Lack, D.: 1948, *Darwin's Finches*, Oxford University Press, Oxford.

Matthews, S.B. and Campbell, A.K.: 2000a, When sugar is not so sweet, *Lancet* **355**, 1309.

Matthews, S.B. and Campbell, A.K.: 2000b, Neuromuscular symptoms associated with lactose intolerance, *Lancet* **356**, 511.

Sala Newby, G.B., Kendall, J.M., Jones, H., Taylor, K.M., Badminton, M.N., Llewellyn, D.H. and Campbell, A.K.: 1999, Bioluminescent and chemiluminescent indicators for molecular signalling and function in living cells, in: *Fluorescent Probes for Biological Function*, 2nd edition, Ed Mason, WT. Academic Press, London, pp. 251–272.

Sala-Newby, G. and Campbell, A.K.: 1994, Stepwise removal of the C-terminal amino acids in firefly luciferase results in graded loss of activity, *Biochim. Biophys. Acta* **1206**, 155–160.

Sala-Newby, G.B., Badminton, M.N., Evans, W.H., George, C.H., Jones, H.E., Kendall, J.M., Ribeiro, A.S. and Campbell, A.K.: 2000, Targetted bioluminescent indicators in living cells, *Methods in Enzymol.* **305**, 478–498.

Wallace, A.R.: 1858, On the tendency of varieties to depart indefinitely from the original type. *Proc. Linn Soc 1st July.*

Wallace, A.R.: 1889, Darwinism. An exposition of the theory of natural selection with some of its applications.

Watkins, N.J. and Campbell, A.K.: 1993, Requirement of the C-terminal proline residue for stability of the Ca^{2+} – activated photoprotein aequorin, *Biochem. J.* **292**, 181–185.

Waud, J., Sala-Newby, G.B. and Campbell, A.K.: 1996, Engineering the C-terminus of firefly luciferase as an indicator of covalent modification of proteins, *Biochim. Biophys. Acta* **1292**, 89–98.

Waud, J., Bermudez-Fajardo, A., Sudadhaharan, T., Jefferies, J., Jones, A. and Campbell, A.K.: 2001, Measurement of caspases using chemiluminescence resonance energy transfer, *Biochem. J.* **357**, 687–697.

COSMIC GENES IN THE CRETACEOUS-TERTIARY TRANSITION

MAX K. WALLIS
Cardiff Centre for Astrobiology, Cardiff University, UK

Abstract. It is proposed that genes coding for Aib-polypeptides arose early on in the K/T transition, presumed from the Earth's accretion of interplanetary (comet) dust. Aib-fungi flourished because of the evolutionary advantage of novel antibiotics. The stress on Cretaceous biology led directly and indirectly to mass species extinctions, including many dinosaur species, in the epoch preceding the Chicxulub impact.

If cosmic genes are to be a driver of species evolution, Hoyle and Wickramasinghe (1981) developed the idea of pathogenic carriers to establish them in terrestrial ecosystems. Carriers could be viruses or more complex organisms. That viable bacteria could reach the Earth within meteorites, such as those known to derive from Mars, is now considered fully plausible. But such material has been reaching the Earth throughout evolutionary history, so it would take material from more unusual sources to cause significant episodic evolutionary changes.

We consider the major evolutionary change of the K/T transition. The K/T boundary as measured by the spike in iridium abundance is now well identified with the asteroid or comet impact that caused the Chicxulub crater. However, the K/T transition as measured by the mass extinctions of Cretaceous species lasted of order 100 kyr or more, right across the 65.0 Myr impact event. The immense outpouring of lava that created the Deccan Traps could well have played a part in the extinctions, but just post-impact if the impact set off volcanic lava flows. The global wild-fires and atmospheric changes consequent on the impact also cannot explain the earlier extinction record.

The iridium data do provide an extraterrestrial signal pre-Chicxulub impact. The iridium abundances in samples above and below the impact spike in the boundary sediments at Stevns Klint (Zhou and Bada, 1989) indicate a strong enhancement of extraterrestrial debris reaching the Earth over some 10^5 years. The same boundary clays are also remarkable for high levels of the exotic aminoacid Aib, which was known from the 1980's to be a major aminoacid in extracts from the Murchison meteorite (Engel and Macko, 2001). By 'exotic' I mean a non-protein aminoacid, not in the standard 21 or 22 making up DNA. It was therefore hypothesised that Aib and a second exotic aminoacid, isovaline, are remnants of extraterrestrial organic material (Zhou and Bada, 1989; Zahnle and Grinspoon, 1990).

It is however, quite unclear how this material could be linked to the pre-Chicxulub mass extinction. We previously investigated if a dust veil could cut down the solar radiation (Wickramasinghe and Wallis, 1994). Alternatively, we suggested that the exotic aminoacids could have been highly poisonous (Ramadurai et al., 1994). Dust veiling was considered to be consequent on the fragmentation of a giant comet, which injects vast amounts of debris down to the smallest sizes into the

Astrophysics and Space Science **285:** 587–592, 2003.

inner solar system over a 100 000 yr period (Napier and Clube, 1979). Quantitative analysis showed dust veiling to be a plausible mechanism, assuming the comet had similar composition in non-volatiles to that of the Murchison meteorite, but it implied that the Aib-carriers had to largely survive degradation processes before being fixed in the sedimentary rocks. The sediment data shows Aib to be comparable in abundance to the ordinary aminoacids in the same samples. Aib also tracks these aminoacids rather than iridium levels. The absence of some compounds like n-alkanes implies extensive microbial reworking of the organic matter (Brisman et al., 2001). Moreover, natural degradation processes exist at the present epoch and Aib is now known as a component of relatively rare biological peptides, which are associated with select microfungi.

Let's hypothesise that Aib is an indicator of an unusual biology – probably with abundant Aib-fungi – that flourished through the K/T boundary. This fits with the concept that the genes coding for the peptaibol-generating enzymes arrived with microorganisms from space – perhaps as actual Aib-fungi. Alternatively the novel genes were incorporated into existing microfungi. The novel species of peptaibol-based fungi were pathogenic to many cretaceous organisms and were particularly virulent because of the novel biochemical properties. Their stress on faunal and floral species caused not only mass extinction but also accelerated evolution and speciation – species that evolved defence mechanisms tended to survive the fungal attack. The competitive evolution of the ordinary and alien organisms resulted in resistant species (including mammals) winning through and symbiotic relationships developing. While the fall-off in Aib in the upper sediments indicates that the fungal invaders lost out in the end, their novel genetic input may have stimulated the evolutionary upsurge of the early tertiary epoch.

Aib Terrestrial and Extraterrestrial

Figure 1 shows the run of Aib and iridium levels in the Stevns Klint sediments. Other Raton Basin sediment data from Brisman et al. (2001) run up against detection limits, having lower aminoacids generally. However, the presence of Aib in the Colorado (Starkville South) but not the Mexico (Raton Pass) sections implies some variability in the occurrence of Aib-fungi and/or peptaibol degrading organisms. The failure to find Aib in pre-impact samples (two each location) may be just a detection limit problem.

We don't know the composition of the organics in Murchison and K/T clays, but do know that hydrolysis releases Aib. It is strikingly abundant in K/T clays – 10^4 times above detection limit and comparable to standard amino acids (10–20 times less than main aminoacids, but similar abundance to minor ones). Isotope and/or isomer analysis might be used to distinguish terrestrial and extraterrestrial origin, but the Brisman et al. work (2001) did not reach that far. In comparison, the iridium does indicate ET material – early in the boundary clays it was evident at

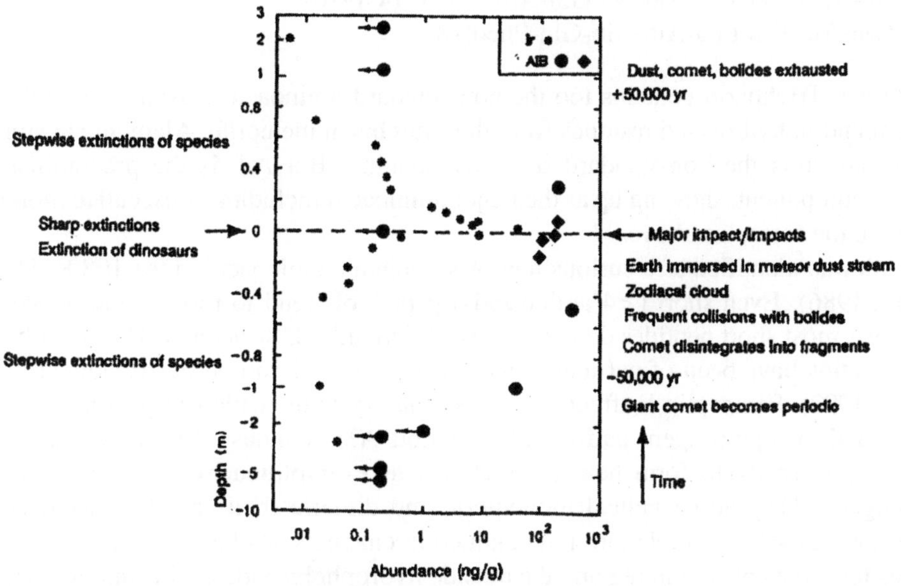

Figure 1. Interpretation of the K/T boundary record after Wallis and Wickramsinghe (1994). Iridium and Aib data in K/T marine boundary clays at Stevns Klint are from Zhau and Bada (1989 – circles: small Ir and large Aib) and Brisman et al. (2001 – diamonds: AIB). Sediments vary around the planet – Brisman et al., 2001 also analysed for aminoacids the non-marine K/T clays of Raton Basin in Colorado and in New Mexico. Aib was found in the Colorado site above the iridium spike, but not below it and not at all in the New Mexico clays. When one takes into account the likely degradation of sedimentary organic matter and the solubility of aminoacids, these new results confirm the significance of Aib found by Zhau and Bada.

10 times the detectable limit compared with 100 times generally and 10^4 times in the peak.

Aib in Structural Biochemistry

Aib (α-aminoisobutyric acid or α-methylalanine) has quite a simple formula:

$NH_2 - C\,(CH_3)_2 - COOH$

Polypeptides that contain Aib are termed *peptaibols*. ***Alamethicin*** is the best studied natural one (Cosette et al., 1999), but all peptaibols have high percentages of Aib. Examples are ***Efrapeptin C*** (Jost et al., 2001):

Ac-Pip-Aib-Pip-Aib-Aib-Leu-βAla-Gly-A ib-Aib-Pip-Aib-Gly-Leu-Aib-X

Trichorzin (Duval-Delphine et al., 1998)

Ac-Aib1-Ser-Ala-Aib-Iva-Gln-Aib-Val-Aib-Gly10-
Leu-Aib-Pro-Leu-Aib-Aib-Gln-Pheol18

Note that Trichorzin contains too the non-standard aminoacid isovaline (*Iva*) that was found in hydrolysed residues from the Murchison meteorite. Alamethicin similarly contains the non-standard α-methyl alanine. But Aib is the predominant exotic component, showing up as the major aminoacid including consecutive monomers in the formulas above.

Aib was identified as a curious non-proteinogenic aminoacid in the 1980s (Das et al., 1986). Even short (<4 aminoacids) peptaibols tend to form stable helices, whereas most short peptides do not have well-defined 3-D structures. The peptaibol Efrapeptins have been of particular interest for antibiotic properties (Gupta et al., 1991, 1992). Efrapeptin F affects insect nervous systems while Efrapeptin tolypin is an anti-mosquito agent and found to inhibit ATP-synthase. Peptaibols have a strong propensity to form helical structures, as exemplified by alamethicin and analogues. They adopt helical monomeric and dimeric (helix-bend-helix) structures, which self-assemble into ion-conducting channels. Probably the hydrophilic faces form the channel lining and the outside hydrophobic side chains interact with phospholipid membranes (Duclohier and Wroblewski, 2001). A channel conducts ions diffusively with little ion selection, implying long-range electrostatic interactions dominate. Forming a channel through a cell membrane enables leakage of the contents.

Anti-microbial peptides in general are the prime mechanism whereby plants and animals repel attacks by the range of bacteria, fungi, viruses and protozoa (Zasloff, 2002). Such peptides can be α-helical, like peptaibols. The above studies indicate that peptaibols could be utilised by Aib-fungi to penetrate microbial membranes. Peptaibols may also constitute part of fungal mechanisms for penetrating plants' cellulose walls. But present-day animals and plants presumably have effective defence mechanisms against peptaibols.

Genes are Cosmic

Fred Hoyle's writings on the theme of cosmic genes envisaged RNA or DNA reaching us from space, but left the question open as to whether it arrived as fragments, as viroids or viruses, or as complete cells (Hoyle and Wickramasinghe, 1981). The cosmic genetic material is ready to function. It could, like HIV, take over existing cell-DNA and set to work generating its own armoury of polypeptides. Genetic engineering has now made this concept familiar, but Fred had the idea well ahead of his time.

In regard to evolution on Earth, terrestrial biology takes the novel cosmic genes into existing biosystems and assimilates or rejects them long-term in accord with any survival advantage. Copying-error or spontaneous mutations do assist with integrating novel genes into the genome, but are seen as contributing only fine-tuning.

The present idea is that cosmic genes drove the K/T species transition. The exotic (peptaibol) biology arose early in the K/T transition, at the start of or early in the extinction period for cretaceous species. It flourished world-wide after 30 000 yr, then dwindled after another 50–70 000 yr (20–40 000 yrs post Chicxulub). The new genetic material not only generated ecological space through extinction of many species, but also caused accelerated evolution and speciation in response to the novel virulence.

One concept is of fungal invaders: Aib-fungi arrived within the extraterrestrial debris alongside the Iridium. Their toxic properties were peptaibol-based and novel to the cretaceous ecosystems. They were immediately pathogenic to bacteria, but may have required some evolutionary adaptation before they were able to effectively attack higher organisms (plants, reptiles etc.) with cellulose walls and polypeptide defences. Maybe the fungi directly attacked the dinosaurs, maybe they indirectly stressed the Cretaceous ecosystems and so provoked faunal mass extinction.

A variant of the concept is of a genetic invasion: irrespective of the carrier organism, the peptaibol-genes gave existing biology the opportunity of adding importantly to the armoury of anti-microbial peptides. Organisms that adopted these genes rapidly, integrating them into their existing genomes, gained a substantial evolutionary advantage. A range of microfungi adopted them for purposes of attack, perhaps other organisms adopted the peptaibol genes for defence against fungi, bacteria and protozoa. For animals and potentially other multi-cellular organisms, the balance between retaining symbiotic bacteria and repelling hostile bacteria might have tipped the scales against peptaibol antibiotics. The anti-microbial peptides of modern biology (Zasloff 2002) can presumably be more sensitively tuned than peptaibols.

Whether fungal or genetic invaders started off the cretaceous extinctions, there followed 100 000 yrs of competitive evolution of alien and ordinary organisms. The invaders were successful for a period, but eventually lost out. Defence mechanisms against peptaibols had to be developed, but evidently resistant species (including mammals) won through. The peptaibol-genes did enrich the kingdom of fungi but in the end proved to be of little overall advantage and retreated to select specialist organisms. They remain, however, in the terrestrial ecosystem, still available for future eventualities.

To test this hypothesis, further data on Aib-peptides in K/T sediments around the world would be useful, with particular emphasis on slightly earlier cretaceous sediments. Identifying the enzyme systems generating peptaibols and the related DNA should also help. If the genes were confined to the specialist microfungi rather than widely distributed but unexpressed, this would support the fungal invader concept.

Where in the evolutionary record might one look for other examples of cosmic genes arriving on Earth? One suggestion (Hoover et al., 1986) was for diatoms, which appeared suddenly at the 112 Myr Cretaceous boundary. These microscopic algae have cell walls of a complex siliceous biopolymer attached to an $Si(OH)_4$ matrix on a protein template. Similar shells of radiolarians are evident in the earlier fossil record but not the distinct diatoms. As production of the intricate cell organisation via an 'evolutionary jump' is implausible, it's worth considering if diatoms provide a second example of the arrival of new cosmic genes.

References

Brisman, K., Engel, M.H. and Macko, S.A.: 2001, *Precambrian Res.* **106**, 59–77.

Cosette, P., Rubuffat, S., Bodo, B. and Molle, G.: 1999, *Biochim. Biophys. Acta* **1461**, 113–122.

Das, M.K., Raghothama, S. and Balaram, P.: 1986, *Biochemistry* **25**, 7110.

Duclohier, H. and Wroblévsky, H.: 2001, *J. Membrane Biol.* **184**, 1–12.

Duval, D., Riddell, F.G., Rebuffat, S., Platzer, N. and Bodo, B.: 1998, *Biochim. Biophys. Acta* **1372**, 370–378.

Engel, M.H. and Macko, S.A.: 2001, *Precambrian Res.* **106**, 35–45.

Gupta, S., Krasnoff, S.B., Roberts, D.W., Renwick, J.A., Brinen, L.S. and Clardy, J.: 1992, *J. Org. Chem.* **57**, 2306.

Hoover, R., Hoyle, F., Wallis, M.K. and Wickramasinghe, N.C., 1986: in: C.-I. Lagerkvist et al. (eds.), *Asteroids Comets Meteors II*, Uppsala University, 359–362.

Hoyle, F. and Wickramasinghe, N.C.: 1981, *Evolution from Space*, New York.

Jost, M., Neumann, B., Stammler, H.-G. and Sewald, N.: 2001, <http://oc3web.chemie.uni-bielefeld.de/5ps/beitrag/mjo01.htm>

Napier, W.M. and Clube, S.V.M.: 1979, *Nature* **282**, 455.

Ramadurai, S., Lloyd, D., Wallis, M.K. and Wickramasinghe, N.C.: 1994, *Adv. Space Res.* **15**(3).

Wickramasinghe, N.C. and Wallis, M.K.: 1994, *MNRAS* **270**, 420–426.

Zahnle, K. and Grinspoon, D.: 1990, *Nature* **348**, 157–160.

Zasloff, M.: 2002, *Nature* **415**, 389–395.

Zhau, M. and Bada, D.L.: 1989, *Nature* **339**, 463–465.

REMEMBERING FRED HOYLE *

On behalf of everyone here tonight, I would like to thank Chandra Wickramasinghe, Geoffrey Burbidge, and Jayant Narlikar for organising, and the Vice-Chancellor of the University for hosting, this splendid meeting in tribute to Fred Hoyle and his inspiring work. Thank you, very much, to you all.

On an occasion like this, so many treasured memories come flooding back. I want to try to give you something of the flavour of working with and around Fred as a student, and then as one of the first five staff members at the Institute, *his* Institute.

I had the very good fortune to become one of Fred's students in 1960, the year when he started off a large group of new students to supplement the one or two he'd previously had. We had great and exhilarating times together, as I'm sure we all recall. Later, in 1964, I went as a post-doc to Willy Fowler's Kellogg Lab at CalTech. Then, around Easter 1966, while Fred was visiting the Burbidges in La Jolla, he invited my former Cambridge flat-mate Peter Strittmatter and myself to take a long walk with him on the beach there. I can heartily recommend this – it's not a bad way to be recruited! Fred outlined his plans and invited us to come back to Cambridge as two of the first five founding staff members of IoTA – his new Institute of Theoretical Astronomy – which was coming into being just then.

This was a wonderful opportunity which Peter and I both seized despite the fact that when we arrived back in Cambridge in September 1966, we found that (a) the Institute was but five courses of bricks high, and (b) little practical details like salaries for the intended staff had yet to be arranged.[1] But in the new year, after about three months or so, we began to be paid a salary so that we weren't so impoverished from then on. (Meanwhile, our bodies and souls were kept together by William Stone Prize Research Fellowships from Peterhouse. Many of the Peterhouse fellows just couldn't distinguish between the two of us – so Peter and I became known collectively as 'the heavenly twins.')

Let me tell you what I feel like doing right now. Having tried to copy Fred in so many things, I'd like to begin the major part of my talk tonight by modelling

* Text of after-dinner speech made by Dr. John Faulkner.

[1] It slowly emerged that, Cambridge University being Cambridge University, it had yet to formally accept the idea of having and hosting the Institute. To decide this momentous question, certain University committees had to pass on it in the order A, B, C. However, by time-honoured tradition, the committees only met once a term, in the order C, B, A. Thus, by design, any substantive decision was dragged out for over a year or more. In fact, Fred grew so tired of this whole procedural log-jam that on his own recognizance, he gave the builders, Rattee and Kett, the go-ahead to start laying the foundations for the Institute before the University had formally decided to host it!

Astrophysics and Space Science **285:** 593–608, 2003.

myself on Fred, in particular on the way he began a specially invited talk at the Royal Society. It went something like this:

'Oohh Helloow.[2] I've been invited here tonight to give you this special lecture on a subject which is extremely interesting; and there's a little bit of a problem with it, which is that I find myself with rather too much to say. The thing is, I've been asked to limit my full talk to an hour and a quarter. Unfortunately, this makes it a critical situation. The trouble is that the subject I want to speak about divides itself naturally into six different topics, and each topic, to fully develop it, will take about 15 minutes. Now I'm sure you all realise that there is a little bit of a problem with that, because six intervals of 15 minutes don't exactly fit into an hour and a quarter. In fact, there's only room for five. So what I'll have to do is to leave aside the sixth topic and only tell you about the first five of them – which is a bit of a pity when you think about it because it turns out that the sixth topic, which I would have told you about had there been time for it, is really the most important and the most interesting of them all. So that really is quite a shame, if I have to leave it out because there's too little time for it. So, in view of this problem, I thought I might start by beginning with the sixth topic – the topic I would have told you about if there were time for it – and after that I'll move on to the other five topics and so complete the story.' You may not believe this, but I swear that this is how Fred did begin to address a special meeting of the Royal Society. We students present were having fits – we'd all heard Fred rationally analyse some scientific problem out loud, but it must have seemed rather odd to his audience that day to hear him thinking through his time limitations so logically and gravely.

Going back to my earliest days with Fred, I was due to start as his student in October 1960. However, this was during one of the many phases in his life when he wasn't going to be around very much. By October he would be off to the United States for an extended period of several months. So, he suggested that we could get a head start on something during the summer, before he left. Could I spend the summer in Cambridge? He showed a surprisingly practical side by pointing out that I had been awarded the St John's College 'Hockin Electricity Prize' for my performance in Part III of the Maths Tripos. That prize (as I recall, the magnificent sum of Forty Pounds!) would nicely take care of my extra expenses. (I'm fairly sure that I owed this award to the combined machinations of Fred and Leon Mestel.)

I was very pleased and flattered at this chance to work with Fred so soon, but already had some astronomically related plans. So, a little hesitantly, I went to see him and said I was thinking of going to Herstmonceux for part of the summer, to spend six weeks as a summer student at the Royal Greenwich Observatory. What did he think about this? Fred looked at me and said, 'Oh, yes. That's a good idea. You should go there. If you *do* go to Herstmonceux and work with Woolley, that

[2] Delivered in a deep-toned, slightly gravelly Yorkshire/mid-Atlantic accent, accompanied by appropriate nose manipulation and finger gesturing – see later.

should *certainly* knock any ideas about becoming an observer out of your head!' This advice turned out to be most prescient.

If memory serves me correctly, I then remained in Cambridge for five or six weeks, reading Landau and Lifshitz's *Statistical Physics* and talking with Fred about the calculation of stellar opacities. Fred thought that stellar opacities were in a simply awful state, and his initial plans were that *I* would be the person to put that right! I confess that I did not warm to this task. I was much relieved when, upon his return to Cambridge almost six months later, he announced, 'Oh, you can forget about doing opacities – it turns out there's a massive undertaking on them at Los Alamos.' (I've always been grateful to Art Cox and his collaborators for rescuing me from a fate worse than death.)

Fred gave his various new students some essential reading assignments during his absence. Three things that I had to read were first, B^2FH, and then two cheek-by-jowl articles: one by Chip Arp on the H-R diagram and an adjacent one by the Burbidges on stellar evolution, in Volume 51 of the 1958 *Handbuch der Physik*. These were massive and superb articles, and I read them all religiously while Fred was abroad.

Fred returned to Cambridge a few months later. Soon after his return, something occurred which I still regard with awe, and quite a miraculous privilege to have experienced. Kumar Chitre and I have talked recently about this intense experience being one of the highlights of our lives as students. Fred wanted to bring three of us quickly up to stellar research speed. So he commandeered one of the big mathematical lecture theatres in Cambridge's old Arts School buildings, and engaged Kumar, Ken Griffiths and myself in an incredible three-day interactive lecturing marathon. In that time he introduced us to many things, things almost completely new to all of us, simultaneously. He taught us the essence of stellar atmospheres and the outer boundary condition for stars, the intricate physics of ionization and opacity (such as which were the major, and which the ignorable electron donors, the mysteries and consequences of the negative hydrogen ion, ...), and a new approach he had developed to convective energy transport. He showed us how to program in FAP – the mnemonic Fortran Assembly Program used with IBM computers – and before our very eyes he developed both an envelope and a basic stellar structure code, all in the course of these three days. It was an absolutely masterly marathon performance, for something like seven or eight hours a day for three days. Fred completely laid out for us how to do the automatic computation of stellar interiors, with reasonably realistic treatment of the outer boundary. And during that time, at first hesitantly and then with more confidence, we students began to see and point out small errors or logical flaws, or to make other suggestions, and were transformed from quiet recipients to contributing junior colleagues.

That experience led to Fred telling us something about his days of working with Brian Haselgrove, which brings me to the Haselgrove/EDSAC stories. In those days when Fred was collaborating with Brian Haselgrove, on the first automatic computations of stellar evolution that were ever done, they worked together on

EDSAC. EDSAC was in effect a Cambridge University home-built computer,[3] built by the people in the Cambridge maths lab, with literally thousands of thermionic valves and mercury delay lines. This made it an intense heat source, and also gave it some major quirks due to unique local conditions. Because of the incredibly untrustworthy electricity supplied by the Cambridge City Electricity Company – which, by the way, had its own idiosyncratic voltage not common to any other place in the United Kingdom[4] – substantial fluctuations would frequently occur in the voltage, and these would wreak havoc with EDSAC. You worked with EDSAC as a hands-on machine, and the voltage fluctuations were a very serious problem, because they could cause EDSAC to blow up, before your very eyes! The most important thing they told you, indeed, the first thing you were taught in the very first class on using EDSAC, was how to turn it *off*. We were told that by the time electrical power reached the machine, there was both an extremely high voltage supply and also a normal, low voltage electricity supply. The problem was, these two voltages were controlled by different switches. One learned that if these power supplies were turned off in the wrong order, every single electronic valve in the building would spectacularly blow. Then, it would take two days of overtime to replace all the thousands of valves – always assuming that enough spares were on hand.

This situation being pretty critical (to adapt one of Fred's favourite phrases), I never did understand why on earth there wasn't one master, mechanically time-delayed trip control that you could turn off! After all, this machine had been designed and built by some of the best brains in Britain, and one would expect them to be able to arrange for something to successfully trip off the power switches in the right order. (Some of these people had been at Bletchley House in the Second World War, cracking the Enigma code and saving the world for democracy. You'd think they could organize a few switches.) Instead, when you were taught how to use EDSAC you were told that there were three switches (I never did understand why the number was three!) and you had to turn them off in a certain order. Now even if you couldn't arrange to have a master switch that simply tripped these off in the desired order, you'd think that any rational person would have put them on the control panel in the order one, two, three; but that was too easy for Cambridge. So they had to be turned off in some odd order – like one, three, two.

What would take place then was this. You'd be happily working with EDSAC and the Cambridge City voltage would start to fluctuate. There was a big pointer/meter displaying the voltage over one of the doors of the large room housing the machine. There was also something illustrating the recent history and trend

[3] I don't intend to belittle EDSAC in any way, despite the idiosyncracies it possessed. Both EDSAC 1 and 2 were firsts in many ways, EDSAC 2 the first full-scale microprogrammed machine. See http://www.cl.cam.ac.uk/UoCCL/misc/EDSAC99/. I still fondly recall EDSAC Autocode, a marvel of compactness and ease of use.

[4] Consequently, from light bulbs to major appliances, manufacturers produced special – and expensive – models geared to this niche market.

in the voltage. You were supposed to keep one eye on that as you worked. But they didn't rely entirely on that display. At some point an automatic procedure would predict that the voltage was about to dip too low, below a previously established critical level. At this moment an awfully loud and insistent alarm klaxon would go off, audible not only in the machine room but throughout the entire Maths Lab. Upon hearing this, absolutely everyone in the room or near vicinity was supposed to dive for the console control and remember to turn it off in the right order, rather than the wrong order. In the latter case you had the high voltage on after the low voltage went off, and everything would blow. This was particularly likely to occur in hot summer weather, and in the unusually hot summer of 1964, Peter Strittmatter and I spent much time hanging around EDSAC waiting for it to come back up, and running each other's codes when we took vacations at different times. (It was strongly rumoured that the non-essential EDSAC people – i.e. those not destined to replace the valves – clandestinely arranged for this to happen at favourable sun-tanning times, to give themselves a nice couple of days off.)

Fred used to tell us that EDSAC, having been designed by people who belonged to the huntin', shootin' and fishin' set, was built rather like an English hunting dog. Thus, if the input occurred at some point here, on this side of the room, naturally the output – like the input, on high speed, chattering, punched paper tape – would emerge at a point diagonally across, at the opposite end of the long rectangular room. This was so in his day, and remained so in mine. Now there weren't enough storage addresses (i.e. valves) to allow for the storage of much intermediate data in a complex and lengthy computation, so the trick was to put the data needed later out on paper tape as intermediate output. This presented one with a small geographical problem. Even though you might reasonably want to take your intermediate results and feed them in for the next stage in the calculation, it didn't matter – the intermediate output was spewing out over there, but the input point was over here. It was a two-person job! It was necessary for Fred to collaborate with Brian Haselgrove, so that while one of them was feeding stuff in at one end, the other was collecting the output at the opposite corner of the room. (Again, Fred used to muse on the fact that EDSAC was designed and built like a dog, and that if you weren't too careful with your inputs then the outputs of both these entities bore a strong resemblance to one another.) So one of them would feed stuff in here, and the other would collect it over there. The paper tape would come zipping at high speed out of the output punching device, ejaculating accompanying confetti blobs with it as it spewed out. The two of them would alternate as they collected this continuous loop of punched tape, rapidly wound it up into a great big circular pin-wheel, and then ran half way around the entire machine to feed it in here. I seem to remember that for some strange reason, in their day there also wasn't a convenient switch to tell the computer to pause and wait until the next intermediate input became necessary. (There either wasn't such a switch, or it wasn't conveniently placed.) So they had to play this waiting, chasing, and feeding in game, running round and round the computer after one another, in order to get their work done. Fred told us that the ultimate limiting factor to the length of their calculations was how long they could do this hectic devil-take-the-hindmost race around EDSAC. They'd be feverishly winding the intermediate tapes up, while taking care not to tear them, or drop and trip over them, rushing around and feeding them in again – and this was what actually limited the length of evolutionary runs they could do. I myself like to think that the term 'a stellar evolutionary run' owes its origin to this incredible footrace of Haselgrove and Hoyle around the EDSAC machine.

Another thing about C.B. Haselgrove – it isn't clear whether he had already died of a brain tumour before or after this story was told – but Fred remarked to us one day that C.B. Haselgrove had been such an excellent collaborator because he was extremely good at computing and at doing it efficiently, something very necessary in those far-off days of extremely limited storage capability. According to Fred, Haselgrove was able to do this because 'he had a very convoluted brain.' Fred would then go off into some flight of psychobiological fancy about the 'firing of 10^{10} neurons' and talk about the convoluted whorls in Haselgrove's cerebral cortex. He once illustrated the consequences of having such a convoluted brain by telling us, during his massive three day instructional marathon, that there was one particularly tricky set of logical branching decisions needed in his and Haselgrove's computer code. Haselgrove had thought carefully about it, and managed to make it possible to do this convoluted set of decisions in only thirteen machine language instructions. Fred emphasised several times how incredible it was to do this in only that many machine instructions. I thought about it that night, came back the next day and showed Fred how to do the same things in nine machine instructions. He then began to peer at me very closely and intently, and I had the uncomfortable impression that he was trying to look inside my head and see how convoluted *my* brain was.

Subsequently I was to learn that Fred really did have great insight into the minds of those working with him, and what *did* go on inside their heads! For me this happened about three years later, after I'd finished my research for my Ph.D. I was working for Fred as a kind of senior research assistant by then.[5] I went to see him one day and discussed with him what I was planning to do next. I'd decided that I wanted to study a problem that he and Martin Schwarzschild had made a first exploratory attempt at, in their seminal giant paper, but not completed. This was the problem of understanding the structure and evolutionary status of horizontal branch stars. So I went to discuss it with him and he gave me his blessing – he said 'You're tackling quite a challenging problem!' (I confess that another attraction of doing this was the following. At an Enrico Fermi summer school in Varenna sul Lago di Como in 1962, Allan Sandage had privately told me that he considered this the major problem holding up progress in understanding the evolution of globular cluster stars and their possible implications for the universal distance scale. He'd

[5] This was the same year in which I did the calculations for the famous 1964 *Nature* paper by Hoyle and Tayler, a fascinating episode whose full and largely unknown story must await another occasion.

declared that this was the kind of problem a young man could make his reputation on – heady advice, indeed, and accurate to boot.)

I went away from my discussions with Fred and began to compute what I hoped would be the first model in a systematic exploration of horizontal branch stars. Some of these models were ultimately extremely sensitive to compute, requiring new techniques for their construction that I won't go into here. However, one of the important physical inputs turned out to be substantially different from the value used in the earlier work by Hoyle and Schwarzschild: the rate of the CNO cycle had reduced by a factor of about 100 since the mid-1950s. Although I didn't realise the full implications of this before I started, it was one of the keys to me getting the solution – something like the observed spread of horizontal branch models for the observed range of metal contents, and this in addition with a marked preference for the previously unheard of 'high' (or 'cosmological') helium content coming out of my slightly earlier calculations for Hoyle and Tayler. That solution had evaded both Hoyle and Schwarzschild *and* Hoyle and Haselgrove.[6]

I started doing those computations in the unusually hot summer of 1964. In that extremely hot weather the Cambridge electricity voltage was even more uncertain than ever, and it was a nightmare nursing things along. I decided to do something which in retrospect was the most ridiculous thing I could have done. For my first attempted horizontal branch model, I chose to consider a metal content of one or two percent by mass. It seemed clear from observations that such a star ought to be up against the giant branch, so I gave it a trial initial luminosity of 100 solar luminosities and a trial surface temperature of 3000 degrees. Only later was I to realise that in this very region of the H-R diagram, where convection makes the inward integrations *so* sensitive, converged computed models had by far the *smallest* circles of convergence of any stellar model I'd ever then, or subsequently, encountered! I was aiming for the craziest possible target.

In those days, waiting to see output from the EDSAC printer was incredibly frustrating – the massive printer slowly went 'clunk, clunk,' and the platten moved the printed output up just one line. Then after an agonisingly long interval it went 'clunk, clunk' again, and another line came up. But initially all that emerged into the visible world were blank lines. At any given time a lot of what had already been printed onto paper lay tantalisingly out of sight, deep within the bowels of the printer. There had to be about six successive sets of 'clunk-clunks' before you could read even the first line of output, or, in my case, of a trial solution. I already suspected that this first solution might take a long time to converge, so I just settled down to await its outcome while an increasingly annoyed queue of waiting potential users began muttering behind me. It was all terribly tedious and it just went on and on but I couldn't drop it – the machine seemed to be hunting around randomly, going absolutely nowhere for an eternity, going clunk, clunk with no apparent end

[6] Their overly large rates for energy generation in the CNO cycle were responsible for all of their immediate post-helium-flash models jamming themselves onto the middle-RGB, instead of coming off into the classical HB region.

in sight. Eventually, I began to notice – as waiting people were saying, 'Come on, turn it off, it's getting nowhere!' – that the summarising parameters of the model began to step away more and more purposefully from where I thought the solution ought to be. Now it was marching across the H-R diagram with ever increasing steps! But eventually the steps taken reduced and reduced still further in size, and ultimately this putative model of a horizontal branch star of supposedly high metal content actually converged with a luminosity of only about 20-30 instead of $100L_{\odot}$ and a surface temperature of 16,000 degrees instead of 3,000. I just couldn't believe it – what on earth was going on? I was absolutely lost.

I went back to Clarkson Close and said to Fred, 'I have a strange problem here. I thought I was going to compute a model of a star jammed up right against the giant branch, but instead it's on the very opposite side of the H-R diagram! It seems to be a perfectly good model, at least it's properly converged.' And Fred looked at it and said 'Oh. Oohh! Oooohhhh!! It looks pretty good. But you know what: it's converged itself just where a star without any metals ought to be, according to the observations. Now then, young Faulkner, I know you pretty well. You've got one of those convoluted brains like C.B. Haselgrove. This time, I wouldn't mind betting you've probably been too clever by half. You probably intended to save as much time as you could with EDSAC because of its limited computational and storage capabilities, right? So you tried to program your way around parts of the computation, like CNO energy generation at too low temperatures, that would just waste unnecessary time. However, I'll bet that if you look into your programme, you'll find that you've been so clever that you've developed a logical loop that completely bypasses the carbon, nitrogen and oxygen cycle! I bet if you look into it you'll find that you've effectively told your model that the CNO cycle doesn't exist *at all*, and it's only using p-p energy in the shell-burning in order to produce the model. So, of course the machine has followed the instructions you really gave it – rather than those you thought you had – and it's converged the model exactly where it's right for a star with no nuclear-energy producing metals whatsoever!' I went away from Fred's house and looked carefully into my code and good heavens, he was right. I *had* completely programmed my way around the CNO cycle! So, although I had ultimately hoped to show the strong dependence of surface temperatures on metal content for these horizontal branch stars, as the observations had indicated, I had reached the desired 'low metal content' (i.e. zero metals) destination via an absolutely ridiculous initial choice of model, being saved only by a most fortuitous computing mistake.[7] Fred, knowing his man, had correctly deduced exactly what kind of problem I had probably created for myself.

Let's get on to other things to do with Fred in those days. It was always very stimulating to be working with Fred, if sometimes a bit bewildering. One frequent problem was, he would take up a conversation you'd been having the day before,

[7] For models on the hot, blue side of the H-R diagram, the circles of convergence were *huge*, stretching all the way to the cool side. This had saved my bacon.

and you'd find yourself utterly bewildered, completely lost. Was this really what you'd been talking about? It seemed to have become completely transformed, overnight. Fred would generally have seen some snag in the thing under discussion the previous evening, thought of a way around it, gone on, seen another snag, etc., etc. By the time you came to talk with him about it the next day, he might have gone through the next three or four such stages in his thinking about the problem. And, since every step he'd gone through was completely logical and obvious in *his* mind, he thought it should be so in *yours*! Surely you, too, had followed the arguments to their inevitable conclusion?

As happens with many inspiring teachers, Fred had certain unique mannerisms and all his students tended to adopt them. He had a particularly characteristic way of saying 'Oohh, Helloow,' and manipulating his nose and then gesturing with his forefinger after he did so. (He would also emphasize critical points in an argument the same way.) We all began meeting one another in the street and going 'Oohh, Helloow,' accompanied by the requisite nose manipulation. (Anyone who saw Jayant Narlikar and me meeting recently again after all these years would have seen that this old habit dies hard with the two of us.) You could always tell when a Fred student was nearby because you'd hear one or more 'Oohh, Helloows' in the vicinity, or coming from around the corner.

(Recalling Fred's characteristic gestures, I remember a much later occasion when he visited Santa Cruz to give an invited lecture. My wife Jeanne and I gave a dinner party in his honour. Among other guests we invited a UCSC colleague, Audrey Stanley, to come and have dinner with Fred. Audrey, an Englishwoman, is a distinguished Professor of Theatre Arts and the original founder of the extremely successful Shakespeare Santa Cruz. At the dinner party Fred engaged in a most stimulating and remarkable conversation in which he explained the environmental and therefore evolutionary importance for the proboscis monkey of having a very large nose. He held everyone spellbound as he explained all this, accompanied by much nose manipulation to emphasize key points. Audrey later told me that she was absolutely beside herself at seeing Fred's 'performance,' and at the illustrative unity between the speaker's subject and his emphatic gestures.)

We students got so much into the habit of impersonating Fred that I'm afraid I began answering the telephone in Fred's voice. But eventually I fell into this trap of my own making. One day the phone rang somewhere in the grad student offices, and I picked it up and said 'Oohh, Helloow.' Then I heard the startled voice of Barbara Hoyle, who knew he was out of town, saying 'Fred, what are you doing there?!' At that, I was so embarrassed I just slammed the phone down and ran out of the room. I have always wondered how exactly Fred, who was at a meeting on the other side of the country, explained to Barbara how his voice had come over the phone when she was calling up the student offices in the Cavendish complex's Phoenix building.

In his own contribution, Cyril Domb mentioned that Fred had deduced something significant about the reflectivity of radar off the sea during the war, while seeing Liberators flying off the coast at Bedruthan Steps, near Newquay in Cornwall. That reminds me of a near-misadventure that took place in 1961. This story involves two Hoyles and Bedruthan Steps, also, but started far away from there, when I went to my first very enjoyable Enrico Fermi summer school on Lake Como. (The Varenna conversation I already mentioned, with Allan Sandage, took place at the next year's summer school.) During the 1961 summer school Barbara Hoyle got into difficulties swimming. Fortunately, I'd obtained the Royal Life Saving Society's Bronze Medallion at school. Because of my training I was able to save her from drowning. (Birmingham's George Isaak, later to become a very prominent helioseismologist, leaped in to help her towards the end as my own strength flagged.) That rescue brought me rather close to Barbara and to the family.[8]

Later that summer of 1961, Barbara and her children Geoff and Liz were going to take a camping vacation very close to Bedruthan Steps, on a farm near Trenance in Cornwall, the place where Barbara and Fred had snatched a holiday break during the wartime. (I think that the very same farmer let us all stay in the very same field.) Jayant Narlikar and I were invited to join the family to share in the vacation. I'm not sure how Jayant travelled down there, but I was an out-of-training former Cambridge racing cyclist. I rather foolishly rode down to Cornwall, without adequate training miles in my legs, and eventually ruined my left knee on those Cornish hills, coming back from this trip. During that holiday at least three of us – Geoff and Barbara and I – made a trip down the Bedruthan Steps into a beautiful bay and walked a considerable way north on the magnificent sand past the now collapsed 'Queen Elizabeth Rock' into another bay, etc., etc. We finished up in a bay rather removed from Bedruthan Steps themselves, beyond a particular and prominent headland. Finally, we noticed rather too late that the sea had come in quite a bit.[9]

It came to the point where it was clear to us that we were trapped – there were high and rugged cliffs all around us and we were not able to make it back on the sand around the headland by which we had walked into this bay. Things looked rather desperate for a time. But then, out of the recesses of her memory, Barbara remembered that exactly the same thing had happened to her and Fred back in the wartime. She suggested that we follow her. Then she made her way into what seemed like an endless dark and echoing cave that was going to be our graveyard as far as I could tell – a cave underneath the headland, quite some way short of the promontory itself. The cave led on and on into near darkness, with just a bit of light shining from behind us. Then it widened at a fairly deep pool. Miraculously, however, although the water came up to high chest height, we waded on through this area and eventually, rounding a dark bend, there was now light ahead of us,

[8] I had the pleasure of visiting Barbara Hoyle, in Bournemouth, just the day after this memorial meeting ended. The Lake Como incident was one of the first things she mentioned.

[9] Liz doesn't remember this so she probably wasn't involved. Jayant has since told me that he was at Herstmonceux for the first part of our joint vacation, arriving after our misadventure. However, he recalls immediately hearing our harrowing descriptions of it.

literally at the end of the tunnel. We made it out alive, much to my surprise. So Barbara repaid the summer's earlier favour, remembering as she did so the route to safety that Fred had pioneered in the later part of the war.

We were Fred's students when it was also one of the heydays for his science fiction. Quite early on, Fred had the idea for what was ultimately a fascinating and popular pair of science fiction TV series. They aired on the BBC in 1961 and 1962 with the titles *A for Andromeda* and *The Andromeda Breakthrough.* Fred ultimately co-authored the series and the subsequent books, as I'll explain. My reason for wanting to tell you this is the following. Incredible though it may seem, during this time Fred discovered a previously unknown star of the very first magnitude. Now you'd think that a bright star like this would have been discovered by the rest of the world, but no, Fred was the first. Here's how it came about:

The BBC had become interested in producing a TV series based on Fred's best-selling book, *The Black Cloud.* However, when they contacted Fred they learned that the rights for that had already been sold to an independent producer.[10] So the BBC could not obtain the rights because they already belonged to someone else. Fred told the BBC not to worry, he had lots of ideas suitable for other possible series. One in particular had been brewing in his mind but he'd not had time to write any of it down. Fred and the BBC then agreed that he would talk his ideas and some key dialogue into a tape recorder. Meanwhile, the BBC chose one of its trusted writers, John Elliot, to come to Cambridge and talk with Fred about it and then help develop the series. Initially, Fred had conceived the whole storyline as just one TV series, to be shot and shown in its entirety. However, Elliot and the BBC found it to contain so much dramatic material, they decided to do one half first. The ending to that could either be a bit enigmatic, or alternatively act as a cliffhanger for the rest of it. And that's how the two successive series originated. (Eventually, following the enormous success of the two TV series, Elliot and Fred co-wrote the books of *A for Andromeda.*)

So here they were, developing an exciting science fiction TV series together. Now, a key figure in *A for Andromeda* was a cool, androgynous humanoid (as Fred had conceived it), who would be built according to blueprint instructions sent by radio signals from an alien civilisation. This humanoid would act as an interface between humanity and the computer, also built to alien instructions. Elliot was having nothing of this: he knew he couldn't have an androgynous interface – 'What you have to have, Fred,' he said, 'is a woman, and a very beautiful woman at that!' – though, in those sexist days I rather suspect he probably said 'a girl.' So the following obvious question became paramount: where would they find the 'girl' who would play the part of this interface? They looked and looked, and still couldn't find an appropriate person. So, somewhat in a spirit of desperation, they went to

[10] Barbara tells me that, despite a well-known hyphenated name and subsequent knighthood, said producer has never done anything with those rights.

the final end-of-year performances at places like RADA, the Royal Academy of Dramatic Art, and London's Central School of Speech and Drama.[11]

At one of these shows they hit gold! On to the stage at some point came a rather cool and beautiful, alluring yet reserved, other-world looking blonde female, the epitome of every red-blooded British lad's dreams in the 1960s. Fred immediately grabbed Elliot's arm and said: '*She's* the one. I want *her*!!' (I can't help feeling that this emphatic declaration could have been misinterpreted.) Guess who 'she' turned out to be?: 'she' was the person who later became known to the world as Julie Christie. So Fred had instantly recognised that this cool, beautiful, very reserved young woman with an impassive but finely chiselled face was *the* person to act the part of Christine (a computer operator who dies and is 'absorbed' by the computer), and Andromeda herself, the subsequent living computer interface in 'A for Andromeda.'

The first series aired from 3 Oct. – 14 Nov. 1961. It was a rapidly growing cult success, and established unprecedented popularity records. From an early small start, the audience doubled with each episode. By the seventh and last episode, 80 per cent of the TV-viewing audience of Great Britain watched it! This is what made Julie Christie well-known and ultimately a star. In fact her own star was to rise quite swiftly. She was already not available (or too expensive) for the second series (which aired 28 June – 2 Aug. 1962), and within only four years she would win a Hollywood Oscar. Thus, she had very rapidly become a pearl with a price far beyond the BBC's reach.[12]

Let me tell you another science fiction story. Fred and his son Geoff wrote *The Fifth Planet*. I'll remind you of the first two sentences. 'Hugh Conway shifted uneasily. An hour before[13] his wife had come to him with such fervour that he knew she must have been unfaithful again.' I was a little bit naughty: having read this, I cheerily remarked to Barbara Hoyle one day that I'd really enjoyed the opening two sentences of Fred's latest science fiction book. She replied somewhat archly: 'Oh, the collaboration between Daddy and Geoff works really well. You see, Daddy provides the science, and Geoff provides the sex – Daddy doesn't know very much about that sort of thing!'

And finally, that brings me to one of my all-time favourite Fred stories. One of the things that always impressed me as a student and a young post-doc in Cambridge and in California was how much Fred, Willy Fowler and the Burbidges

[11] This tactic had often enabled the BBC to get hold of very promising young actors at rates it could afford before their services cost too much.

[12] But see www.btinternet.com/~screeny/guides/andromedabreak.htm; according to this web-site, Fred thought the second production was a near disaster, and expressed dissatisfaction with the BBC's refusal to give Julie Christie a 300 Pound option payment to retain her services for that sequel. Cheap! Clearly she should have been retained, and Auntie Beeb blew it. No wonder Christie tends to downplay her early BBC role. The viewing audience's consolation prize was Susan Hampshire, instead.

[13] Sic; the rhythmically needed comma is missing.

really enjoyed their science. They got one heck of a kick out of doing it together. I had the pleasure of seeing the four of them working together, of Willy Fowler writing things down on a blackboard, of some calculation being needed and Willy whipping out his slide-rule, and Fred doing it faster in his head because he knew the logarithm tables intimately. They had this friendly competition in which Fred would often beat Willy – who was an absolute whiz with a slide-rule – to the answer. Meanwhile, the Burbidges would be adding in ideas and information from their expertise, everything from optical spectroscopy, radio astronomy and cosmic ray physics being thrown into the hopper. It was a true four-person collaboration among equals – a wonderful thing to see in operation. I have never seen anything so exciting in scientific collaboration as the one that had produced B^2FH. They always enjoyed their science so much, and it was a wonderful thrill to see how much they got out of it. The particular story that follows involves Fred and Margaret.

We now come to early 1968, when something quite remarkable in the annals of Cavendish radio astronomy occurred. At that time there were four astronomy groups in Cambridge, and certain members of them would get together for lunch on Fridays. Now here was an amazing thing: three of these groups completely engaged in a two-way exchange of information, but the fourth group, which came from the Cavendish, would absorb information but never emit any. This was the Ryle group. Let's face it: Martin Ryle seemed utterly paranoid because of the bitter fight that the Cambridge radio astronomy group had had with the Australians over the notorious 2C survey. Australian radio astronomers had shown that their Cambridge competitors had had a real problem, not appreciating how their own side lobes were responsible for their over-enthusiastic interpretation of data from the 2C survey. Ryle was extremely upset, and subsequently gave instructions that no member of the radio group should give the slightest information to, or talk with, anyone outside the group.[14] (It was said that his personal graduate students were not even allowed to talk to anyone in the rest of the radio group about what they were doing, that they were told they'd be cut off without support if they did so.) It was a very bizarre situation.

Whether or not everything that was said about the reasons for the tight-lipped nature of the radio group was strictly true, the everyday behaviour of its members lent credence to these widespread assertions. Back to our story. We used to have regular astronomers' lunches on Fridays, the 1968 lunches taking place in a spacious private dining lounge at the relatively new Graduate Centre. At these lunches there would be very engaged conversations between three of the groups, who were always fairly well represented. A few members of Ryle's group would come as well. Peter Scheuer, John Baldwin and John Shakeshaft were regulars, and I seem to recall Malcolm Longair coming, too. If the topic were general astronomical science or gossip, they'd be as involved as the rest of us, but they'd get a bit tight-

[14] This was a rather odd thing to decide, if you think about it. It was through holding their cards too close to their collective chests, that they'd fallen into this problem in the first place.

lipped and sit there shifting uncomfortably should the conversation show signs of taking even the slightest turn towards what might be going on in their group.

Then came a Friday in February 1968 when a change occurred, but only modestly as you will understand. The meal had long ended, and the conversation had wound down. People started to leave. Peter Strittmatter and I were still there by the main table. As the radio astronomers exited, John Shakeshaft paused at the door, turned, and said, essentially in afterthought but in his very precise way, 'Oh, by the way, I really think you'll want to pay attention to a very important seminar that will be given next week - it will be extremely significant and you won't want to miss it!' Peter Strittmatter said, 'That sounds great John, but what's it about?' And Shakeshaft said, a bit uncomfortably, as though he might already have said too much, 'Well, I'm afraid I'm not at liberty to say; but please, *do* watch out for the announcement – you'll definitely recognise it when you see it and you won't want to miss it!'

So, on Monday or Tuesday there appeared a notice at the Institute and on other astronomical noticeboards around Cambridge saying 'Discovery of a new type of pulsing radio source.' Tony Hewish was going to give a hastily announced seminar, either on that very Tuesday or shortly thereafter, about something that had not yet been published (!); this latter aspect itself was quite noteworthy, a first in the Cavendish group's history as far as most of us were concerned. It turned out that they had learned that their paper on the discovery of pulsars would definitely appear in that Friday's *Nature*. As a very special case, therefore, Ryle had relaxed his otherwise invariable rule that nothing should be talked about prior to it appearing in hard print. He was actually allowing a seminar to be given a day or two early, about something that wasn't going to appear in *Nature* until Friday. Amazing!! We really should have all been so grateful.

So, we all go and hear about this truly remarkable breakthrough, in a packed Maxwell Lecture Theatre. The circumstances of the discoveries themselves are discussed, the first four 'rapidly pulsing radio sources' (the term 'pulsars' still lay in the future), the fact that they had been known maybe half tongue in cheek as LGM1, 2, 3 and 4, 'Little green men,' because they weren't sure if they were really cosmic sources or signals from alien civilizations and so on. We learn about this and Fred starts speculating and others start speculating. It was a very exciting and frenzied time.

Only a day or two later Fred had to go to a committee meeting in London of either the DSIR or the SRC or whatever it was called in 1968, and he found himself sitting next to Bernard Lovell. Fred later told me that the topic of the exciting new radio sources came up, and that the following conversation then ensued. Fred said to Lovell, 'Oh, Ooh, Bernard, have you heard about these exciting new pulsing radio sources that the Cavendish group has discovered?' 'Oh yes, I know about them, Fred.' And Fred carried on, 'What a pity it is that neither in their colloquium nor in their paper in *Nature* have they given the positions. They're holding the positions pretty close to their chest.' He was a bit surprised when Lovell responded,

'Oh, but *I* know the positions Fred!' Fred replied, 'How can that be?!' Lovell then said: 'Ryle told me that they'd exhausted the means they have at Cambridge for exploring these things, and so they've given me the positions so that we can now use Jodrell to look at them.' He added, 'And I happen to know as well that they phoned them through to Allan Sandage in Pasadena, so that he could look for them optically at Palomar, but they told him to keep the positions secret.' Fred then asked, 'Oh, is that right? Well Bernard, did they ask *you* to keep them secret?' And Lovell replied 'Well, no Fred, as a matter of fact they *didn't*!' So Fred said, 'Well then Bernard, would you mind letting me know what these positions are?!' Lovell thought only a moment and said 'Not at all, Fred, I'll phone you tonight or tomorrow.' So he called Fred and told him what the positions of these new objects, the future 'pulsars,' were.

Then Fred decided to have a bit of fun. He called up Margaret Burbidge in La Jolla, and said 'Oh, Oohh, Margaret, helloow! No doubt you've heard about these amazing new pulsing radio sources that have been discovered in Cambridge?' And Margaret of course replied, 'Oh yes, Fred, I certainly have heard about them. They sound extremely exciting, but what a pity they're keeping the positions secret!' (Almost exactly what Fred had said to Lovell.) Fred then told her, 'That's what *they* think, but Margaret, it turns out that I've managed to get the positions from Bernard Lovell. And I'll tell you something else. I've also learned that Allan Sandage has been told the positions, so that he can look for them at Palomar, and he thinks he's the only optical observer who's got them. Now, I tell you what: why don't *you* call Allan up, and in the spirit of friendly international scientific collaboration, why don't you ask him whether he has the positions, and when he tells you he hasn't, why don't you tell him what they are?!' Margaret then replied, 'Oh Fred, what a good idea!' And so, a day or two later, Margaret made a return phone-call to Fred, and said, 'Well, Fred, I did as you suggested. I called Allan up, and first of all we talked about something inconsequential. Then I brought the subject around to these wonderful new pulsing radio sources, and I said to Allan, "Isn't it a pity that they haven't publicly released the positions?" And Allan said, "Oh yes, Margaret, we would all love to know them, I'm sure, but I understand that they're keeping them under wraps for the moment." And I then replied "Well Allan, as it happens, I know what the positions of these radio sources are, so please take a piece of paper and a pencil, and I will tell you the positions so that you too can check them out!" So I then dictated the positions of the pulsars to him. There was a lo-o-o-o-ng silence, and then he thanked me, *most* profusely!'

That story brings me to the end of my reminiscences of Fred Hoyle, a man always so much larger than life. And now, as we think about Fred, and what knowing him meant to each of us, let me ask you all to take your glasses, to raise them, and to drink a toast to the memory of Fred Hoyle, one of the most creative, most inspiring, most gifted and accomplished of giants in the history of science, a man who truly dominated the astrophysics of the last half of the twentieth century. To Fred Hoyle!

Acknowledgements

I would like to thank Dr Jane Gregory of UCL's Department of Science and Technology Studies for recording my after-dinner speech in Cardiff, and for generously providing me with a transcript on which this expanded account is based.

MIX
Papier aus verantwortungsvollen Quellen
Paper from responsible sources
FSC® C105338

If you have any concerns about our products,
you can contact us on
ProductSafety@springernature.com

In case Publisher is established outside the EU,
the EU authorized representative is:
Springer Nature Customer Service Center GmbH
Europaplatz 3, 69115 Heidelberg, Germany

Printed by Libri Plureos GmbH
in Hamburg, Germany